An Introduction to the History of Structural Mechanics

Part I

Edoardo Benvenuto

An Introduction to the History of Structural Mechanics

Part I: Statics and Resistance of Solids

With 112 Illustrations

Springer-Verlag
New York Berlin Heidelberg London
Paris Tokyo Hong Kong Barcelona

Edoardo Benvenuto
Università di Genova
Ordinario di Scienza delle Costruzioni
Facoltà di Architettura di Genova
Genova, Italy

Mathematics Subject Classification: 01-xx, 73xx, 82xx

Library of Congress Cataloging-in-Publication Data
Benvenuto, Edoardo.
An introduction to the history of structural mechanics / Edoardo Benvenuto.
p. cm.
Contents: v. 1 Statics and resistance of solids—v. 2. Vaulted structures and elastic systems.
ISBN 0-387-96227-1 (v. 1: alk. paper).—ISBN 0-387-97187-4 (v. 2: alk. paper)
1. Structural analysis (Engineering)—History. I. Title.
TA646.B46 1990
624.1'71'09—dc20 89-26230
CIP

Printed on acid-free paper.
This work was originally published in Italian by G.C. Sansoni, 1981.

Text photocomposed using the LATEX system.
Printed and bound by R.R. Donnelley & Sons, Harrisonburg, Virginia
Printed in the United States of America.

9 8 7 6 5 4 3 2 1

ISBN 0-387-96227-1 Springer-Verlag New York Berlin Heidelberg
ISBN 3-540-96227-1 Springer-Verlag Berlin Heidelberg New York

To my mother, Giovanna

Acknowledgments

This book draws its origin from my textbook on the science of structures and its historical development which I published with the Sansoni Publishers in 1981 (*La Scienza delle Costruzioni e il suo Sviluppo Storico*, Florence, 1981). I am indebted to Professor Clifford Truesdell, who kindly appreciated my attempt to outline a history of the relation between rational mechanics and structural engineering and presented my work to the late Mr. W. Kaufmann-Bühler of Springer-Verlag for an English edition. In fact, this book is not a translation of the original. Mr. Kaufmann-Bühler suggested that I transform the text into a general introduction to the history of structural mechanics concentrating on some specific topics and expanding their historical references. I wrote the new text in Italian. I am very grateful to my colleague and friend Prof. Aurelia V. von Germela for having carefully interpreted my complex academic style and transformed it into fluent English. I thank very much also Mrs. Molly Wolf for her clever and precious final copy-editing, which made my manuscript more dry and light for American preferences. Very useful to me was Dr. Peter Barrington Jones' kind collaboration, and I am particularly grateful to my assistant, Arch. Massimo Corradi, for his beautiful drawings, hand-made in "old style". Finally, I am glad to thank my colleagues Prof. Gianpietro Del Piero and Prof. Paolo Podio Guidugli for their useful suggestions regarding the topics in the first volume.

Foreword

This book is one of the finest I have ever read. To write a foreword for it is an honor, difficult to accept.

Everyone knows that architects and master masons, long before there were mathematical theories, erected structures of astonishing originality, strength, and beauty. Many of these still stand. Were it not for our now acid atmosphere, we could expect them to stand for centuries more. We admire early architects' visible success in the distribution and balance of thrusts, and we presume that master masons had rules, perhaps held secret, that enabled them to turn architects' bold designs into reality. Everyone knows that rational theories of strength and elasticity, created centuries later, were influenced by the wondrous buildings that men of the sixteenth, seventeenth, and eighteenth centuries saw daily. Theorists know that when, at last, theories began to appear, architects distrusted them, partly because they often disregarded details of importance in actual construction, partly because nobody but a mathematician could understand the aim and function of a mathematical theory designed to represent an aspect of nature.

This book is the first to show how statics, strength of materials, and elasticity grew alongside existing architecture with its millenial traditions, its host of successes, its ever-renewing styles, and its numerous problems of maintenance and repair.

In connection with studies toward repair of the dome of St. Peter's by Poleni in 1743, on p. 372 of Volume 2 Benvenuto writes

> This may be the first case in this history of architecture where statics and structural mechanics are successfully applied to a real problem with maturity and full consciousness of their implications. It marks a turning point between two eras: one in which tradition and prejudice ruled the art of building, and another in which the mathematicians' and physicists' new theories, elaborated in academies and laboratories, were allowed to make their contribution. It is somehow pleasant to realize that this anticipation of the great nineteenth-century synthesis of

> science and technology came not from an ordinary bit of building but from one of the most daring and beautiful creations of the Renaissance at the height of its splendor.

On p. xx of the introduction

> The division between inspiration and technique is of very recent origin and is largely artificial. In building, science and art have always been united in the creative act. Not even the most narrow-minded aesthete or engineer can part the two without losing something. To see Brunelleschi, Michelangelo, Guarini, Wren, Mansart, Soufflot, a hundred others, *merely* as great artists is to deprive them of credit for their brilliant engineering. Their wonderful technical innovations, their perfect determination of the weights that had to be balanced and the mechanisms of collapse that had to be opposed—these give coherence and splendor to their works.

The two paragraphs just quoted provide a kind of summary, indeed partial, of what Benvenuto wishes to tell us and to let us learn, step by step, not as philosophy or by journalistic simplisms, but by reading expert observations upon a gradual, not always direct history of the science of construction. The last paragraphs of his book read in part as follows:

> The long, stormy commotion [about the ideas of Menabrea, Castigliano, Crotti, and Mohr] enlivened scientific literature for more than a century. Persuasive hypotheses, even more persuasive confutations, fruitful but fallacious intuitions, sterile but unexceptionable verdicts, agreements reached unexpectedly—all have been forgotten. What we remember today are the instruments of engineers, the formulae in daily use. If we asked an engineer about the origins of the equations he or she uses constantly, the reply would be disappointing. They exist; nothing else matters. Why be curious about their derivation?
>
> True, the authors with whom we conclude our historical outline were able to supply such effective technical solutions that, in their hands, the real meaning of the questions they tackled seems to have been lost. But history has its uses....

Indeed it does, as the reader will learn.

Not only is Benvenuto a man of astonishing erudition and breadth, but also he loves his science and is humble before it. He thinks clearly, clearly organizes his material, difficult and complicated as it seems, and writes clearly with direct and masterly expression. In leafing over or reading his book, we recognize a great work, one doubtlessly flawed by many small errors among several grand truths. Parts of his matter, bit by bit or lacuna by lacuna, may well be corrected or filled by historians in coming decades, but his book can never be replaced as a general, pioneering treatise, a survey of a great field heretofore seen only dimly, from a distance, but never trodden. Never before have I learned so much about the history of mechanics from a single book.

As is often the case with books that start from the foundations of a subject, the beginning of Benvenuto's is the part hardest to understand. The reader accustomed to scientific works could well begin with Chapter 5 of Volume 1, "Galileo and his 'Problem'", or with Chapter 8, "Early Theories of the Strength of Materials". Perhaps, even, he might begin with Volume 2, which opens with "Knowledge and Prejudice before the Eighteenth Century". Above all, to get an idea of the spread of the work, every reader should study first of all and carefully the two tables of contents, for the titles of the subsections are fascinating. He who is not already expert in both architecture and mechanics will see there some names he has never before encountered, associated to problems or structures or theories he is unlikely to know. In fact, Benvenuto's clarity and directness are such that a reader might start by fishing out some subsections. Any place you open this book and read in it, you will be fascinated by what is there. Wherever you start, for example at the passage first quoted above, I wager you will end by studying the whole book.

Part I of Volume I, although some may profit best from reading it last, is of great value. Very few readers will know already all of the contents of §1.2, "The Enigma of Force and the Foundations of Mechanics". It begins with a résumé of what should now be regarded as vague meandering, impotent struggles, foolish attempts at reduction, and justified doubt regarding the nature of force, the first problem "against which science finds itself powerless." It ends with "one of the most important *events* in the history of mechanics," namely, Walter Noll's organization of the mechanics of continua as a mathematical science. There not only is "system of forces" taken as a primitive term, but also it is clarified by a list of its *mathematical properties.* The theory of systems of forces makes mathematical sense, just as Hilbert's axiomatization of Euclidean geometry in terms of the undefined objects "point", "line", and "plane" makes mathematical sense. That will not stop philosophers from musing about force and historians of science from dilating upon old, obscure, unmathematical ideas about force, but it does make "force" something a modern scientist, be he mathematician or be he architect, can use as he does "point", "line", and "plane". The intuitive notions, both in geometry and in mechanics, remain; not only that, they help both in applications and in creative thought; but the precise concepts stand behind both.

Of course, Benvenuto makes use of secondary works, but also he studies carefully and analyses meticulously the originals to which they refer. It is not unusual—as I can vouch through reading his treatment of some sources that I described too hastily some thirty years ago—not unusual, I say, that in the end he silently corrects the secondary work he has studied.

Benvenuto rightly refers to many Italian sources which are largely unmentioned in the general literature. As in many other fields, Italians were the great leaders in architecture, structures, and remedies for the apparent beginnings of failure. Architects from other countries studied in Italy, and Italian architects designed castles and palaces from Russia to Spain. The Italians were also second to none in theoretical analyses of architectural members and assemblies. Failure to study Italian sources directly is a general malady of the precise history of science.

Occasionally Benvenuto refers to a rule or solution of a problem as "correct" or "incorrect". Even the sociological historians, with their belief that the sciences are no more than ephemeral fads, much as history was called by a famous and once powerful man "the lies that men agree to believe," can not justly cavil here, for in architecture the correctly designed arch is one that does not fall except under conditions it was not intended to withstand.

C. Truesdell

Contents of Part I

Foreword vii
Introduction xvii

I The Principles of Statics 1

1 Methodological Preliminaries 3
1.1 The Special Objects That Gave Rise to Mechanics 3
1.2 The Enigma of Force and the Foundations of Mechanics . . 7
1.3 Statics as "Science Subordinated to Geometry as Well as to Natural Philosophy" 14
1.4 Momentum: Fixed Word, Fluid Concept 16
1.5 The Aristotelian Roots of a Vocabulary for Mechanics . . . 20
1.6 A Short Outline of Aristotle's Physical Principles 25
1.7 Modern Metamorphoses of the Immobile Mover: Towards the Principle of Conservation 30
1.8 The "Mechanical Problems": The Peripatetic Explanation of the Law of the Lever and the Parallelogram Rule 34

2 The Law of the Lever 43
2.1 Archimedes' Demonstrations 43
2.2 Interpretations (and Improvements) of Archimedes' Proof . 48
2.3 An Alternative Approach: Pseudo-Euclid and Huygens . . . 56
2.4 Marchetti's New Approach and Daviet de Foncenex's Improvements 61
2.5 De la Hire's Proof, Lagrange's Remarks and Fourier's Contribution 64
2.6 Towards the "Dethronement" of the Law of the Lever: Saccheri and de Maupertuis 67

3 The Principle of Virtual Velocities 77
3.1 Medieval Roots 77
3.2 Guidobaldo del Monte, Galileo, and the Principle of Virtual Velocities 80
3.3 Descartes: "Explicatio Machinarum Unico Tantum Principio" 85
3.4 Bernoulli and Varignon 88
3.5 Riccati's "Universal Principle of Statics" 91
3.6 Lagrange's First Demonstration 95
3.7 The Approaches of Fossombroni and Fourier 98
3.8 The Principle of Virtual Velocities and Constraints: Poinsot's and Ampère's Contributions and Lagrange's Second Proof 105

4 The Parallelogram of Forces 116
4.1 Daniel Bernoulli's Claim 116
4.2 Daniel Bernoulli's First Geometrical Demonstration 119
4.3 Bülffinger's Paradox 122
4.4 Riccati's Solution 123
4.5 Foncenex's Memoir and Lagrange's Criticism 126
4.6 Foncenex's Fundamental Lemma 127
4.7 Foncenex's and D'Alembert's Functional Equation 130
4.8 D'Alembert's Memoir of 1769 134
4.9 Further Developments: D'Alembert, Poisson, Cauchy, Dorna and Darboux 136
4.10 Conclusion 141

II De Resistentia Solidorum 143

5 Galileo and His "Problem" 145
5.1 Introduction 145
5.2 Galileo: A Short Account 147
5.3 The Subtext: Galileo's Atomism 152
5.4 The Primacy of Geometry over Logic in the Discorsi 154
5.5 The First Day of the Discorsi 158
5.6 Attempts to Explain the Cause of Resistance 163
5.7 For and Against the Power of the Vacuum 166
5.8 First Intimations of an Atomistic Theory of Resistance . . 169
5.9 Democritus or Plato? 173
5.10 The Second Day 176
5.11 Opening Remarks 179

5.12 Corollaries 183
5.13 The Problem of Solids of Ultimate Dimensions 188
5.14 The Problem of Solids of Equal Resistance 194

6 First Studies on the Causes of Resistance 198
6.1 Experimental Confutations: The *Horror Vacui* 198
6.2 Mersenne and the Problem of Resistance 203
6.3 Descartes' Concept: Stasis as the Best Adhesive 206
6.4 The Atomist Rossetti and His Explanation of Resistance . 209
6.5 Atomism and Vacuum: Newton, Leibniz and Clarke 217
6.6 Newton's "vis interna attrahens": Elasticity and Resistance 221
6.7 Boscovich's Reformation of the Old Atomism 223
6.8 Developments of Boscovich's Theory: Early Nineteenth-Century Research on Elasticity 227

7 The Initial Growth of Galileo's Problem 233
7.1 Introduction 233
7.2 First Steps in the Controversy about Solids of Equal Resistance: Blondel's "Evidence" 235
7.3 Marchetti's "Evidence" on Solids of Equal Resistance . . . 241
7.4 Marchetti's Axiomatic Approach to the Resistance of Solids 244
7.5 Viviani's "Evidence" 246
7.6 Antony Terill and Solids of Ultimate Dimensions 252
7.7 Fabri: Elasticity as an "Intermediate Force" 254
7.8 Pardies' Statics 257

8 Early Theories of the Strength of Materials 262
8.1 Elasticity Enters the Theory of Resistance 262
8.2 Mariotte's Contribution 265
8.3 Leibniz's *New Demonstrations* 268
8.4 New Problems: Catenaries and Elastic Curves 271
8.5 Jakob Bernoulli's Fundamental Work 274
8.6 Varignon and the Galileo–Mariotte Dichotomy 277
8.7 Musschenbroek and the Imperfections of Matter 280
8.8 The Last of the Eighteenth-Century Treatises on Resistance 284

Author Index 294
Subject Index 299

Contents of Part II

III Arches, Domes and Vaults 307

9 Knowledge and Prejudice before the Eighteenth Century 309
9.1 "A Strength Caused by Two Weaknesses" 309
9.2 Viviani's "On the Formation and Size" of Vaults 311
9.3 Fr. Derand's Rule 313
9.4 The First "Scientific" Treatment of the Statics of Arches . 315

10 First Theories about the Statics of Arches and Domes 321
10.1 Philippe de la Hire 321
10.2 Arches and Catenaries: David Gregory and Jakob Bernoulli 326
10.3 Philippe de la Hire's Memoir of 1712 331
10.4 Belidor's Variant 336
10.5 Couplet's Two Memoirs 338
10.6 Bouguer's First Static Theory of Domes 344

11 Architectonic Debates 349
11.1 The Italians: An Introduction 349
11.2 The Case of S. Maria del Fiore in Florence 349
11.3 St. Peter's Dome and the Three Mathematicians 351
11.4 Giovanni Poleni's "*Historical Memoirs*" 358
11.5 Poleni's Theoretical and Experimental Work 359
11.6 Boscovich and the Cathedral of Milan 371

12 Later Research 375
12.1 The "Best Figure of Vaults": Abbé Bossut 375
12.2 Coulomb's Theory of Frictionless Vaults 386
12.3 Coulomb's Theory: Friction and Cohesion 394

12.4 Italian Studies on Vaults in the Late Eighteenth Century . 399
12.5 Lorgna's Essays 404
12.6 Fontana's Treatise 407
12.7 Mascheroni's "*New Researches*": The Limit Analysis of Arches 412
12.8 Mascheroni and Domes of Finite Thickness 420
12.9 Salimbeni's Treatise 425
12.10 The Nineteenth Century: Further Developments 428

IV The Theory of Elastic Systems 439

13 The Eighteenth-century Debate on the Supports Problem 441
13.1 Introduction 441
13.2 The Birth of the Question 442
13.3 Discussion in Eighteenth-century Italy 447
13.4 Volume 8 of the *Memorie della Società Italiana* 455

14 The Path Towards Energetical Principles 461
14.1 The Debate Continues 461
14.2 The Nineteenth Century: An Introduction 466
14.3 The Philosopher Who Understood Everything 470
14.4 From Cournot to Dorna 476
14.5 Clapeyron and the Case of the Continuous Beam 479
14.6 Menabrea's Elasticity Principle 488

15 The Discovery of General Methods for the Calculation of Elastic Systems 492
15.1 Clebsch's Treatise and the "Method of Deformations" . . . 492
15.2 Maxwell's Fundamental Memoir on Frames 499
15.3 Maxwell and the "Method of Forces" 504
15.4 The Goal Attained 507

16 From the Theory of Elastic Systems to Structural Engineering . 513
16.1 Alberto Castigliano 513
16.2 Some Aspects of Castigliano's Work 516
16.3 Francesco Crotti's Clarification 523
16.4 Mohr's "Beiträge": Statically Determinate Trusses 530
16.5 Mohr's Solution for Statically Indeterminate Trusses 537
16.6 German Disputes about Castligliano's and Mohr's Methods 542

Author Index 544
Subject Index 548

Introduction

The battle between weight and rigidity constitutes, in itself, the single aesthetic theme of art in architecture: and to bring out this conflict in the most varied and clearest way is its office. Architecture accomplishes such a task, barring the direct route of free expansion to those indestructible forces, slowing them up by deflecting them; thus the battle continues and shows, in manifold forms, the unceasing efforts of the two opposing forces. Left to its own devices, a whole building would [collapse into] a compact mass, pressing by its mass upon the ground, on which the weight inexorably pushes....

Rigidity, on the other hand ... opposes such an effort with vigorous resistance.

The immediate manifestation of the natural tendency [of gravity] is hampered by architecture, permitting only a mediated manifestation, in tortuous ways. For example, scaffolding can exert pressure on the ground only by means of a column mass; the vault has to hold itself up and the pillars are the only means that satisfy the downward tendency, and so forth. But by virtue of these forced and contorted ways, by virtue of the obstacles, the forces immanent in these rude masses of stone have a way of revealing themselves in the clearest and most varied forms.... Therefore the beauty of a building lies in the final visible suitability of every part; to a finality not external and arbitrarily fixed by man (the work in this case would belong to practical and applied architecture) but rather concerned with the consistency of the whole, for which the place, the size and the form of every part must be in such an essential relationship that taking away any part from any place would plunge the building into ruin.[1]

---

[1] A. Schopenhauer, *Die Welt als Wille und Vorstellung*, book 3, section 44, included in *Arthur Schopenhauers sämtliche Werke*, P. Deussen ed., (Munich, 1911) Vol. 1, pp. 252–253.

This odd fragment was written not by a technician dedicated to structural calculation, nor by a historian of architecture looking for an interpretive key to give coherence and meaning to his exposition, but by a philosopher whose real interests were elsewhere. Schopenhauer's intention was to describe the "world as will and representation"; his observations on architecture are merely an aside, one small stone in the mosaic of his speculative system. But for this reason, the passage quoted above is of interest. It shows, in stripped-down images, an aspect of architecture that is easily accessible to non-specialists. Moreover, Schopenhauer's passing comments reflect an ancient perception, perhaps forgotten now but still implicit: the admiration and marvel aroused by the great architectural works that seem both to vie with God's work in nature and symbolically to renew it, subjugating the hostile forces of nature.

The sense of wonder that stimulates us to explore the phenomena that seem to contradict nature's laws is at the origin, in part at least, of every technical artifice. And this sense of wonder is, perhaps, a primordial dimension of meaning in architecture; it expresses not so much the mystery of the unknown as the mystery of man's dominion over it. Look at the myth of Daedalus the very archetype of this domination. His creation is a combination of art and play, a manifestation of unexpressed rationality which succeeds in clearing up all the dark tangles. He creates the labyrinth that binds and imprisons that somber, telluric god-beast, the Minotaur; and he makes the wings with wax and feathers, the mechanical tool which, in the imagination of the Ancients, would finally defeat gravity.

This sense of wonder appears in a work long attributed to Aristotle, the *Mechanical Problems.* This is the first text which deals with some standard problems of statics and the science of structures. "Miraculously," the treatise begins,

> some facts occur in physics whose causes are unknown; that is, those artifices that appear to transgress Nature in favor of man. In many cases, in fact, Nature works against man's needs, because it always takes its own course. Thus, when it is necessary to do something that goes beyond Nature, the difficulties can be overcome with the assistance of art. Mechanics is the name of the art that helps us over these difficulties; as the poet Antiphon put it, "Art brings the victory that Nature impedes."

Early works on architecture frequently express this sense of man's marvel over his victory over Nature. The architect not only constructs useful, beautiful buildings; he also exerts dominion over the natural laws, endowing his creation with "vigorous resistance." The three terms *firmitas, utilitas, venustas*, "strength," "utility," "beauty" or "charm"—these imbue the first treatise on architecture, Vitruvius' *De architectura.*

It is hardly surprising that, for the ancients, the image of the architect has demiurgic connotations. In a famous dialogue of Paul Valéry[2] this "nearly divine" aspect is expressed in the following words of Phaedro to Socrates when speaking of his friend the architect Eupalinos: "How marvellous, when he spoke to the workmen! There was no trace of his difficult nightly meditation. He just gave them orders and numbers." To which Socrates responds, "God does just that".

Orders and numbers: orders signify the design and technical decisions, and numbers symbolize the harmony and coherence that only the universal language of mathematics can express. These two, combined with an appropriate technology and perfect communication among those involved, would, according to the story of the Tower of Babel, be so successful as to link heaven and earth and threaten God Himself. "And the Lord said, Behold, the people is one, and they have all one language; and this they begin to do: and now nothing will be restrained from them, which they have imagined to do" (Genesis 11:6).

Although it is hazardous to try to draw too much from myth and poetic figures of speech, these glimpses from the remote past tell us something about the relations that originally united statics (the mechanics of materials and structures) with technique and art. Today we think of the themes and problems that rational mechanics applies to structures and materials as rather marginal to the objectives, design problems and outcomes of architecture. Architects study stress in building elements and learn the rules for correct dimensions, but as merely technical formulas—sophisticated formulas, of course, and obviously essential for structural purposes, but (with rare exceptions) somehow peripheral to the real essence and meaning of architecture itself. And to science, these formulas are only one feature of a far broader horizon. Architecture and science meet only on the periphery of each.

But it was not always so; the historical reality is far different. Consider the evidences quoted above: from Babel to Daedalus, from Schopenhauer's words to Valéry's thought. We should look for a different understanding, one which adheres better to history itself. Mechanics and architecture do not meet on the fringes of each discipline but at their very hearts.

The journey we are about to undertake will explore extraordinarily interesting questions about technical mechanics. We will explore not only the history of science but the history of ideas and architecture—for the separation of these branches is by no means so certain as we may think and is of comparatively recent vintage. The history of construction shows us how debatable is the subordination of technique to theory; in fact, in most of the cases we will look at, the theory arose from technique, not the other

[2] P. Valéry, "Eupalinos ou l'architecte," in *Œuvres*, Vol. 2, Bibliothèque de la Pléiade, Éditions Gallimard 1960, p. 83.

way around. The scientific explanation often came to the fore at the end of a long journey—that of constructive techniques—whose origins are lost, as Koyré puts it, "in the mists of time"; and whereof it was presented in the form of rational acknowledgment of what was known but not understood, as Minerva's latch which, according to Hegel, lifts at nightfall.

We find, in fact, that ancient techniques slowly arrived at satisfying levels of complexity and perfection long before theory caught up with them. And theory evolved not so much because the techniques needed an intellectual underpinning as because of individuals' curiosity. The theorists wanted living proof of the excellence of their theories; the technicians knew what worked and were often not much interested in *why* it worked. The Dome of St. Peter's Basilica sprang heavenward without the benefit of theory. It not only preceded mathematical analysis but begot it.

The division between inspiration and technique is of very recent origin and is largely artificial. In buiding, science and art have always been united in the creative act. Not even the most narrow-minded aesthete or engineer can part the two without losing something. To see Brunelleschi, Michelangelo, Guarini, Wren, Mansart, Soufflot, a hundred others, *merely* as great artists is to deprive them of credit for their brilliant engineering. Their wonderful technical innovations, their perfect determination of the weights that had to be balanced and the mechanisms of collapse that had to be opposed—these give coherence and splendor to their works.

Returning to Schopenhauer in seeing "the battle between weight and rigidity" as "the single theme of architecture," we find him discerning in statics not a means but an end, not an instrument but a meaning. Of course he goes too far; his language tries to fix the concepts with an overlay of terms: "weight," "indestructible force," "effort," "rigidity," "resistance," all taken from ordinary language but all ultimately indefinable. Schopenhauer, who was anything but a scientist, may be excused a certain excessiveness and imprecision. In this, however, he joins a large and respectable company; part of the history of statics is the process of compacting and defining its vocabulary, as we shall see.

Look at his images. The first is of the "battle between weight and rigidity"; we can translate this into the *resistance of solids*, a theme first broached by Galileo and subsequently expanded, enriched and transformed into both reasons and models. The second image is of the "tortuous ways" by which gravity has to express itself when it is "hampered by architecture." We now express these in terms of the composition and decomposition of forces, a theme we shall discuss in some detail, and we shall see their practical application to the *statics of arches, domes and vaults*. Schopenhauer's third image is of the intrinsic "finality," the "consistency of the whole," the "suitability of every part," such that the loss of a single part "would plunge the building into ruin" —the whole field of *structural mechanics*, in short. The definition of the laws of equilibrium is seminal; it lies at the basis of all these applications to architecture. As we shall show, since ancient times, the *principles of statics* were interpreted as a consequence of geometric and metaphysical axioms, rather than as merely physical laws. Their truth was located beyond the range of empirical knowledge, almost the trace of that "one language" which pervades every rational explanation of reality.

During the eighteenth century, under the influence of rational optimism, the conviction arose that the laws "of the repose and movement of bodies" were in turn subordinate to a great universal design, one that manifested the beauty, harmony and perfection of Nature as the best work of the Supreme Architect. This was hardly a novel notion; "final causes" were a legacy from Aristotelianism until they were ousted, after long and vigorous combat, by the "efficient causes" beloved of post-Renaissance science. The novelty lay in the translation of "final cause" into a minimum principle, innocent of teleonomic intention.

These four themes form the subject of this book. We shall trace them from their origins to the threshold of the modern age, in which much is still under discussion and still enlivens scientific research. Our intention is to go by distant and almost-forgotten routes, some of them hardly more than footpaths. We will examine forgotten premisses and ancient errors—gently, we hope, and without prejudice or blame. As Leonard Woolf put it, "the journey not the arrival matters."

Today the science of structures has a formidable air of perfection. Everything is logical; all is related to the great deductive systems of rational mechanics and mathematical physics. Nothing, apart from the name, seems to have much to do with architectural applications. Of course the science of structures *does* apply to real buildings, it does have an empirical basis, but this seems rather limited compared to its luxuriant, unbelievable theoretical refinement—a refinement that has given coherence, harmony, order to the parts and the whole of the discipline, revealed new formal analogies, widened the range of problems resolved, and pared down the languaged used to treat them. Socrates' "Order" and "number," the Biblical "one language": we have these now, and they demonstrate the power of the original architectural act.

But how did we come by them? We shall see.

Part I

The Principles of Statics

1

Methodological Preliminaries

1.1 The Special Objects That Gave Rise to Mechanics

The history of many branches of science follows the same generalized pattern: knowledge grows by the expansion of experimental techniques and the refinement of mathematical (or non-mathematical) formalization. Someone asks a question, which leads to the development of the techniques needed to approach the question. We harvest the data and come up with a rule, a relation, a law, a generalization. Then we get on with the next question. This pattern allows historians to work neatly from the earliest confused and incomplete discoveries to the later, better defined ones, finally reaching the present day when the old premisses have found their conclusions and the new questions have been stated.

There are, however, numerous instances in the history of thought which do not fit this pattern. Such is the case of metaphysics; its evolution is based not on a series of new discoveries and questions, but on the perennial conflict of interpretations which always concern the same original queries.

The history of mechanics is similar. It, too, will not fit tidily into a narrative model based on the growth of empirical knowledge. The first data establishing the science of mechanics did not derive from pure experimentation, but from practical experience, and have changed little over the centuries. But their immediacy does not simplify matters—quite the contrary. As Carnot observed, "these facts are too familiar to permit us to know up to what point, without them, reason alone could state its definitions."[1]

To clarify these "elementary truths" is the very aim of mechanics. Its goal is not so much to ascertain the facts of experience as it is to interpret them in the light of evident principles. It is, however, difficult to know whether such principles are simply based on the observation of facts, as physical laws are claimed to, or rather whether they involve interpretive assumptions.

For instance, a body stays in equilibrium when the forces acting on it balance each other. This is a proposition whose logical role is uncertain: does it really describe a "fact"? Common observation shows us innumerable situations in which a body is at rest, but description soon gives way

[1] L.M.N. Carnot, *Principes fondamentaux de l'équilibre et du mouvement* (Paris, 1803), p. 5.

to interpretation. We analyze the various circumstances which accompany the resting position of the body, reducing them all to the action of one single, protean entity: force. The presence of a weight, the tension of a wire, the point bearing on a surface, the cohesion among the particles of a solid, the effects of a deforming stress, etc.—all these can be interpreted as manifestations of active and reactive forces, internal or external, and could therefore be included in the mathematical formulation of equilibrium. But the physical reasons which could relate them through the concept of force are not independent of a principle of balance; they are a consequence of it, and perhaps an implicit statement of it.

Experience, therefore, is not of much use in discovering new facts which could confirm, or at least coordinate, laws. It does come in handy, though, as a means of recognizing, in the most commonly known facts, a means towards interpretation from principles. This is precisely why early theoretical thinking about statics and mechanics took as its references particular *objects*, things like the lever, used since ancient times as necessary tools. Such implements began to be thought of as special, carrying a new meaning, a hidden *fragwürdig/denkwürdig* (questionable/memorable) message, as though they were specially labelled.

The entire history of mechanics up to the close of the eighteenth century can probably be told in terms of particular objects which embody a concept—which give a physical image to a line of thought and testify to a principle [Figure 1.1]. From Hellenic times to the present day, the balance and the lever have symbolized momentum and all the changing interpretations of it, as well as early intuitions about the principle of virtual work (or velocities) and the equilibrium laws.

During the seventeenth century the range of such objects widened; at the same time, the criteria for including them in the range became clearer and their special aspects were better recognized. The language describing them became richer and more selective, with stricter rules which began to anticipate theory. Stevin's work, for example, shed a new light on the inclined plane, which became the focus for deep mechanical truths and, above all, for the decomposition of forces according to the parallelogram rule. From this, it was only a short step to the recognition that a weight suspended by a system of ropes embodies the principle of the composition of forces. We can see how the funicular polygon studied by Stevin and Roberval leads to today's concept of statics, which was concluded by Varignon with his *Nouvelle mécanique*.

Guidobaldo's able treatment of the pulley opened the door to a wider and more definite knowledge of the virtual work principle (or the principle of virtual velocities) finally stated by Johann Bernoulli. Lagrange mentions a system of pulleys in his *Mécanique analytique* in order to give a geometrical demonstration of this principle.

If we look at dynamics, we see other examples from practical experience—for example, the falling weight or the missile from a crossbow or cannon,

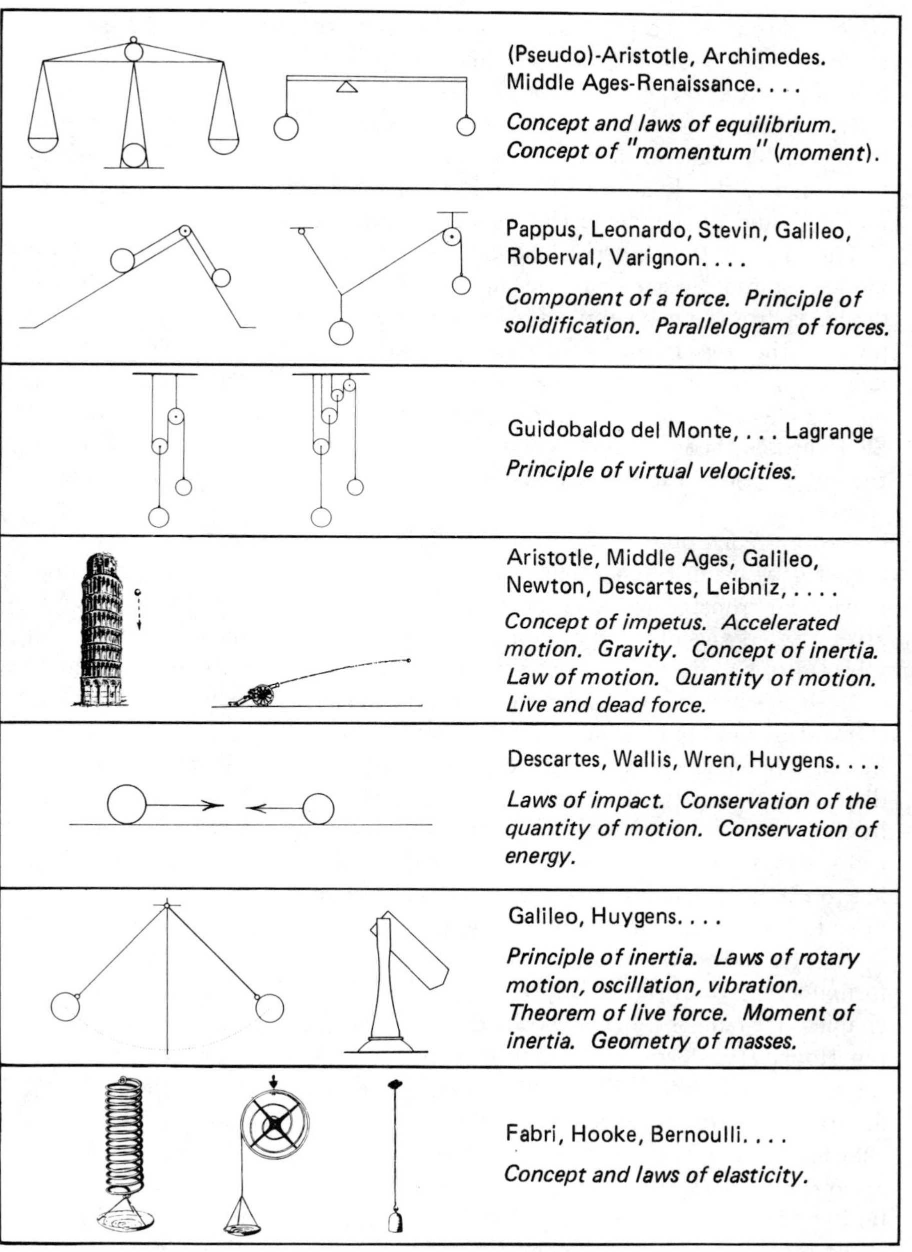

FIGURE 1.1. The special objects that gave rise to mechanics.

which led to the concept of impetus. These physical phenomena were recognized as early as the Middle Ages, and later became Galileo's model for the principle of inertia and for studying accelerated motion. The falling body initiated one of the famous disputes which stirred the scientific community in the eighteenth century: Leibniz's disagreement with Descartes, which finally led to the definition of "live force" as distinct from the "quantity of motion," and to the law of the conservation of energy.

The case of two spheres knocking against each other on a horizontal plane inspired Descartes to the outline of that sovereign principle which descends from the heights of Aristotelian metaphysics to commonplace physics, the conservation of motion. In London, in 1668, the Royal Society held a contest for the decoding of the laws of impact. Wallis' contribution on non-elastic impact and those of Huygens and Wren (the architect of St. Paul's Cathedral) on elastic impact were later to prove fundamental to the clarification of the subtle distinctions which permeate the principles of conservation.

Another such object: the pendulum. Galileo changed a lamp rocked by a gust of wind in the cathedral of his native Pisa into a source of primal truths; he transformed a crude fact into a scientific one. The pendulum gives rise to the discovery of the laws of rotary motion, oscillation and vibration, as well as to concepts of acceleration and centripetal force which, in Newton's hands, were radically to change the face of mechanics. Newton's third principle, that of action and reaction, came to light in an experiment in which he used a weight connected to a suspension point by a thread. By displacing a suspended weight to a given level and direction, and observing its movement to the same level in the opposite direction, Huygens derived the theorem of live forces. In Huygens' hands, the compound pendulum led to the introduction of new mechano-geometrical quantities which were later to be clarified by the geometry of masses.

Finally, consider the coil or helicoidal spring or (more simply) the deformable rod or wire. In Hooke's work, it acquires a primary role in mechanics. It had long been known that a spring lengthens as it is pulled, but Hooke transformed this mundane observation into a proportion which implies measurability. The *ut tensio sic vis* (extension varies directly with force) law is now a part of our physical alphabet, as is Hooke's observation that the force involved can be measured by means of a corresponding elongation. Today we interpret Hooke's law as the empirical description of a limited class of bodies which possess a particular type of elasticity, and we make practical use of dinanometers and spring balances. But we have not fulfilled what was probably Hooke's original intention: to derive the definition of static force from measuring elongation according to an operative concept.[2]

---

[2] P.W. Bridgman, *The Logic of Modern Physics* (New York, 1927).

To "rescue the phenomena" (σῴξειν τὰ φαινόμενα) —that is, to give Hooke's law an empirical meaning—we must find other ways to define and measure the force itself. To do this presents no problems; there are many such examples. But none of them can fully explain the single most important puzzle in mechanics, the concept we call "force".

1.2 The Enigma of Force and the Foundations of Mechanics

What is force? Leonardo's definition is of little help to a strict scientific formulation: "I say that force is a spiritual virtue, an invisible power, which, through accidental exterior violence, is caused by motion and placed and infused into bodies which are [thus] removed and deviated from their natural use, giving to such virtue an active life of marvelous power."[3] Even today, after an extraordinary development of scientific thought, Leonardo's vision is probably the best we can manage. The notion of an "invisible power," present in bodies but separate from them, and somehow causally connected to motion, is the basis of the first general agreement about the meaning of the word "force". This intuition is imprecise and may well be wrong, and all later thought has tried to discredit it, but it is important all the same.

The history of mechanics is shot through with such vague intuitions. Sometimes they precede and sustain it; sometimes they fetter it; sometimes they provide safe ground for further progress. During its development, mechanics used this "burden of concept" (Hegel's term) in order to give rational consistency to the intuition of force and turn it into something more intelligible.

In fact, the essential feature of mechanics is not the constant and inevitable use of the concept of force, but the recurrent conflict about its interpretation. This confusion was expressed by a profusion of terms—*dynamis*, *vis*, *potentia*, *momentum*, *action*, *energy*, *stress*, and so forth—that at first seemed interchangeable and were only later differentiated before being finally brought together in the light of clarifying laws.

Not even the introduction of these laws eliminated the confusion, since they were themselves conceived intuitively and to fulfill the double intention generally accompanying the enunciation of a physical law: to delineate a concept of pre-existing reality, and at the same time to *connote* that concept by means of the phenomenological properties assigned to it—that is, both to describe the phenomenon and to stand for it.

Many scientists of the last century tried to reduce the concept of force to a purely conventional device, deprived of any physical relevance. Lazare Carnot was, perhaps, the first to transform his criticisms of "force" into

---

[3] Institut de France, manuscript A 34v.

a new understanding of the principles of mechanics. According to Carnot, the subject of mechanics is motion, while force must be rejected as "a metaphysical and obscure notion".[4] A. Barré de Saint-Venant considered the concept of force to be an "occult intermediary" which should be omitted from empirical sciences.[5] A number of solutions, suggested by (for example) E. Mach, G. Kirchhoff and H. Hertz in Germany, by de Saint-Venant and H. Poincaré in France, and by W.K. Clifford, P.G. Tait and K. Pearson in England, featured the common purpose of reforming the principles of mechanics so as to transform the notion of force into a purely intellectual construct.[6] On the other hand, the breadth of the synthesis needed to make such a reduction created serious reservations, and led such scientists as F. Reech and J.F.C. Andrade to make contrary proposals, while others such as E. Du Bois-Reymond took skeptical views. In a famous booklet of 1880, Du Bois-Reymond succeeded in distinguishing the enigma of force as the first problem against which science finds itself powerless.[7]

The dispute has never been resolved at this philosophical level. A well known essay by Max Jammer, *Concepts of Force*, presents numerous arguments and quotations to show that any attempt to abolish the notion of force by reductions will fail to eliminate the problematic nature of the concept. The "ontological *status* of 'force'"—he says—"remains an open question even today," especially with reference to the new views on physics (such as the theory of relativity or the theory of interactions in atomic and nuclear systems) that emerged in our century.[8]

---

[4] Carnot, *Principes fondamentaux*, p. 47.

[5] A. Barré de Saint-Venant, "Mémoire sur les sommes et les différences géometriques, et sur leur usage pour simplifier la mécanique," *Comptes rendus*, Vol. 20 (1845), p. 624.

[6] E. Mach, "Über die Definition der Masse," *Repertorium der Experimentalphysik*, Vol. 4 (1868); G. Kirchhoff, *Vorlesungen über* Mechanik (Leipzig, 1876); H. Hertz, *Die Prinzipien der Mechanik in neuem Zusammenhang dargestellt* (Leipzig, 1894); A. Barré de Saint-Venant, *Principes de Mécanique fondés sur la Cinématique*, (Paris, 1851); H. Poincaré, *La Science et l'Hypothèse* (Paris, 1902); P.G. Tait, "On Force," *Nature*, Vol. 17 (1869); K. Pearson, *The Grammar of Science* (London, 1869); W.K. Clifford, *The Common Sense of the Exact Sciences* (London, 1885).

[7] F. Reech, *Cours de mécanique d'après la nature généralement flexible et élastique des corps* (Paris, 1852); J.F.C. Andrade, *Leçons de Mécanique Physique* (Paris, 1897); E. Du Bois-Reymond, *Ueber die Grenzen des Naturerkennens: Die Sieben Welträthsel* (Leipzig, 1881).

[8] M. Jammer, *Concepts of Force. A Study in the Foundations of Dynamics* (Cambridge, Mass.:Harvard, 1957), preface to the Italian edition (Milan, 1971), p. 7. In fact, relativity and quantum mechanics—the favorite resort of philosophers and historians who welcome "crises" and "revolutions" as an excuse for failure to face the basic problems that beset students of classical mechanics—have nothing to do with the matter. The history of mechanics is the history of classical mechanics, to witness some recent, additional conceptual difficulties.

This opinion is not surprising: it corresponds to an interpretative key of history which today dominates the "philosophy of science." We may summarize this "philosophy" as follows: science grows according to *two laws* which lead to different directions, but in history are somehow interlaced. The first law is represented by the image of *Apollo's arrow.*[9] The arrow hits the quarry and kills it: in its practical development, science succeeds in solving a lot of particular questions and therefore these questions disappear as such. The second law is represented by Paul Valéry's simile of a *land of ancients' Inferno*,[10] where Danaides, Ixions and Sisyphes eternally labour, filling bottomless hogsheads and lifting weights which always fall again: in its theoretical development throughout the centuries, scientific work seems to be concerned with defining again and again a small list of basic words which give form to the foundations of theories.

A two-fold task is assigned to a general "history of science": on one side, to record in chronological order the experimental discoveries and the solved theoretical questions which provide modern science and technology with answers useful to man's needs; on the other side, to demonstrate that these particular achievements have not put an end to the work of Sisyphes and Danaides, but made their work more and more tiring, burdening the same basic vocabulary of science with ever new claims and aims.

A similar way of thinking can be found in many works of contemporary historians and philosophers of science. It also permeates the above mentioned essay by Jammer: "force" would belong to that small list of words about which questions cannot be answered but merely shifted. Thus—as Clifford Truesdell points out, with reference to another of Jammer's books—the author "seems to be content with quiet juxtaposition of conflicting opinions."[11] We may wonder however what connection there may be between the *real* history of mechanics and this philosophical representation. It is true—as R. Dugas put it—that "few sciences required such an effort of human spirit as mechanics did: the conquest of some axioms has lasted more than two thousand years."[12] But to deny that this goal has finally been attained in the present century, would ignore one of the most important *events* in the history of mechanics. This event, developed over a long period, throws light on the innumerable attempts of the past so as to give the entire theory of mechanics a rigorous and abstract formulation.

We refer to the line of thought which began at the end of the nineteenth century especially in the field of pure mathematics, in view of a deeper clarification of the fundamental concepts by axiomatic expression of the formal

---

[9] See G. Colli, *La Nascita della Filosofia* (Milan, 1975), pp. 39–46.

[10] P. Valéry, "Poésie et Pensée Abstraite," in *Œuvres*, Vol. 1, Bibliothèque de la Pléiade (Paris, 1957), pp. 1316–1317.

[11] C. Truesdell, "Max Jammer's *Concepts of Mass in Classical and Modern Physics*," *Isis*, Vol. 54 (1963), pp. 290–291.

[12] R. Dugas, *Histoire de la Mécanique* (Neuchâtel, 1950), p. 11.

principles that govern them. The prominent contributions of David Hilbert to the axiomatisation of geometry are well known: in his essay of 1900, the *Grundlagen der Geometrie*, the primitive concepts, "point," "straight line" and "plane," relinquish the vagueness of their intuitive roots and become rigorously related through a set of axioms that "*describe some basic facts homogeneous with our intuition.*" In fact, we must refer to Hilbert's famous lecture delivered the same year at the International Congress of Mathematicians held at Paris. This lecture traced, with admirable insight, the new goals that should be pursued by mathematical research from the beginning of the century onwards; it was later published in the *Archiv der Mathematik und Physik* with the title "Mathematische Probleme."[13]

The *Sixth Problem* proposed by Hilbert concerns our subject: "The investigations on the foundations of geometry—he said—suggest the problem: *To treat in the same manner, by means of axioms, those physical sciences in which mathematics already plays an important part; in the first rank are the theory of probabilities and mechanics.*" Hilbert was well acquainted with earlier studies of the foundations of mechanics. In his lecture he mentioned the treatises of Mach, Hertz, Boltzmann, and Volkmann: "[they are] meaningful in investigations—he added—from a physical point of view; it would be very desirable that also the mathematician would discuss the foundations of mechanics.... If geometry is to serve as a model for the treatment of physical axioms, we shall try first by a small number of axioms to include as large a class as possible of physical phenomena, and then by adjoining new axioms one after another to arrive gradually at the more special theories.... The mathematician will have to take account not only of those theories coming near to reality, but also, as in geometry, of all logically possible theories, so as to reach a general and all-comprehensive view on the consequences which an assumed system of axioms properly entails."[14]

This challenging program was to be followed by numerous mathematicians during our century, even if some of them did not make a specific reference to Hilbert's approach. Worth mentioning are the contributions of such authors as G. Hamel, P. Painlevé, M. Brelot, A. Bressan, and others whose works are carefully listed in the "Additional Bibliography P: Principles of Mechanics," at the end of C. Truesdell and R. Toupin's treatise of 1960, *The Classical Field Theories.*[15]

---

[13] D. Hilbert, "Mathematische Probleme," *Archiv der Mathematik und Physik*, 3rd Ser., Vol. 1 (1901), pp. 44–63, 213–237; (reprinted in Hilbert's *Gesammelte Abhandlungen*, Vol. 3 (Berlin, 1935), pp. 290–329).

[14] *Ibid.*, pp. 62–63.

[15] C. Truesdell, R. Toupin, "The Classical Field Theories," pages 226–793 of Volume III/1 of Flügge's *Encyclopedia of Physics* (Berlin-Göttingen-Heidelberg, 1960), pp. 788–790.

But the true turning point in the history of the axiomatisation of mechanics, according to the line traced by Hilbert, must be located in 1957, when Walter Noll wrote his (unpublished) report *On the Foundations of the Mechanics of Continuous Media*," at the Carnegie Institute of Technology (Rep. No. 17, U.S. Air Force Office of Scientific Research). Some of the main results there obtained by Noll—especially on the subject of forces: mutual forces and Newton's third law—were first published by Truesdell and Toupin in their outstanding treatise mentioned above.[16]

This was only the first step. The same year (1957), Noll presented his new approach (again relative to continuum mechanics) at the Symposium held at Berkeley on *The Axiomatic Method, with Special Reference to Geometry and Physics*, whose Proceedings were published in 1959.[17] "I want—he wrote in the "Introduction"—to give here a brief outline of an axiomatic scheme for continuum mechanics, and I shall attempt to introduce the same level of rigor and clarity as is now customary in pure mathematics." The basic terms of mechanics ("body," "motion," "force," "dynamical process," etc.) form the subject of the various sections. Noll succeeds in freeing the corresponding concepts from their imaginative or metaphysical obscurities: they become *mathematical objects* and therefore perfectly defined and clear, for—as Hilbert put it in his "Mathematische Probleme"—"*in mathematics there is no ignorabimus!*"[18]

Later on, in his Bressanone lectures of 1965,[19] Noll perfected his system of axioms for classical mechanics: essentially the same, but cleared of restriction to continua. As Truesdell comments on this text:

> There the set of all bodies B, C, etc., called a *material universe*, is simply a Boolean algebra. B and C are *separate* if their meet is the null body. The axioms of forces concern a vector-valued function f of pairs of separate bodies: $f(B, C)$ is the force exerted by C on B. The mass $M(B)$ of B is the value of an abstract nonnegative measure defined on the material universe. Noll's method and results follow Hilbert's prescription: by "a small number of axioms to include as large a class as possible of physical phenomena." Special choices of material

[16] *Ibid.*, §196A, pp. 533–534.

[17] W. Noll, "The Foundations of Classical Mechanics in the Light of Recent Advances in Continuum Mechanics," in *The Axiomatic Method, with Special Reference to Geometry and Physics* (Amsterdam, 1959), pp. 266–281; reprinted in W. Noll, *The Foundations of Mechanics and Thermodynamics—Selected Papers* (Berlin-Heidelberg-New York, 1974), pp. 32–47.

[18] D. Hilbert, *op. cit.*, p. 52.

[19] W. Noll, "The Foundations of Mechanics," in G. Grioli and C. Truesdell (ed.), *Non-Linear Continuum Theories* (Rome, 1966), pp. 159–200; reprinted with some improvements in W. Noll, "Lectures on the Foundations of Continuum Mechanics and Thermodynamics," *Archive for Rational Mechanics and Analysis*, Vol. 52 (1973), pp. 62–92; see also W. Noll, *The Foundations...*, *cit.*, pp. 294–324.

> universe correspond to different domains of mechanics. Again, this is Hilbert's prescription: "by adjoining new axioms" we arrive at "the more special theories".... There can be no doubt that the treatment there provides *a solution of Hilbert's Sixth Problem* in reference to classical mechanics.[20]

Noll's approach to the foundations of mechanics is now widely known: in several countries it is the basis for university teaching of rational mechanics. For this reason, and because of the level of mathematical abstraction which is required for a deep understanding of this matter, we shall not dwell on the subject. The reader who is not acquainted with these new advances of rational mechanics may find a rigorous and very clear treatment in Truesdell's textbook *A First Course in Rational Continuum Mechanics.*[21]

Let us now go back to the "enigma of force." It remains true, perhaps, that if we wish to grasp the "ontological status" of this physical entity, we must repeat what G. Hamel wrote to Truesdell in a letter of 1952: "in the concept of force lies the chief difficulty in the whole of mechanics."[22] On the other hand, the axiomatic scheme we have mentioned above allows us to leave aside any inquiry into "ontology." Thus, to the questions: *What are forces? How are forces determined?* a suitable answer can be given. "The answer desired—Truesdell observes with reference to certain attempts at handling the questions in a philosophical way—is not epistemological or semantic. It is mathematical. Forces are undefined objects like points and lines in geometry. Their properties are mathematically stated. Using those properties, mathematicians can prove theorems about forces."[23]

In these words a clear relation emerges between natural sciences and formal systems: mathematics plays the role of the *terminus medius* in that "great syllogism" (Hegel's words) that allows man to reach a scientific knowledge of nature, having *Experience* as the guide and *Thought* as the creator. In fact,

> [n]ot only does any theory reduce and abstract experience, but also it overreaches it by extra assumptions made for definiteness. Theory, in its turn, predicts the results of certain specific experiments. The body of theory furnishes the concepts and formulae by means of which experiment can be interpreted as in accord or disaccord with it. To *overturn* a theory by the

---

[20] C. Truesdell, *An Idiot's Fugitive Essays on Science* (New York-Berlin-Heidelberg etc., 1984), Essay No. 39, pp. 538–539.

[21] C. Truesdell, *A First Course in Rational Continuum Mechanics*, Part 1, *General Concepts* (New York-San Francisco-London, 1977); first published in French (Paris, 1973) and in Russian (Moscow, 1975); second edition, revised, corrected, and expanded, in press (1990).

[22] C. Truesdell, *An Idiot's Fugitive Essays...*, *cit.*, pp. 523–524.

[23] *Ibid.*, p. 553.

> results of experiment, we seek the aid of the theory itself; in terms of the theory, from experiment we may find agreement which develops confidence in the theory, but *establish* a theory by experiment we never can. Experiment, indeed, is a necessary adjunct to a physical theory; but it is an adjunct, not the master.[24]

This means that the *mathematical model* embodied by a formal theory holds primacy in the development of empirical knowledge:

> Granted that the model represents but a part of nature, we are to find what such an ideal picture implies.... The oversimplification or extension afforded by the model is not error: The model, if well made, shows at least how the universe *might* behave, but logical errors bring us no closer to the reality of *any* universe. *In physical theory, mathematical rigor is of the essence.*[25]

The present "introduction to the history of structural mechanics" cannot dwell upon these last achievements of rational mechanics. Our narrative is concerned with the past, and its major aim is to explain the fascinating development of architecture (structural engineering) and rational mechanics in alternating stimulation: in this field, as well, the primacy of mathematical models and the role played by construction technology as "an adjunct, not the master" will find meaningful evidences.

Nevertheless, in the first part of this volume, we shall direct our attention to the history of the millenary debate that accompanied and surrounded the concept of force—in reference to the principles of statics. As we shall see, this history will allow us to discern the different steps of scientific thought for freeing force from its vaguely intuitive or philosophical background and establishing it as a strictly mathematical object. Turning back to Valéry's simile, we shall acknowledge that Danaides, Ixions, and Sisyphes have been useful for this purpose: the different, though often deceptive, attempts to give the profusion of words related to force an exact meaning, paved the way for the modern construction of mechanics as a new branch of mathematics, in the light of Lagrange's program.

In the above cited "Additional Bibliography P" put at the end of their treatise of 1960, Truesdell and Toupin observed that "[a]ny partial bibliography of work on the concepts and axioms of mechanics from the origins through the time of Lagrange would be misleading. No adequate critical history has ever been written. The remarks on this subject given in treatises or general histories of physics are often mendacious and usually so

---

[24] C. Truesdell, R. Toupin, *op. cit.*, §3, pp. 228–229.

[25] *Ibid.*, §4, p. 231.

incomplete and inaccurate as to be totally misinformative."[26] Our history here aims to contribute (as a very small stone) to filling this lacuna, at least in part.

1.3 Statics as "Science Subordinated to Geometry as Well as to Natural Philosophy"

In this chapter—as, in fact, throughout this book—we will focus our attention on statics, with occasional mention of the principles of mechanics in general. From the time of its Hellenic inception, statics has given rise to some of the most stimulating questions of natural philosophy and exact sciences. "The science of weights is subordinate to geometry as well as to natural philosophy" ("Cum scientia de ponderibus sit subalternata tam geometriae quam philosophiae naturalit, ...") wrote the anonymous medieval author of a commentary on Jordanus de Nemore's *Elementa* (which P. Duhem calls a "*commentaire péripaticien*").[27] The same thesis, similarly expressed, was to be accepted by Biagio Pelacani, Nicola Tartaglia and other scientists at the beginning of the modern age. Statics was placed on a distinguished level with respect to other sciences—somewhere between physical research and pure mathematics.[28]

In the eighteenth and nineteenth centuries, the language changed but the concept did not. The most famous authors gave statics, and even the whole of mechanics, a high place, last among the branches of mathematics and first among the natural sciences. It is in this light that the "grand metaphysical problem" (d'Alembert's words) put forward by the Prussian Academy of Sciences in the eighteenth century must be viewed. The question "whether the laws of statics and mechanics are actually necessary or contingent" was one about which Daniel Bernoulli, Euler and d'Alembert argued, from different viewpoints but with substantially uniform intent. The same spirit pervades the works of Lagrange, Laplace, Poinsot and even Carnot, as well as dozens of others. We have therefore a long line of authorities all inviting us to grapple with the ancient question of the nature of static principles, and, if necessary, to shake off the restriction of scientific empiricism.

Historically, the principles of statics were generally thought to have two attributes: they were simultaneously propositions with empirical relevance and theorems of a deductive system whose axioms were so immediate as to require no specific confirmation. Nor was this the only dualism: either statics could be seen as an extreme and anomalous case of dynamics and

---

[26] *Ibid.*, p. 788.

[27] P. Duhem, *Les Origines de la statique* (Paris, 1905), Vol. 1, pp. 128–134.

[28] N. Tartaglia, *Quesiti et inventioni diverse* (Venice, 1554), p. 82v.

subordinate to it, or alternatively "it is in the nature of things that dynamics is based upon statics".[29]

Before looking at the history of the various principles of statics, it may be useful to try to set up an interpretive hypothesis to guide us through the maze. The crux of the question, as you might expect, is the uncertainty surrounding the nature of force. We can only resolve this uncertainty (or at least set it to one side) if we admit that the principles and equations of statics contribute to the *connotation* of the concept of force. But this is just why these equations and principles are ambiguous; they both determine the identity of force and describe its particular properties. Here, then, is what we can derive: there exists an entity t whose nature is just slightly sketched by intuitive pre-scientific imagination. We can say something accurate about t only if its identity is drawn from given propositions which concern it, and which constitute the principles of the theory ϑ of which t is the object.

Consider the three principles of statics in their simplest formulation. Let n_1 and n_2 be the distances between the fulcrum of a lever and weights t_1 and t_2, placed at the ends of the arms. Then the relevant principle is expressed by the equation $t_1 n_1 = t_2 n_2$. Using $n_1, n_2, \ldots, n_k$ to indicate the virtual "relative" displacements (or velocities) of the points on which forces $t_1, t_2, \ldots, t_k$ are applied in equilibrium, the corresponding principle is given by $\sum_1^k t_i n_i = 0$. Finally, using n_1, n_2 and n_3 to give the cosines of the angles included between a given straight line and the straight lines by which the concurrent forces t_1, t_2 and their resultant t_3 act, the principle of the composition implies that $t_3 n_3 = t_1 n_1 + t_2 n_2$. As may be seen, a term n of a geometric nature and an equation of balance is associated with the term t in every case. To be precise, given n as the geometric quantity, tn represents the entity on which the balance rests. The three principles of statics can therefore be reduced to:

$$\sum_1^k t_i n_i = 0. \tag{1.1}$$

Statics is subordinate to geometry in the sense that, for a system S in which forces t_i $(i = 1, 2, \ldots, k)$ act, the law $i \mapsto n(i)$ is known. It is subordinate to philosophy in the sense that the balance equation, like the principles of statics in general, has an ambiguous role, both rational and factual. A simple general definition of the system S may be obtained by supposing that it is characterized by two parameters ξ, η, such that by varying ξ, the quantities t_i change according to a known law $t_i(\xi)$. Similarly, varying η changes the quantities n_i following the law $n_i(\eta)$. Such a system might include the set of levers for which the ratios between the arm lengths and

[29] L. Poinsot, "Théorie générale de l'équilibre et du mouvement des systèmes," *Journal de l'École Polytechnique*, Vol. 6, No. 13 (1806), p. 234.

between the magnitudes of the weights are identical, or any example of two concurrent forces, in which the ratio of the two forces and the difference between the angles of the straight lines of action are identical. This decision is highly relevant, for if we decide that equation (1.1) is valid for the system so defined (for all values of parameters ξ, η), then the principles of statics are unified in the functional equation

$$\sum_1^k t_i(\xi) n_i(\eta) = 0 \tag{1.2}$$

introduced by L.J. Magnus in 1830 and later studied by other authors.[30] We know that solutions to equation (1.2), whose variables belong to an arbitrary field, may be represented as

$$t_i(\xi) = \sum_1^r a_{ij}\varphi_j(\xi), \quad n_i(\eta) = \sum_{r+1}^k b_{il}\psi_l(\eta), \tag{1.3}$$

where r is an integer between 0 and k, and $\varphi_1, \varphi_2, \ldots, \varphi_r, \psi_{r+1}, \psi_{r+2}, \ldots, \psi_k$ are arbitrary systems of mutually and linearly independent functions. Furthermore, the constants a_{ij}, b_{il} $(i = 1, 2, \ldots, k; j = 1, 2, \ldots, r; l = r + 1, r + 2, \ldots, k)$ must satisfy the condition

$$\sum_1^k a_{ij} b_{il} = 0 \qquad (j = 1, 2, \ldots, r; l = r + 1, r + 2, \ldots, k). \tag{1.4}$$

Conversely, every system of functions in the form (1.3) with the condition (1.4) satisfies (1.2). It is easy to find a geometric representation of the solution (1.3)–(1.4). In fact, it requires that the vectors of the sets $\{t_i(\xi)\}$ and $\{n_i(\eta)\}$, each depending on one parameter, be orthogonal pairwise. It is also easy to recognize that most of the formal properties which statics expresses about forces are contained in the solution (1.3)–(1.4), in the form of its particular specifications.

1.4 Momentum: Fixed Word, Fluid Concept

The quantity *tn* mentioned above has a distinguished name: in Greek, ῥοπή, and in Latin, *momentum*. A recent book by Paolo Galluzzi[31] outlines the

---

[30] L.J. Magnus, "Über die Relationen der Functionen, welche der Gleichung $F_1y\varphi_1x + F_2y\varphi_2x + \cdots + F_ny\varphi_nx = F_1x\varphi_1y + F_2x\varphi_2y + \cdots + F_nx\varphi_ny$ genug tun," *Journal für die reine und angewandte Mathematik*, Vol. 5 (1830), pp. 365–373; cf. J. Aczél, *Lectures on Functional Equations and their Applications* (New York, 1966), p. 160.

[31] P. Galluzzi, *Momento: Studi galileiani* (Rome, 1978).

extraordinary and eventful history of this concept. His analysis goes back to biblical times, and discusses the Hebrew term *šaḥaq*, which means the fine "dust" added to the weight on a scale, so light that it fails to disturb the scale's balance (see Isaiah 40:15). In Latin, *momentum ponderis* signified the tiny weight that just offsets the scale, however slightly, that next-to-nothing which causes a slight inclination, while *momentum temporis* suggested the temporal instant, the indivisible unit of time. Passages in works by Albertus Magnus and Bonaventura da Bagnoregio also refer to these usages. In his *Mecaniche*, Galileo refers to momentum as "minimum weight."

But there is another tradition as well. Archimedes' *Equilibrium of Planes* and Eutochius' *Commentary* (in its various medieval and early modern editions) use the term ῥοπή in a more modern sense. It is understood to be a tendency for objects to move by natural motion, with a velocity proportional to their weight. With one exception (a 13th-century Latin text in the Vatican, ascribed to William of Moerbeke, where ῥοπή is transliterated as *reptio*) Latin authors generally translated Archimedes' term as *momentum*. The two concepts—momentum as the effect or efficacy of a weight, and the infinitesimal element or instantaneous change—are not, even now, completely divorced: Italian preserves such words as *repente*, *repentino* (sudden) which derive from the Greek ῥοπή.

Consider the weight. Whatever its real effects, it may be strictly defined only by considering its position—for instance, its distance from the fulcrum. This adds another element to the definition of momentum, one which leads to the definition of the mechanical moment. Francesco Maurolico's *De quantitate sermo noster* (1554) incorporates the additional shading:

> The body therefore acquires its own weight according to quantity and quality: the weight then receives its *momentum* from the space from which it hangs. So, when spaces are reciprocal to weights, the *momenta* are identical, as was demonstrated by Archimedes in his book on the equilibrium of planes.[32]

In Vitruvius' book *De architectura*, the word "momentum" occurs only twice, both times with this meaning. Vitruvius uses the concept to explain why a given weight of a steelyard can balance heavier counterweights, depending on its distance from the fulcrum.[33]

The *oscillazioni di momentum* (as Galluzzi calls them) between minimal weight, instant of time, instantaneous change, natural inclination, and the efficacy of weight are also to be found in the Peripatetic tradition, which remained rigidly faithful to textual analysis of Aristotle, as well as in the

---

32 In *Maurolici Abbatis Prologi sive sermones, quidam de divisione artium, de quantitate, de proportione*, G. Bellifemine, ed. (Melphicti, 1968), p. 46.

33 M. Vitruvius Pollio, *De architectura Vitruvii*, F. Krohn, ed. (Leipzig, 1912), p. 234.

writings of the most progressive scientists of the sixteenth and seventeenth centuries. Intention is added to the word "momentum" to indicate potentiality of motion, a "virtuality" which (for that reason) is called *virtus*. For example, Alexander Piccolomini, in his *Paraphrasis* of the Μηχανικὰ Προβλήματα (*Mechanical problems*) gives ῥοπή the meaning of *virtus*. From here, it is a short step to the development of words whose provenance may vary, but whose meaning incorporates the two notions of immanent possibility and immediate capability with regard to motion. Momentum begins to be connected with the "force of the motive soul," "impetuosity," "impulsion" (see for example Giuseppe Ceredi, 1567).[34] It may also be linked to the terms τό κίνημα, and τὸ νῦν, which Aristotle uses to denote motion in its elementary "indivisible" appearance (*Physics*). In Galileo's mechanical works this extraordinary range of meanings reaches its widest. Galluzzi, in his book, follows the metamorphoses of the term throughout Galileo's work. The ambiguity persists right up to the close of the eighteenth century, so much so that Lagrange mentions it in the first chapter of *Méchanique analitique* (Paris, 1788), suggesting that we go back to the "more natural and more general" notion of momentum/moment as virtual work—Galileo's idea, according to Lagrange.

This presents an epistemological case, often overlooked, but of great interest. The word "momentum" goes through several stages of meanings and interpretations, and becomes in itself a puzzle to be solved and a source of connections and analogies. Usually technical terms stay put; they are simple instruments of communication, subordinate to concepts. If the concept changes, terminology shifts to accommodate it. But in the case of "momentum," the *word* persists, a fixed term around which different concepts revolve. This lexical stability deeply influences thought because it makes the term a part of history, welds it to tradition, and at the same time stimulates debate to clarify the definition. This, as it happens, is one of the major routes of scientific research. The subordination of word to concept and the strict, arbitrary assignment of the word to its meaning—a process which characterizes scientific theories, the explanation of phenomena and definition of laws—are, in this case, subject to limitation. The word itself plays a hidden role because of its sheer persistence. It is the custodian of manifold intentions, an object of hermeneutic research, and a spur to historiographic reflection about the *nexus rationum inter se*—the connections between ideas. It takes on both an explanatory and a lawgiving role.

In statics, one can represent the manifold meanings of momentum by the term tn, to which a "judgment" J, in the form of the balance equation (1.1) belongs. This equation, however, only has empirical meaning if we refer to a "pre-judicial" notion of t—that is, to a previous intuitive knowledge of

---

[34] G. Ceredi, *Tre discorsi sopra il modo d'alzar acque da' luoghi bassi* (Parma, 1567).

force. But t is not fully connoted independently from the judgment J on tn; therefore J cannot be recognized as a true physical law. On the contrary; the identity of t is reached by way of $J(tn)$, so that this becomes (at least in part) a definition of t. The puzzle presented by the couple $[t, J(tn)]$ may be outlined in this way: On one side we have the term t, whose meaning is given vaguely by a pre-judicial understanding. On the other, we have the judgment $J(tn)$, asserting an essential property of that term through momentum, tn. To interpret t, we have to elaborate the term tn. The latter represents force in its efficacy, its action—we could even say, its being as it is. The proposition J may be understood in two ways: as a definition of t and as a description of a physical fact by a previous notion of t.

There are similar epistemological situations elsewhere. The binomial *force/momentum* can, for example, be compared to the ontological difference between "Being" and "being as it is," according to Heidegger's *Sein und Zeit*, in which he poses the question of the meaning of Being.[35] A still closer analogy exists between our discussion of t and n and the stimulating treatment proposed by J.D. Sneed in his analysis of the logical structure of physico-mathematical theories. For a given theory ϑ, according to Sneed, it is necessary to distinguish two kinds of terms: the theoretical terms, which correspond to t, and the non-theoretical ones (n, in our discussion). The former express the quantities or entities to whose definition the theory contributes; the latter are concerned with the quantities or entities whose definition derives from theories independent of ϑ. Sneed focuses on the question we presented above: what empirical relevance may be assigned to a theory in which theoretical terms appear?[36]

Unfortunately, the complex answer to this question outlined by the author involves a mainly philosophical interest, but does not seem to correspond "to anything that theorists of mechanics do or ever have done."[37] Furthermore, the application of Sneed's solution to "classical particle mechanics" lays itself open to serious criticisms. However, our narrative of the historical development of the principles of statics will keep an eye on the question posed by Sneed, so as to ascertain its real relevance in the history of scientific inquiry into the foundations of mechanics.

The law of the lever, the principle of virtual velocities and the parallelogram rule—all derive from previous intuitive knowledge of the concepts involved, as received from ancient tradition. We must therefore return to the roots of this tradition, to the first attempts at scientific explanation of nature and its laws. And here we meet Aristotle. It was he, perhaps more

---

[35] M. Heidegger, *Sein und Zeit* (Tübingen, 1927).

[36] J.D. Sneed, *The Logical Structure of Mathematical Physics* (Dordrecht, 1971).

[37] C. Truesdell, *An Idiot's Fugitive Essays...*, *cit.*, p. 574

than any other, who determined the "prejudicial" concepts and the linguistic instruments to which even current scientific thought and epistemology are, at least indirectly, indebted.

1.5 The Aristotelian Roots of a Vocabulary for Mechanics

Very little remains today of Aristotelian mechanics. From the seventeenth century on, mechanics went through a long, difficult struggle to shake itself free from those ponderous axioms, that rigidity, that net of metaphysical argument, which the commentators on Aristotle's works had laid upon it. But the commentators had at least created a common way of thinking and of looking at facts. For this reason, and because of the great influence of Aristotelian *Weltanschauung* on scientific thought for more than a thousand years, we should at least review its characteristic traits.

Aristotle (384–322 BC) approached the fundamental concepts and themes of physics in a manner far removed from the spirit of current research. The real difference lies, not in his hypotheses or the solutions he arrived at, but in the questions he posed and the problems which he found worthy of attention. His physics has no present-day equivalent—no development, no integration into current knowledge, not even an opposite. If it has any place at all, it is in the "philosophy of nature" which still has a place (if a secondary one) in contemporary thought. This philosophy has

> as its object just movement, a mutable being as far as it is mutable; the *being*, then ... but not the *being* as being, or a *being* according to its mystery of intelligibility, which is the object of metaphysics; the object of philosophy of nature is the *being* considered according to the conditions that imprison it in this universe of poverty and division which is the material universe, the *being* according to the true mystery associated with becoming and with mutability of the movement in the space in which bodies are in interrelation, of the movement of generation and of substantial corruption which is the deepest reign of their ontological structure, of the movement of vegetative growing in which the ascent of matter to the order of life reveals itself.[38]

This "definition" of a philosophy of nature (such as it is) comes from a famous essay by Maritain, and derives from Cajetan's *De subjecto naturalis philosophiae*. It reflects, quite faithfully after two thousand years, the general sense of the Aristotelian treatment of nature, as expressed in the

[38] J. Maritain, *La philosophie de la nature: essai critique sur ses frontières* (Paris, 1935), p. 113.

Physics, *On Heaven*, *On Generation and Corruption* and the meteorological and biological works. Today's scientific thought finds it completely irrelevant. Whatever was vital in the problems Aristotle studied has been transferred to other, more appropriate realms of knowledge, especially to epistemology and the philosophy of science.

But we should remember this: that at its very beginnings, using rudimentary experimental and interpretive techniques, science was shaped by a reckless attempt to grasp the whole in a single stroke. Instead of trying to describe and explain physical reality in all its diversity, it aimed at a synthesis so all-embracing as to consider every detail beforehand.

This is evident from the pages of Aristotle's principal treatise, the *Physics* (Φυσικὴ ἀκρόασις) . The subjects discussed in it include the nature of a physical being and the doctrine of the four causes (efficient, final, formal and material). The critics of the Eleatic school of thought and their successors are firmly set "on the path of Metaphysics, which is the study of the first reality."[39] Aristotle distinguishes physics from metaphysics as follows:

A natural being has only the *actuality* that a mobile being can achieve. Nature is composed of form and matter, but the form which interests a physicist is not the pure form which the metaphysicist seeks. In Aristotle's physics, the form acquires the specific meaning of mover and end of what is capable of movement in nature. Only at the end of his speculative path can a physicist arrive at the peak from which he can view the Prime Mover (VIII, 5)—unique, unextended, set at the outermost sphere of the universe, indivisible (VIII, 10), the generator of a perpetual, uniform movement, circular from the beginning of time (VIII, 9). The Prime Mover is the pure form which renders the physical universe intelligible. At this summit, metaphysics permeates physics, gaining roots in reality as well, since the theory of substance developed by metaphysics cannot ignore movement. Thus matter and form are, for Aristotle, reciprocal, eternal and uncreated.

From a scientific point of view, the Aristotelian fusion of metaphysics and physics was, as we know now, a disaster. For example, for centuries its assertion of the priority of rotating motion over rectilinear motion hindered the discovery of the mechanical laws of uniform rectilinear motion with which the Galileian principle of inertia is concerned. The effects of the Aristotelian synthesis on theology have been equally insidious. Even today, the best theologians know well how difficult it is to free theology from illusive but comforting syllogisms "proving" God's existence—syllogisms which reduce the Deity to a super-cosmic predicate.

If we confine ourselves to the text of the *Physics*, we can bypass or even reject the theological tones of book 12 of the *Metaphysics*. In the *Physics* Aristotle concentrates his efforts on the search for what does not change

---

[39] Aristotle, *Physique*, ed. and tr. by H. Carteron (Paris, 1926), Vol. 1, introduction, p. 17.

in natural phenomena, where, at first sight, everything seems mutable. Behind such research there is the ancient enigma of "becoming" which, since Parmenides and Zeno, had obsessed philosophical thought. As a matter of fact, the pre-Socratic school of Elea professed the paradoxical thesis that becoming is a self-contradiction, an absurd exchange of being and non-being. According to this school, all changes—including the movement of bodies, from the flight of an arrow to the race between Achilles and the tortoise—force us to consider that impossible moment in which "what is" is the same as "what is not" and vice versa: that is, the nullification of being and the embodiment of nothing.

Aristotle's great solution to this troubling sophism is splendidly set forth in book 7(Z) of the *Metaphysics*, in chapters 7, 8, and 9. To summarize: Aristotle affirms that the foundation and intelligibility of "becoming" derive from the concept of *substance as a union of matter and form.* Everything which becomes has an efficient cause which is the starting point for becoming. If something becomes, the form (the fulfillment of becoming) derives from matter, which is not a simple absence of form but rather its possibility or potentiality. For instance, a craftsman making a bronze sphere does not produce either the form in itself (the roundness) or the matter in itself (the bronze); he only unites existing form with existing matter. This union is expressed in the substance of the bronze sphere, in the reality by which "becoming" is accomplished. Substance is therefore an act, an activity, completion of an action, the goal of becoming. The union of matter and form into substance thus acquires a dynamic value. At this point, we can identify matter with power and form with act. In this determined and concrete reality of substance, becoming is no longer a nullification of being, an embodiment of nothing, but is instead the union of what is possible (active or passive power or potentiality, in its capacity to produce or undergo change) and what really is (the act, an object's existence).

It might, at first sight, seem that this theoretical buildup of the first concepts of metaphysics has nothing whatsoever to do with "real" physics, as we know it now. But this is not true. Mechanics owes much more to the duality *power/act* than to empirical research or to the technical employment of "simple machines" which was its first specific object. Think, for example, of these simple linguistic facts: "power" is δύναμις (*dynamis*) in Greek and *potentia* in Latin. Act, in Greek, is ἐνέργεια (*energeia*) or sometimes ἐντελέχεια (*entelecheia*); in Latin it is *actus. Potentia*, *dynamis*, *actus*—all terms correspond closely to those of the language of mechanics as it exists today. Through time, they have metamorphosed and acquired new (and sometimes unpredictable) meanings. But they form a sort of connective tissue between past and present.

This is not all. According to Aristotelian concepts, "motion" loses its most striking aspect of the difference between before and after, marked by an addition or subtraction. As Aristotle affirms in book 3 of the *Physics*, motion is the act of a power insofar as it is a power (ἡ τοῦ δυνάμει ὄντος ἐντελέχεια, ᾗ τοιοῦτον) (II, 1, 201a, 10). In other words, motion is what a mobile thing must essentially make real in order to be mobile. Nothing new happens; nothing really changes. If we consider motion, all we do is to recognize, in every given circumstance, a definite relation between the potential and the actual, inferring a perfect congruence. What is actual is always limited by what is possible, but what is possible happens only if there is an efficient cause for such a realization. The passage from potentiality (or power) to act always requires a cause. Something must make it happen. From this derives the axiom which governs the last two books of the *Physics*: "If a thing is in motion, it is of necessity being kept in motion by something" (Ἅπαν τὸ κινούμενον ὑπό τινος ἀνάγκη κινεῖσθαι).[40]

This comes close to being a statical concept of movement. Movement is not a delusion, as the Eleatic school asserted; instead, Aristotle's explanation affirms the perfect correspondence between before and after. Power and an efficient cause presume their product, and act; such an act in turn embraces that power and that cause. The equivalence between mover (cause acting on movable matter) and actual motion is always perfectly verified. Nonetheless, the explanation of movement drawn from this equivalence may not be conclusive. The mover might itself be subject to movement and therefore require another mover. The explanation becomes perfect only if one finds a mover which is not subject to change, an Immobile Mover.

Aristotle's real goal—the last stop on his great journey through the *Physics*—is to demonstrate the existence of such an Immobile Mover. Only in this way can movement be completely and rationally clarified. His crucial assumption for his proof of the existence of one or more Immobile Movers (Metaph. XII, 1073–4) is expressed in the thesis that *the whole existing movement in the universe is eternal and indestructible.* He sets forth three proofs for this thesis in book VIII of the *Physics*:

1. If all things may be described as being either movers or mobile, and if at a certain moment one of them becomes the first mover and another the first moving thing (while in the previous moment there was absolute immobility) then it is necessary to assume a preceding change.

> There must have been some cause for that stopping short of actual motion which constitutes being at rest; so before motion could take place, there must have been some change which prevented that cause from hindering motion. Therefore, before the supposed first change there must have been [this is absurd] another change (VIII, 1, 251a, 25–27).

[40] This and subsequent quotations from the *Physics* are taken from *Aristotle: The Physics*, tr. Philip H. Wickstead and Francis M. Cornford (Cambridge, Mass.: Harvard University Press, 1929).

2. The second proof is similar: If the mover and the mobile thing were first inactive and immobile, and from a certain instant they came into the condition of producing a movement, this implies that before that instant a change must have occurred which allowed them to pass from one condition to the other (VIII, 1, 251b, 1–9).

3. If it is true that "time is the numerical aspect [i.e. the measure] of movement," it follows that a movement cannot have a beginning, just as time has no beginning (VIII, 1, 251b, 10–27).

Aristotle's *Physics*, therefore, presents a principle of conservation which foreshadows by almost two thousand years the debate between Cartesians and Leibnizians. The immobility of the first mover is a tautological reaffirmation of the principle. If movement is eternally conserved in its entirety, then its cause, a mover, must be equally conserved; it is "immobile" (VIII, 6, 258b, 10). Can we give Aristotle credit for prophesying the principle of conservation of energy? The answer is not obvious; it is difficult to decipher the intentions which guided the Stagirite in his analysis of movement.

For example, chapter 5 of book 8 is dedicated to the study of three terms which exist in every movement: that which is moved, that which is moving, and that by which a mover moves. Obviously that which is moved does not necessarily move something else in turn. Therefore the reverse is also true; the mover is not necessarily moved by something else. This suggests the existence, or at least the possibility, of something that does not change—that is, it is immobile—during a movement. It still follows that every "isolated system" (in Aristotle's terminology, "every self-moving mover") cannot change completely. It always has a self-conserving element, its own immobile mover. Aristotle demonstrates this using the above mentioned definition of motion as an act of moving insofar as it is moving, and the thesis that a mover is an act. Since the mover is isolated, its mover cannot be moved in turn by external causes. The mover, therefore, cannot change, and is conserved in its actual state, already existing before the motion (VIII, 5, 257b, 6–20).

The text is obscure, and this interpretation is tendentious. It claims too much, playing a little with the original text. But the words themselves are certain: power and act, embodying a principle of conservation. In the centuries which followed, men continued to make use of the realization that something exists which is self-conserving during motion, and that this something has to do with hypotheses regarding power or force (*dynamis*) or relative to the conservation of an act, or of energy, or of "actual" motion.

1.6 A Short Outline of Aristotle's Physical Principles

Aristotle's huge fresco includes all nature, and his concept of movement applies to all fields of becoming. "Whenever anything changes," he says,

> it always changes either from one thing to another or from one magnitude to another, or from one quality to another, or from one place to another .... Again, in each of these four cases, there are two poles between which the change moves; in substantive existence, for example, form and formlessness; in quality, white and black; in quantity, the perfectly normal and an achievement short of perfection; and so, too, in the case of vection, up and down, or the action of levity and gravity. So there are as many kinds of changes as there are categories of existence. (III, 1, 200b, 33–35; 201a, 3–8)

It is difficult to base science on so protean a concept, to apply the same ideas to the motion of planets and the growth of a plant.

This difficulty is resolved in book 7 of the *Physics*. The first chapter aims at demonstrating the existence of a Prime Mover—at showing, that is, that every chain of cause related to a given motion includes a finite number of cause-and-effect transactions. This demonstration essentially concerns the case of a "local motion," developing in a finite period of time. But this does not reduce its generality, as we shall see. The argument he puts forward is this: if one admits an endless chain of movers for a given local motion, one must also admit that the infinite thing (the chain itself) could carry out, in a finite time, infinite motions. This, Aristotle says, is absurd, because it is impossible that "an unlimited movement is gone through in a finite time by something which is either limited or unlimited" (VII, 1, 242b, 32–33).

We shall not dwell on the value of such a thesis. We shall, instead, note that Aristotle himself does not seem to be fully convinced by it. What is the sense of expressing as a "thing"—a single physical body—the concatenation of causes which explains the motion of a given body? Couldn't each mover be a separate body? An infinite number of bodies moving simultaneously for a finite time—this is not absurd at all.

Here, then, is Aristotle's proposition for a positive demonstration of the Prime Mover: We are to admit, as experience shows us everywhere, that "our series of movers and moved must be either continuously in contact with one another so as to form one single thing composed of them all": ἀνάγκη τὰ κινούμενα καὶ τὰ κινοῦντα συνεχῆ ἢ ἅπτεσθαι ἀλλήλων, ὥστ'εἶναι τι ἐξ ἁπάντον ἕν (VII, 1, 242b, 27–29). By means of this postulate, Aristotle fixes a vision of the physical world which was to last through centuries and is still alive today: concept and physical reality intermingle to such an extent that the logical connections connoting the concept immediately correspond to the material links between the parts of the physical object.

But unlike Aristotle, we are familiar with the notion of *action at a distance*. He finds it mysterious while we find it Newtonian. Not that we understand it; the reasons for the reciprocal physical influence between two bodies separated by an empty space remains an enigma. Mary B. Hesse[41] presents a vivid narrative of the historical evolution of the concept of action at a distance. She traces the repeated attempts either to grasp it or to get rid of it, starting with the "mechanicalism" of Hellenic science, through the medieval tradition, to arguments by Descartes, Newton and Leibniz. She deals with nineteenth-century field theories, as interpreted by Faraday, Maxwell and Hertz, progressing on to Einstein's general relativity theory and finally to the new outlooks of quantum mechanics. Hesse's conclusion is that physics itself may be unable to solve the question definitively, since the problem of action at a distance is not so much one of a particular theory as it is one of "metaphysical structure." This nearly "skeptical" conclusion, however, is not really satisfactory: the concept of action at a distance would rather require a *mathematical* clarification, and this can be achieved in the light of the axiomatic approach to the foundations of mechanics that we have briefly presented in section 1.2.

Returning to Aristotle's text: the second chapter of book 7 resumes his discussion of the previous thesis of contact action and develops it further. His aim is to prove by experience that all types of movers and all kinds of movements have some sort of propinquity or union between mover and moved. Without going into great detail, it is enough to note the type of mover that intervenes in "local motion," the "transporting mover" (the other types are "altering," "accreting" and "diminishing"). According to Aristotle there are four kinds of transport: pulling, pushing (or thrust), carrying (or transfer) and turning (or rolling). "To these four all possible ways of moving can be reduced" (VII, 2, 243a, 17–18). For instance, a collision is only a sharp push. Expansion and contraction, inspiration and expiration and other motions which make parts come closer or draw further apart—all are forms of traction and thrust, and traction and thrust incorporate straight transfer and rolling. Traction and thrust, therefore, require contact (VII, 2, 244a, 4).

These concepts seem almost prescient. They anticipate the principle of local action which modern rational mechanics puts at the basis of the theory of constitutive laws. In the distinction between straight transfer and rolling, we see the germ of the similar distinction in the cardinal equations of motion and equilibrium. Furthermore, Aristotle states that "turning can be resolved into pulling and pushing, for the agent that turns the subject does so by pulling one part towards itself and pushing another part away from itself" (VII, 2, 244a, 1–3). In spite of the vagueness of this statement,

---

[41] M.B. Hesse, *Forces and Fields: The Concept of Action at a Distance in the History of Physics* (Edinburgh, 1961).

we have to admire his perspicacity, his power of abstraction and, in the final analysis, the physical importance of his observation, which seems to anticipate the idea of centripetal and centrifugal forces.

We should, however, keep in mind the fact that the paths of history are tortuous in the extreme. What seems a bright beginning may turn into darkness, and good ideas may be corrupted. In fact, the Aristotelian postulate of contact action created insurmountable difficulties which came to light during the Middle Ages and for centuries hampered the development of modern mechanics. Aristotle himself recognized some of the problems: "If everything that is in motion is being moved by something," he asks himself, "how comes it that certain things, missiles for example, that are not self-movable, nevertheless continue their motion without a break when no longer in contact with the agent that gave them motion?" (VIII, 10, 266b, 28–30). He was stumped by inertia. The solution he found is famous for its sheer wrongheadedness: in the case of projectiles, motion would be maintained by contact with air or water or "other such intermediary as is naturally capable of moving and carrying motion" (VIII, 10, 267a, 4–5). Only John Buridan, in the 14th century, managed to demolish this thesis.

The rest of the second chapter and all of the third chapter of book 7 deal with types of movement alien to mechanics. Only later, in chapter 7 of book 8, does Aristotle clarify and tidy up the plethora of terms contained in the concept of motion, assigning a logical, chronological and ontological primacy to *local motion*—that is, to the motion which concerns mechanics and is its specific object. (Remember the title of Euler's famous work, *Mechanica sive motus scientia.*)

The fourth chapter of book 7 deserves more detailed treatment. In it, Aristotle presents the first systematic examination of criteria for comparing the various types of motion. He declares that direct comparison between motions belonging to different "types" is impossible. It is necessary, therefore, to establish a definite comparison rule, based on identity of form and subject, not on the equivocal similarity of terms. When used to compare various cases of local motion, this rule claims that "equal velocity means passing the same distance in equal time" (τὰ ἐν ἴσῳ χρόνῳ ταὐτὸ μέγεθος κινούμενα ἰσοταχῆ) (VII, 4, 249a, 19–20). Every other aspect of movement can be safely ignored (ἀδιάφορον) since the difference between two movements is defined only by the distances covered and by the time required to cover them. This passage is critical to the meaning of Aristotelian mechanics and its interpretation over the centuries until after the reforms of Galileo and Newton. As a methodological decision, it whisks away every circumstance affecting a local motion in order to concentrate on two intervals, one in space, one in time. In this extreme abstraction, velocity becomes the *only* available term to represent motion. The duality *power/act* can be solved only in terms of velocity. Power (*dynamis*) relates only to the possible velocity, and act (*energeia*) measures only the velocity reached by a moving body. This derives not from experimental observation, but from the structure of the linguistic and conceptual model which Aristotle constructed for the purpose of making motion intelligible.

The fifth chapter of book 7 forms the true basis of all western science. Here Aristotle completes his treatise in mathematical terms, proposing the first physico-mathematical equation ever to appear in the history of science. Once again, he makes a methodological decision in order to define a given level of abstraction in the phenomenological description of motion: "That which is causing motion is always moving something in something and as far as somewhere" (᾿Επεὶ δὲ τὸ κινοῦν κινεῖ ἀεί τι καὶ ἔν τινι καὶ μέχρι του) (VII, 5, 249b, 27). We should consider nothing but the mover, the movable body, time, and (in the case of local motion) the distance travelled. To translate literally:

> by the expression "in something" I mean "within time," and by the expression "as far as somewhere" I mean "according to the measure of the distance traversed" .... Thus, let A be the moving agent, B the mobile, C the measure according to which it is moved, and D the time taken. In an equal time an equal force [*dynamis*], i.e. A, will move half of B for twice C, but for C in half of D; for so the proportion will be observed (οὔτο γὰρ ἀνάλογον ἔσται) . Again, if the same force moves the same body in a given time and over a given distance, then it will move over half the distance in half the time; and half a force will move half a body over an equal distance in an equal time. Let E be a force of half A, and Z a body of half B; they have the same relation of force A to load B, so that they will move over the same distance in the same time [VII, 5, 249b, 28–31; 250a, 1–9].

These famous lines express what P. Duhem calls "Aristotle's axiom." We find them condensed in the *On the Heavens* (book 3):[42] under the action of an equal force (*dynamis*) "the smaller [lighter] body will be moved farther by the same force ... and in fact the velocity of a lesser body is to that of a greater body as the greater body is to the lesser." (III, 2, 301b, 4–5, 11–14) The whole enormous effort of post-Aristotelian exegesis may, perhaps, be represented by this relation. It is at once exceptionally clear and absolutely obscure, persuasive and deceptive, right and wrong, depending on the interpretation.

Aristotle's axiom states two theses:

1. For every possible displacement, the force A and weight B are inversely proportional to the distance covered in a given time.

---

[42] This and subsequent quotations from *On the Heavens* are taken from *Aristotle: On the Heavens*, tr. W.K.C. Guthrie (Cambridge, Mass.: Harvard University Press, 1939).

2. The distances covered in any two motions are to each other as the times or the forces are to each other.

The first of these theses is fundamentally correct, and anticipates the principle of virtual velocity. The second is misleading, not so much because its meaning is obscured by the lack of definition for the term "force," but because of the prejudices that it brought into the field of dynamics, impeding the discovery of the laws of motion.

What were Aristotle's real intentions? It is difficult to say. He refers his solution for local motion to the general explanation he proposes for the problem of "becoming" in nature. (At the end of chapter 5, he claims to extend his formulae to non-local alterations and accretions.) We have to believe that his formulae are a mathematical metaphor for a metaphysical concept, not a way of calculating or predicting local motion. They give mathematical expression to the necessary *proportions* of cause and effect underlying all motion. Aristotle himself emphasises this: "for so the *proportion* will be observed" (emphasis mine). He sees proportion as evidence of conservation, a sign of perfect balance between "before" and "after." It allows us, he believes, to recognize something permanent in things which are mutable, and renders the mutation intelligible.

Aristotle concludes his itinerary in book 8 with the demonstration that the Prime Mover cannot be subject to mutation and remains eternally immobile (VIII, 5). In this way, the principle of conservation finds confirmation in the existence of a suitable entity, one which both causes and sustains the conservation. The physical law comes close to being personified in the form of the first Immobile Mover, whose super-cosmic—perhaps divine—attributes Aristotle outlines. And this was to cause major problems in the evolution of physics, since post-Aristotelian scholars had first to dispose of the notion that no general proposition about reality could be made without inferring the existence of an eternal sustainer of it. This prejudice is perfectly understandable. Think of the Ancients' world, a universe rife with gods, godlings, powers, intelligences and anything else imaginable as a sustainer of universal laws. Modern science was to arise from another cultural context, one in which Christianity had accustomed people to a quite different concept of divine intervention and which rejected mythological explanations of natural phenomena. Given this change, the existence of a necessary and universal law may be asserted *in itself*, without needing someone or something to be responsible for its continuance.

The development of *mechanica sive motus scientia* in relation to the principle of conservation is important from this point of view. Aristotle's First Mover comes very close to God, and medieval thought seized gratefully upon the concept. The modern mechanics of the seventeenth and eighteenth centuries secularized the idea; instead of a sustainer, we have a mechanical entity for which the thesis "something is conserved" is valid (energy, perhaps, or motion). Finally, from the nineteenth century onward, the principle of conservation itself has become a means for describing the features of the something which is conserved. Instead of assigning that something to the field of mechanics, we have learned to accept that we have to search outside this science for the terms appropriate to the unfalsifiable thesis of conservation. We have the rules, the mathematical relations, but the heart of the problem has slipped out of our field of inquiry.

1.7 Modern Metamorphoses of the Immobile Mover: Towards the Principle of Conservation

Let us examine this historical evolution. The mechanical inquiry into a principle of conservation starts with the Aristotelian binomial *power/act*. Surely one of these terms must contain the element which is conserved, or is the basis for conservation. As applied to local motion, the binomial lost its ontological attributes and became secularized in the force/actual motion couple. We can trace this line of thought back to Galileo's formulation of the principle of inertia (in *Discorsi e dimostrazioni matematiche*, 1638) and particularly to Descartes. Descartes proposes that God is the primary cause of motion, and that He always maintains an equal quantity of it in the universe.[43] With reference to statics, he formulates "the only one principle of all machines": that "with the same forces by which a 100-pound weight can be lifted up by two feet, a 200-pound weight can be lifted one foot."[44] Both statics and dynamics have conservation principles of their own. Attempts to explain the phenomenon of impact (Wallis and Wren in 1668, Huygens in 1669 and 1700) are based on the same perception.

Then Leibniz comes upon the scene. He disagrees with "Descartes' memorable error" and proposes his own thesis.[45] He does not quarrel with the concept, only with its mathematical expression, replacing the quantity of motion (mass times velocity) with mass times the square of velocity (*vis viva cum motu actuali conjuncta*).[46] This difference was to divide the scientific community for the better part of a century.

Today it is hard to see why. A good many historians are ready to ridicule the quarrel, seeing it as a tempest in a teapot. But this is wrong; the eighteenth century saw the uniqueness of the principle of conservation as a

---

[43] R. Descartes, *Principles of Philosophy* (first published 1644), tr. V.R. Miller and R.R. Miller (Dordrecht, 1983), part 2, no. 36.

[44] R. Descartes, *Opuscula posthuma*, (Amsterdam, 1704), p. 13.

[45] G.W. Leibniz, "Brevis demonstratio erroris memorabilis Cartesii et aliorum circa legem naturae in conservatione quantitatis motus," *Acta Eruditorum Lipsiensibus*, Anni 1686, pp. 161–163.

[46] G. Leibniz, "Specimen dynamicum," *Acta Eruditorum Lipsiensibus*, Anni 1695, pp. 145–157.

necessary premiss which gave consistency to a unified concept of force and thus to the entire system of mechanics. The Leibnizians, in distinguishing "live" forces (*vis viva*) from "dead" ones (*vis mortua*), resurrected the dichotomy between power and act, under the common denominator *vis*. *Viva* and *mortua*, life and death—the terms are reminiscent of the ancient enigma of "becoming," shifting from being to non-being. But the contradiction between them is nullified by *vis*, which conserves and presides over compensation.

But what law describes this compensation? Christiaan Huygens, in studying the compound pendulum in 1658, derived an extraordinary equation:

$$\text{force} \times \text{displacement} = \frac{1}{2}\text{mass} \times \text{velocity}^2.$$

Lagrange came to the conclusion that Huygens' equation was the first expression of a fundamental law of live and dead forces (*Méchanique analitique*, 1788). In 1748, Daniel Bernoulli presented the same law "in a general sense," later applying it to the study of hydrodynamics.[47]

Note, though, that the word "force" is still ambiguous. If live and dead forces are related, and if dead forces are mistaken for those that predominate in statics, then trying to establish a parallelogram rule for the composition of forces only yields inextricable paradoxes. Hermann and (more importantly) Bülffinger —faithful Leibnizians both—show this in the Acts of the Academy of St. Petersburg.[48] The solution, put forward by V. Riccati (another Leibnizian) is impeccable, but in fact it derives from his distinction between force and the action exerted by a force to produce a certain slight displacement.[49] It is not force *per se* but the action performed by the force for an infinitesimal displacement, or velocity, that is subject to a principle of conservation. Johann Bernoulli wrote in a letter to Varignon of 26 January 1717 of such "action" by force, calling it energy—the first independent appearance of this word in the modern vocabulary of statics. Previously this science was thought to be concerned only with potential motion and therefore with "powers;" but Bernoulli's statement of the principle of virtual work gave "energy" a new meaning that includes the concept of potentiality.[50]

---

[47] D. Bernoulli, Rémarques sur le principe de la conservation des forces vives pris dans un sens général, *Histoire de l'Académie Royale de Berlin*, 1748; 2d ed., *Die Werke von Daniel Bernoulli*, Vol. 3, Birkhäuser-Verlag, Basel-Boston-Stuttgart, 1987.

[48] J. Hermann, "De mensura virium corporum," *Commentarii Academiae Scientiarum Imperialis Petropolitanae*, Vol. 1 (1726); J. Bülffinger, "De viribus corpori moto insitis et illarum mensura," *Commentarii Academiae Scientiarum Imperialis Petropolitanae*, Vol. 1 (1726).

[49] V. Riccati, *De caussa [sic!] physica compositionis ac resolutionis virium* (Bologna, 1744).

[50] In P. Varignon, *Nouvelle mécanique* (Paris, 1725), Vol. 2, p. 174.

The topic of actual motion needed, first, a suitable measure if conservation were to be proved, and second, a modern term. The evolution of the latter was slow, and led to some dispute. In the English school, the first thinker to revive the ancient term "energy" was Thomas Young, in his *Lectures on Natural Philosophy* (1807). According to him, energy could be expressed as the product of the mass or weight of a body and the square of its velocity. This product had been called the "live" or "ascendent" force, and some had thought it a suitable gauge for measuring quantities associated with motion. Young, however, was of the opinion that the force assessed in this manner needed a distinct determination.[51]

Note that energy has begun to take on a certain consistency. It is no longer simply the second term of the Aristotelian binomial *dynamis/energeia*, nor is it immediately connected with actual motion. It is made to measure for the conservation principle, expressing its true mechanical importance.

But we have not reached the end of our road. The original division of power and act turns up again, this time in the division of the concept of energy into two aspects. As early as 1828, G. Green[52] had introduced the notion of potential function of a system with reference to: "the effect of all the forces acting on a point according to a given direction." In 1853, W.J.M. Rankine termed this "potential energy." Between 1870 and 1880, W. Thomson (Lord Kelvin) came up with what proved to be the accepted expression, "kinetic energy," which he and P.G. Tait call "actual energy" in their great *Treatise on Natural Philosophy* (1867).

History had come full circle. The eighteenth century had fled from Aristotle's dualism and from his assignment of the conservative attribute of the Immobile Mover to *energeia*. The nineteenth century reversed this, accepting energy as conserved and as twofold: potential (concerning forces, usually independent of velocity) and kinetic, or actual (concerning motion, proportional to squared velocity).

Of course the conservation law is not valid for all mechanical systems, since energy can be dissipated in a number of ways. In two letters to Euler (24 May and 8 November 1738) Daniel Bernoulli recognized one such in the non-elastic impact, foreseeing the existence of non-mechanical forms of energy.[53]

The nineteenth century saw the unravelling of the puzzle, coming to the conclusion that the sum of potential and kinetic energy is conserved only

---

[51] T. Young, *Lectures on Natural Philosophy* (London, 1807).

[52] G. Green, *An Essay on the Application of Mathematical Analysis to the Theories of Electricity and Magnetism* (Nottingham, 1828) included in N.M. Ferrers, ed., *Mathematical Papers of the Late G. Green* (London, 1871), pp. 1–82.

[53] Cf. P.H. Fuss, *Correspondance mathématique et physique de quelques célébres géométres du XVIII siècle* (St. Petersburg, 1843), Vol. 2.

in systems subject to special conditions. This is the theory held by such authors as G.G. Coriolis, J.V. Poncelet, S.D. Poisson, L. Carnot and so forth. But these limitations do not invalidate the principle of conservation; on the contrary, it is firmly re-established, overstepping the bounds of mechanics. What seemed to be a law founded on the principles of the discipline became, in turn, a principle for judging and defining the discipline's boundaries.

The history of this change would be superfluous; it is well known, and outside the scope of this book. Poincaré made an interesting epistemological observation in his famous essay *Science et hypothèse* (1902): the more we extend the range of the conservation principle, the more its empirical meaning fades away. In other words, the more general it becomes, the less immediately ascertainable it is. In the case of a simple mechanical system, where the potential energy U is a function only of the coordinates, and the kinetic energy T is proportional to the square of the velocities of the material points, the principle is simple and clear: $U + T =$ constant. "It is true," states Poincaré,

> that if $T+U$ is constant, so is any function of $T+U$, $\phi(T+U)$. But this function will not be the same as the sum of two terms, one independent of the velocities and the other proportional to their squares. Of all the functions which remain constant, only one has this property—$T+U$ itself, or its linear functions (the same thing, in effect). This, then, is what we call energy.[54]

When the system is more complex or involves non-mechanical forms of energy, then the conservation principle adds another term: $T + U + Q =$ constant. T and U have the same meaning as above, and Q represents the internal molecular energy in, for example, thermal, chemical or electrical forms. "All would be well," Poincaré writes, "if these three terms were absolutely distinct."

> But this is not the case. Let us consider electrified bodies. The electrostatic energy ... will evidently depend on their charge—that is, on their state; but it will equally depend on their position. If these bodies are in motion, they will act electrodynamically on one another, and the electro-dynamic energy will depend not only on their state and their position, but on their velocities. We have therefore no means of making the selection of the terms which should form part of T, U and Q, and of separating the three parts of the energy. If $T+U+Q$ is a constant, the same is true of every function whatever $\phi(T+U+Q)$. If $T + U + Q$ were of the particular form that I have considered above, no ambiguity would ensue. Among the functions

[54] H. Poincaré, *Science and Hypothesis* (tr. of *Science et Hypothèse*) (New York, 1952), p. 125.

> $\phi(T + U + Q)$ which remain constant, there is only one that would be of this particular form, namely the one which I would agree to call energy. But I have said this is not rigorously the case. Among the functions that remain constant there is not one which can rigorously be placed in this particular form. How then can one choose from them that which should be called energy? ... there is nothing then but an enunciation [of the principle of conservation of energy]:—*There is something which remains constant.* In this form it, in its turn, is outside the bounds of experiment and reduced to a kind of tautology.[55]

A sad conclusion? Not at all. When science trembles on the brink of tautology—where its verdicts approach the unfalsifiable—then naive empiricism fails. But new prospects open up for the role of language, the pre-eminence of the word compared to the conceptual construct in the growth of knowledge.[56] These themes are difficult and fascinating, but all we can do is to dangle them before the reader, to tease his curiosity, since even to touch on them would involve a lengthy argument.

1.8 The "Mechanical Problems": The Peripatetic Explanation of the Law of the Lever and the Parallelogram Rule

Our analysis of the Aristotelian source would not be complete unless we mentioned the first text in which the philosopher's principles are directly applied to the solution of mechanical problems.

Until the nineteenth century, Aristotle was thought to have been the author of the *Mechanical Problems* (Μηχανικὰ Προβλήματα),[57] and naturally this belief increased the work's influence throughout the Middle Ages. Today, however, this attribution has been discounted as a result of studies carried out by Heiberg, Wohlwill, Zeller and others at the end of the nineteenth century. All the same, the work is of considerable importance because of its technical orientation. It tries to explain in general terms how the simple machines of that period worked, and to condense their laws into a single principle.

Its basis is a sort of wonder. How can mechanical art apparently contradict the laws of nature?

---

[55] *Ibid.*, pp. 125–127.

[56] Cf. E. Benvenuto, *La Scienza delle costruzioni e il suo sviluppo storico* (Florence, 1981), pp. 893 ff.

[57] Cf. Aristotle, *Minor Works*, tr. W.S. Hett (Cambridge, Mass., 1936), pp. 326–441 (hereafter cited as *Mechanical Problems*).

> Among the problems included in this class are included those concerned with the lever. For it is strange that a great weight can be moved by a small force, and that, too, when a greater weight is involved. For the very same weight, which a man cannot move without a lever, he quickly moves by applying the weight of the lever.
>
> Now the original cause of all such phenomena is the circle; and this is natural, for it is in no way strange that something remarkable should result from something more remarkable, and the most remarkable fact is the combination of opposites with each other. The circle is made up of such opposites. ...
>
> Therefore, as has been said before, there is nothing strange in the circle being the first of all marvels. The facts about the balance depend upon the circle, and those about the lever upon the balance, while nearly all other problems of mechanical movement can depend upon the lever.[58]

This approach is obscure and inconclusive, but it is still worth reading. The attention paid to the balance and lever is interesting. Not only are they two technical instruments whose effects are immediately obvious, and their application is extremely wide, but their behavior casts light on the fundamental laws of equilibrium. In fact, the balance and lever were fundamental to the understanding of statics right up to the eighteenth century. Galileo and his followers treated the problem of the bent beam by considering—more or less fruitfully—the principle of the lever. Eighteenth-century thinkers interpreted the behavior of vaults and arches by means of a suitable system of levers. The principles were also used in considering the problem of retaining walls, according to indications in the works of B. Forest de Belidor, C. Bossut, C. Coulomb, etc.

Let us examine—however briefly—the two propositions "demonstrated" by the Peripatetic author of the *Mechanical Problems*: the explanation of the law of the lever, and the composition of movement according to the parallelogram rule.

The author explains the law of the lever in terms of kinematic analysis. The motion analyzed is that of the "mover" and the "moved body" situated at the ends of the lever; not the actual motion (because, in fact, at equilibrium this motion is excluded), but the potential or virtual motion. In terms of the binomial power/act, power has the main role. The equilibrium must be concerned with the powers (i.e., the forces) in which the possibility of motion expresses itself, even if it is only a possibility. Therefore, if the equilibrium has set conditions to which preclude motion ("act of the mobile as such"), it must also include the widest field of possible motions to which the actual motion belongs.

[58] *Mechanical Problems*, 1, 847b, 10 to 848a, 15.

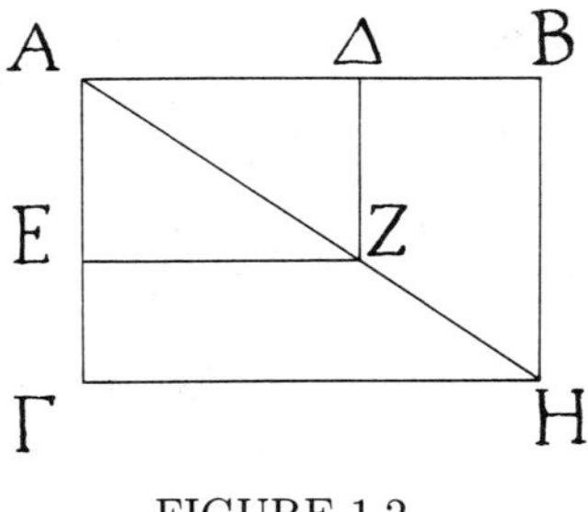

FIGURE 1.2.

The first and twenty-third problems in the *Mechanical Problems* contain references to actual motion, and the author demonstrates the parallelogram rule by means of kinematic considerations:[59]

> Whenever a body is moved in two directions in a fixed ratio it necessarily travels in a straight line, which is the diagonal of the figure which the lines arranged in this ratio describe. Let the ratio according to which the body moves be represented by the ratio of AB to $A\Gamma$ [Figure 1.2]. Let $A\Gamma$ move towards B while AB be moved towards the position $H\Gamma$; now let A travel to Δ, and let AB travel a distance determined by the point E. Then if the ratio of the movement is that of AB to $A\Gamma$, then the line $A\Delta$ must bear the same ratio to AE. Then the small parallelogram has the same proportions as the larger, so that its diagonal is the same, and the body will move to Z. It can be shown that it will behave in the same way at whatever point its movement be interrupted; it will always be on the diagonal. Conversely it is obvious that an object travelling with two movements along a diagonal will always move in the ratio of the sides of the parallelogram.

Note that this proof concerns only movement, not force. From the context, however, we can infer that the kinematic analysis is not an end in itself.

The first question begins by asking "why larger balances are more accurate than smaller ones" and goes on to examine the "double movement" described by the end of the radius of a circle. As we shall see, this examination leads to the explanation of the law of the lever proposed in the third question. Similarly the parallelogram rule appears parenthetically in the first question, referring to the particular case of a fixed ratio between two movements acting on a body. In the context of the author's arguments at the beginning of the book, such a case becomes the natural reference which allows us to evaluate what happens when a body travels around a circle.

---

[59] *Ibid.*, 1, 848b, 10–25.

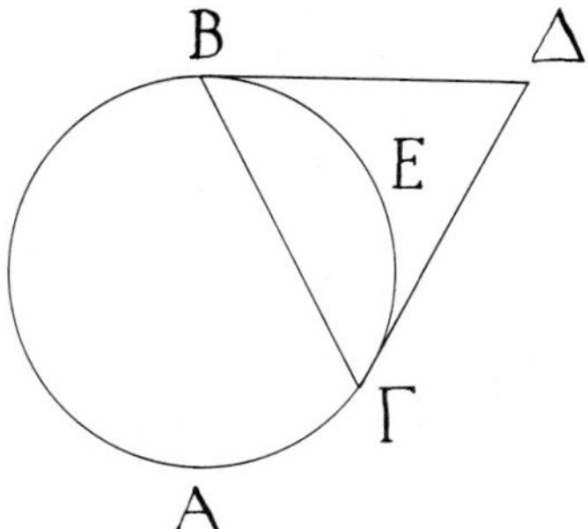

FIGURE 1.3.

> If [a body, placed on point B of the circle $AB\Gamma$], travelled with velocities in the ratio of $B\Delta$ to $\Delta\Gamma$ [Figure 1.3], it would move along the diagonal $B\Gamma$. But, as it is, seeing that it is in no such proportion it travels along the arc $BE\Gamma$. Now if of two objects moving under the influence of the same force (ἀπὸ τῆς αὐτῆς ἰσχύος) one suffers more interference (ἐκκρούοιτο πλεῖον), and the other less; it is reasonable to suppose that the one suffering the greater interference should move more slowly than that suffering less, which seems to take place in the case of the greater and the less of those radii which describe circles from the centre.[60]

In this passage, the author suggests a "reasonable" connection between velocity and interference—an interpretation which leads kinematic analysis into the domain of dynamics. In Duhem's opinion, this tacit transition derives from the proportionality of force and velocity which characterizes Peripatetic dynamics.[61] This reference to "Aristotle's axiom" which Duhem maintains in spite of G. Vailati's criticism,[62] may or may not be justified. But there is no doubt that the *Mechanical Problems* relates the idea of motion to other concepts belonging to the Aristotelian philosophy of nature.

This is evident in a passage in which the author sets forth his understanding of circular motion. Let us consider, he says, any radius which describes a circle; it moves along a curve naturally (κατὰ φύσιν) in the direction of the tangent, but is attracted to the center contrary to nature (παρὰ φύσιν). The lesser radius always moves in its unnatural direction; for because it is nearer the center which attracts it (τοῦ κέντρου τοῦ ἀντισπῶντος) it is

---

60 *Ibid.*, 1, 849a, 3–11.

61 P. Duhem, *Les origines de la statique* (Paris, 1905), Vol. 1, pp. 109–110.

62 G. Vailati, "P. Duhem. Les origines de la statique," *Boll. di Bibliografia e Storia delle Scienze Matematiche* (Jan.–March 1905).

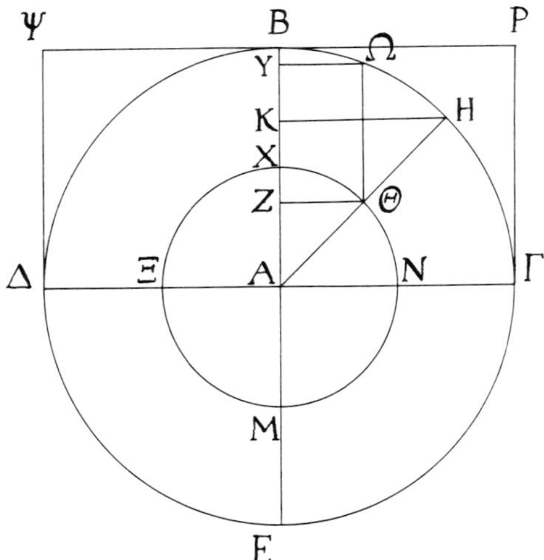

FIGURE 1.4.

the more influenced (κρατεῖται μᾶλλον)." [63] The author has adapted the Peripatetic distinction between natural and unnatural motion to infer an attraction to the center and an influence exerted on the rotating body. The meaning of these terms is unclear; the Greek word which expresses the action of attracting (ἀντισπάω) also means "to pull in a contrary direction," "to deviate," while the term κρατέω for the action of exerting an influence also connotes "to have power," "to dominate," "to impede." In both cases, as we see, there is no reference to the cause of motion—to force (ἰσχύς).On the contrary, the idea of unnatural motion implies a bar to the natural connection between the cause of a motion and the motion itself. We can leave undefined what this connection is if we concentrate on the comparison of two motions, and explain the influences which hinder the natural effects of that cause.

And this is exactly what the author does; he compares the motion along the circles $B\Gamma E\Delta$ and $XNM\Xi$, both with the center A (Figure 1.4). Suppose that the point B travels over $B\Omega$, and the point X over $X\Theta$, with $Y\Omega = Z\Theta$. The natural motions of B and X are equal, but the unnatural motion of B (i.e. BY) is less than the unnatural motion of X (XZ). This explains why X travels more slowly than B when it and B are linked by the rotating radius AB. The explanation is not only geometrical but also physical: "because the intereference [with X] is greater."[64] Moreover, as X moves to Θ, B moves to H. From geometry, it follows that KH is to BK as

[63] *Mechanical Problems*, 1, 849a, 14–20.

[64] *Ibid.*, 1, 849a, 30–31: διὰ τὸ γίνεσθαι μείζονα τὴν ἔκκρουσιν.

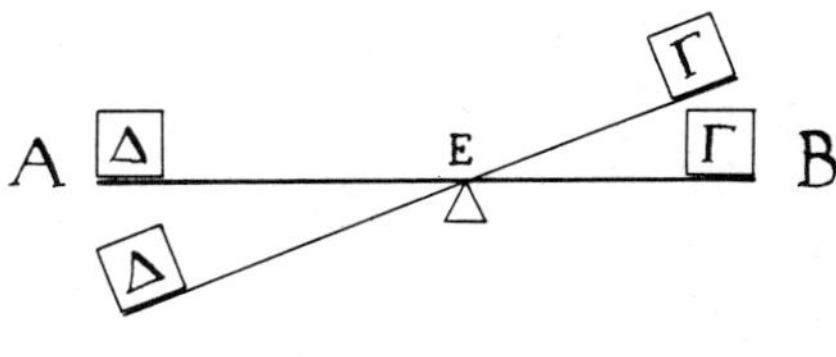

FIGURE 1.5.

$Z\Theta$ is to XZ. This geometrical relation is charged with physical meaning, for B will move to reach its "true" position H

> when the proportion between the natural and unnatural movements is true. If then, the natural movement is greater in the greater circle, the unnatural movement would at that point have the same proportion only in the sense that the point B would travel along the arc BH in the same time as the point X would travel along the arc $X\Theta$.[65]

Thus, since the ratio of natural to unnatural motions is the same in X and B, we may say that the cause of motion is equally hindered in both cases. Therefore the natural effect of the same force (τῆς αὐτῆς ἰσχύος) applied to X and B causes them to traverse the arcs $X\Theta$ and BH, respectively, in the same time. This, then, is the reason "why the point more distant from the center travels more quickly than the nearer point, though impelled by the same force"[66] And this, too, is the reason why greater balances are more accurate than smaller ones, for "the extremity of the balance scale must move at a greater rate under the influence of the same weight, inasmuch as it is further from the cord."[67]

At this point, the explanation of the lever is obvious. The fulcrum E (Figure 1.5) acts as the center of a circle, while the arms AE and EB may be seen as radii.

> Since under the impulse of the same weight the greater radius from the centre moves the more rapidly, and there are three elements in the lever, the fulcrum ... and the two weights, the one which causes the movement, and the one that is moved; now the ratio of the weight moved to the weight moving it is the inverse ratio of the distances from the centre. ... The reason has been given before that the point further from the centre describes the greater circle, so that by the use of the same force, when the motive force is farther from the lever, it will cause a greater movement.[68]

---

65 *Ibid.*, 1, 849b, 8–15.
66 *Ibid.*, 1, 849b, 19–22.
67 *Ibid.*, 1, 849b, 24–26.
68 *Ibid.*, 3, 850a, 36–850b, 6.

Duhem[69] traces this considerable proposition directly to the Aristotelian axiom expressed in book 7 of the *Physics* and book 3 of *On the Heavens*. The method of virtual velocities applied by the author of *Mechanical Problems* to the equilibrium of the lever would be a mere corollary of the law of dynamics. Duhem's thesis is based more on possibility than on certainty. As we have noted, in all the arguments presented above the author makes only one reference to an (obscure) law of motion, and that reference concerns the relation between velocity and "interference" with the cause of motion. But the author never treats the cause itself, only those factors which impede or redirect motion from its natural course.

Nonetheless, Duhem is certainly correct in affirming that the method of virtual velocities introduced in the *Mechanical Problems* has traditionally been considered a consequence of Aristotle's axiom. The Peripatetic tradition was characterized by its assumption of a direct relation between velocity and force, and this law of motion became the basis for the transition from kinematics to statics and dynamics. With reference to the law of the lever, Duhem quotes Simplicius' *Commentaria in physicorum* and the Arabic and Latin medieval commentaries, before coming to the Renaissance and Galileo. This parallels the evolution of the object of mechanics, from an attempt to understand the hidden marvels of machines to the study of motion and its causes.

From this point of view, the parallelogram rule presented in the *Mechanical Problems* is perhaps more important. Even in more modern times, numerous authors have used the law of motion as an obvious means of transition from the composition of movements to that of forces; witness the "demonstrations" offered by Newton and Varignon at the end of the seventeenth century. However the two differed in their treatment of the law of motion, their proofs of the composition of forces are quite similar. Both Newton's *Principia* (corollaries 1 and 2 of the introductory chapter, "Axiomata sive leges motus") and Varignon's *Nouvelle mécanique ou statique* (lemmas 1 and 2 of "Lemmes pour l'intelligence des sections suivantes") deduce the parallelogram rule by showing that a mass point acted upon by "joined" forces in different directions runs along the diagonal of the parallelogram at the same time that it would take to describe the sides under the action of "separate" forces.[70] The ideal force which would correspond to compound motion is defined as the "resultant" of these separate forces.

The validity of the parallelogram rule depends neither on the Aristotelian law of dynamics nor on that which underlies the modern mechanics of Galileo and Newton. Even if the fundamental axioms differ, the result is the same. We should, however, note that the classical tradition dominated

---

[69] P. Duhem, *op. cit.*, Vol. 1, pp. 5–8, 108–110, Vol. 2, pp. 291–293.

[70] I. Newton, *Philosophiae naturalis principia mathematica* (London, 1687); P. Varignon, *Nouvelle mécanique ou statique, don le projet fut donné en MDCLXXXVII* (Paris, 1725).

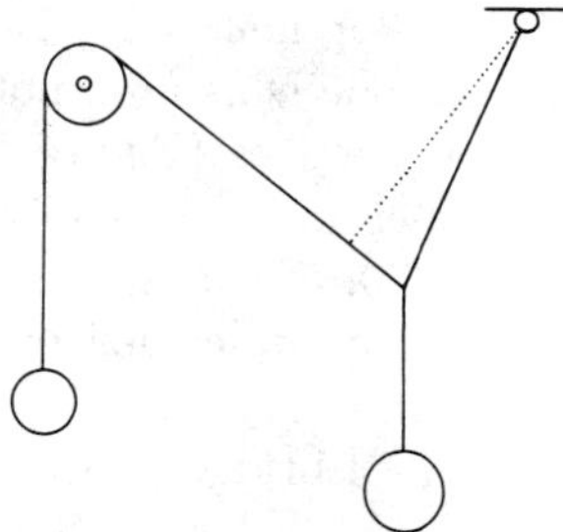

FIGURE 1.6.

almost the whole of the seventeenth century, and the parallelogram rule was invoked to confirm this tradition. Gilles Personne de Roberval's *Observations* (1675), which provide perhaps the first correct statical demonstration of the composition of forces, may be interpreted in this light.[71] So may the discussion in 1688 of Henry Basnage, the author of *Histoire des ouvrages des sçavans*, and the Oratorian priest Bernard Lamy, who had defined the parallelogram rule in "a new way" in 1687, in a letter to Dieulamant annexed to his *Traité de méchanique*.[72] Basnage accused Lamy of plagiarizing Varignon, not understanding that Varignon's "demonstration" was based on Peripatetic dynamics while Lamy's depended on the new post-Galilean law of motion.

The mechanics of the sixteenth and seventeenth centuries witnessed the evolution of a trend which both paralleled and took off from the evolution of the laws of motion, without depending on them. It was composed of developments in that most traditional field of statics, the explanation of simple machines. Leonardo da Vinci's intuitions; the esoteric arguments of Girolamo Fracastoro (to whom some authors attribute the first formal statement of the parallelogram rule); the major contribution of Simon Stevin, who connected the parallelogram rule to the study of inclined planes and the statics of wires; the concepts in Galileo's *Discorsi*; all of these, and others, helped to establish a lexicon and syntax for the concept of force and for the laws of equilibrium.[73]

---

[71] G. Personne de Roberval, *Observations sur la composition des mouvements et sur le moyen de trouver les touchantes des lignes courbes* (Paris, 1693), reprinted in *Mémoires de l'Académie Royale des Sciences depuis 1666 jusqu'à 1699*, Vol. 6 (1730), pp. 1–93.

[72] B. Lamy, *Nouvelle manière de démontrer les principaux théorèmes des élémens des méchaniques pour servir d'addition au Traité de Méchanique du R.P. Lamy* ... (Paris, 1687).

[73] G. Fracastoro, *De sympathia et antipathia rerum* (Venice, 1546); S. Stevin, *De beghinselen der weeghconst* (Leiden, 1586), latin translation in *Hypomnemata mathematica. Mathematicorum hypomnematum de statica* (Leiden, 1608).

One consequence of this development was that the parallelogram rule acquired a different foundation, one which did not consider composite motion at all. One of Leonardo's sketches (Figure 1.6) provides a foretaste, later developed by G.P. de Roberval in his *Traité de méchanique* and in a letter to Johann Hevelius of 1650, of the importance of funicular equilibrium. Funicular equilibrium can be described by means of an application of the law of the lever, but it also expresses the rule of composition and decomposition of forces. It forms a bridge between the balance of the lever and the parallelogram rule—and the law of the lever, since Archimedes' time, was held to be demonstrable in a strictly deductive and rational way.

To understand the theory behind this, we must turn again to the beginning —to the classical origins of the program of mathematical and deductive thought which today underlie rational mechanics.

2

The Law of the Lever

2.1 Archimedes' Demonstrations

Archimedes (287–212 BC), in his works on statics, first attempted to give the science of equilibrium an axiomatic, deductive structure similar to Euclid's for geometry. In our discussion of his work, we shall confine our attention to a few passages in his treatise *On the Equilibrium of Planes* (Περί 'Επιπέδων 'Ισορροπιῶν). While Aristotle relates mechanics to a physical theory, aiming for a universal synthesis, Archimedes thinks of statics as a rational and autonomous science, founded on almost self-evident postulates and built upon rigorous mathematical demonstrations. Both his intention and the source of his empirical data are also different. He is interested not in motion, but in equilibrium; not in the causes of velocity, but in the effects of a weight suspended at the end of a balance arm, that is, at a distance from a fixed point.

The notion of weight that underlies Archimedes' treatment is strictly connected with that of a material body; the weight is a measure of its matter, just as extension is a measure of its form. As geometry is concerned only with the form of a body, ignoring its matter, so the "science of weights" (as it was to be called in the Middle Ages) deals only with weight, ignoring form. Two properties describe a weight: it can be suspended from a point; and one weight is heavier, equal to, or lighter than another.

In the text which we are to consider, however, Archimedes does not explain all this. He assumes that we understand weight; the concept is implicit in this work. This assumption creates some serious lacunae in his argument, gaps which successive authors criticized and tried to fill in. For example, the demonstration of the law of the lever requires the use of a hypothesis which Archimedes takes for granted when he substitutes for one suspended body another of the same weight but of different shape. When Galileo reformulated the Archimedean demonstration, he noted the absence of this hypothesis, assuming that Archimedes had simply taken it for granted on the basis of immediate evidence. Concerning two prisms, AD and DB "which are in equilibrium around point C," Galileo observes that they,

> reduced to cubes or balls or any other shapes whatever, provided only that they keep the same suspensions, G and F, will continue to be in equilibrium around point C. I believe no one

> can doubt this, for it is quite evident that shape does not change a weight so long as the same quantity of material is preserved.[1]

We will return to this point later, in considering the interpretive proposal put forth by Giovanni Vailati at the close of the last century.

For the moment, let us examine the postulates which Archimedes sets at the foundation of his statics, and his demonstration of the law of the lever in propositions 6 and 7 of his first book.[2] There are seven postulates, which can be divided into two groups. The first three concern the lever. They are elementary observations, indicating the existence or nonexistence of equilibrium, but not establishing units or quantities. The first states that "equal weights at equal distances are in equilibrium (Αἰτούμεθα τὰ ἴσα βάρεα ἀπὸ ἴσων μακέων ἰσορροπεῖν)." The obverse is confirmed in a corollary: "equal weights at unequal distances are not in equilibrium, but incline towards the weight which is at the greater distance." Furthermore, according to the second postulate, "if, when weights at certain distances are in equilibrium, something be added to one of the weights, they are not in equilibrium but incline towards that weight to which the addition was made." The third postulate states, "Similarly, if anything be taken away from one of the weights, they are not in equilibrium but incline towards the weight from which nothing was taken."

The principal postulate is the first. Its validity seems to transcend experimental data, since it relies on considerations of symmetry and can be derived from the principle of "sufficient reason." The next three statements express the fact that the equilibrium is governed by two properties: the equivalence (or lack of it) of the weights at the ends, and their respective distance from the fulcrum. The first postulate is the only one to deduce equilibrium as a consequence of hypotheses about distances and weights. The other two take equilibrium itself for granted and state only that weights and distances are responsible for it. Archimedes' aim is to deduce the law of the lever for the more general case, starting from the simplest case in which symmetry ensures *a priori* the existence of equilibrium. This "demonstrative" procedure was to be a constant and powerful attraction for later scientists, in spite of its limited usefulness.

The other four postulates concern the concept of the center of gravity for plane figures:

4. When equal and similar plane figures coincide if applied to one another, their centers of gravity similarly coincide.

---

[1] G. Galilei, *Discorsi e dimostrazioni matematiche intorno a due nuove scienze* (Leiden, 1638), p. 83; reprinted in the National Edition, *Le opere di Galileo Galilei* (Florence, 1933), Vol. 8, p. 154; G. Galilei, *Two New Sciences*, tr. Stillman Drake (Univ. Wisconsin Press, 1974), p. 112

[2] Archimedes, *On the Equilibrium of Planes*, ed. J.L. Heiberg, 2d ed. (Leipzig: 1910–1915), Vol. 2, 124.3–126.3, 132.13–138.8.

5. In figures which are unequal but similar the centers of gravity will be similarly situated. By points similarly situated in relation to similar figures I mean points such that, if straight lines be drawn from them to the equal angles, they make equal angles with the corresponding sides.

6. If magnitudes at certain distances be in equilibrium, [other] magnitudes equal to them will also be in equilibrium at the same distances.

7. In any figure whose perimeter is concave in the same direction the center of gravity must be within the figure.

Vailati[3] observes that these postulates do not properly define the center of gravity, but only cite it as if the term already had a specific meaning.

> [But] neither the work under consideration [*On the Equilibrium of Planes*] nor any other work by Archimedes handed down to us, features an explicit definition of this *technical term*, and this is all the more remarkable inasmuch as the concept corresponding to it, and the general considerations referred thereto, constitute, so to speak, the nucleus or the backbone to which all Archimedes' mechanical speculations are connected, from his demonstration of the principle of the lever to his research about equilibrium conditions of floating bodies.[4]

The prevailing opinion among historians (Vailati included) is that the treatment of the center of gravity in *On the Equilibrium of Planes* depends on a lost work by the same author. Pappus bears witness to this; in his references to Archimedes' and Hero's theories of the center of gravity (κέντρα τῶν βαρέων) at the beginning of book 8 of *Mathematicae collectiones* (Συναγωγή), he presents arguments which, although fragmentary, embody a well-constructed organic treatment. Probably Archimedes set forth the principles of the doctrine in his book *About Balances* (Περί ξυγῶν). This is the only work of Archimedes which is known not to survive; it is quoted by Pappus elsewhere in his *Collectiones*.

In propositions 6 and 7 of the first book of *On the Equilibrium of Planes*, Archimedes demonstrates the law of the lever, first in proposition 6, which hypothesizes that the quantities considered (weights and distances) must be commensurable, and second, in the more general case (proposition 7). Ivor Thomas' translation of this memorable and important text is worth quoting:

---

[3] G. Vailati, "Del concetto di centro di gravità nella statica d'Archimede," *Atti Accademia Reale delle Scienze di Torino*, Vol. 32 (1896–1897) (hereafter cited as Vailati, "Del concetto"); reprinted in G. Vailati, *Scritti* (Florence, 1911) (hereafter Vailati, *Scritti*), pp. 79–90.

[4] *Ibid.*, p. 79.

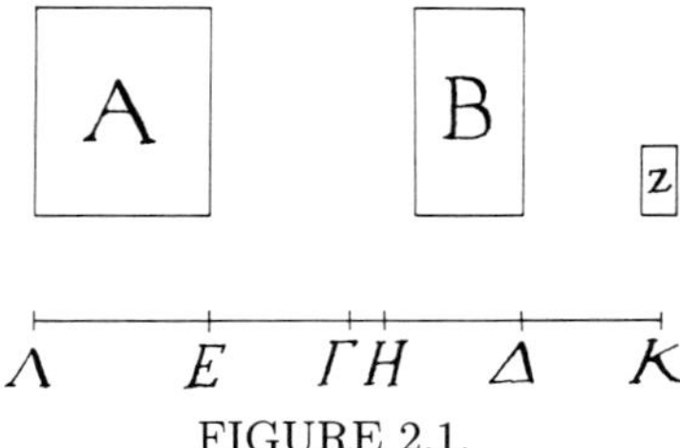

FIGURE 2.1.

Proposition 6. *Commensurable magnitudes balance at distances reciprocally proportional to their weights.*

Let A, B [Figure 2.1] be commensurable magnitudes with centres [of gravity] A, B, and let $E\Delta$ be any distance, and let

$$A : B = \Delta\Gamma : \Gamma E;$$

it is required to prove that the centre of gravity of the magnitude composed of both A, B, is Γ. Since

$$A : B = \Delta\Gamma : \Gamma E$$

and A is commensurable with B, therefore $\Gamma\Delta$ is commensurable with ΓE, that is, a straight line with a straight line [Eucl. X.11]; so that $E\Gamma$, $\Gamma\Delta$ have a common measure. Let it be N, and let ΔH, ΔK be each equal to $E\Gamma$, and let $E\Lambda$ be equal to $\Delta\Gamma$. Then since $\Delta H = \Gamma E$, it follows that $\Delta\Gamma = EH$; so that $\Lambda E = EH$. Therefore $\Lambda H = 2\Delta\Gamma$ and $HK = 2\Gamma E$; so that N measures both ΛH and HK, since it measures their halves [Eucl. X.12]. And since

$$A : B = \Delta\Gamma : \Gamma E,$$

while

$$\Delta\Gamma : \Gamma E = \Lambda H : HK$$

—for each is double of the other—therefore

$$A : B = \Lambda H : HK.$$

Now let Z be the same part of A as N is of ΛH; then

$$\Lambda H : N = A : Z$$

[Eucl. V. Def. 2]. and

$$KH : \Lambda H = B : A$$

[Eucl. V.7, Coroll.]; therefore *ex aequo*

$$KH : N = B : Z$$

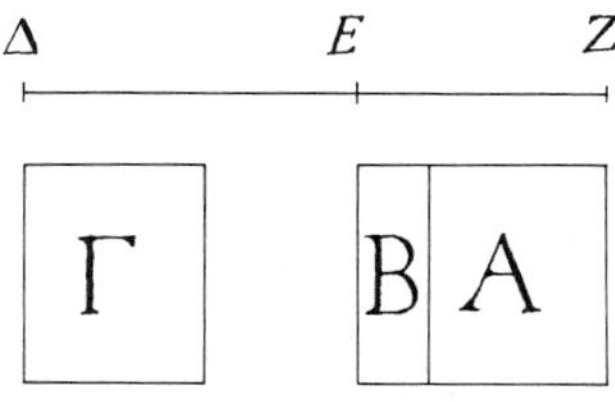

FIGURE 2.2.

[Eucl. V.22]; therefore Z is the same part of B as N is of KH. Now A was proved to be a multiple of Z; therefore Z is a common measure of A, B. Therefore, if ΛH is divided into segments equal to N and A into segments equal to Z, the segments in ΛH equal in magnitude to N will be equal in number to the segments of A equal to Z. It follows that, if there be placed on each of the segments in ΛH a magnitude equal to Z, having its centre of gravity at the middle of the segment, the sum of the magnitudes will be equal to A, and the centre of gravity of the figure compounded of them all will be E; for they are even in number, and the numbers of either side of E will be equal because $\Lambda E = HE$ [Prop. 5, coroll. 2]. Similarly it may be proved that, if a magnitude equal to Z be placed on each of the segments [equal to N] in KH, having its centre of gravity at the middle of the segment, the sum of the magnitudes will be equal to B, and the centre of gravity of the figure compounded of them all will be Δ [Prop. 5, coroll. 2]. Therefore A may be regarded as placed at E, and B at Δ. But they will be a set of magnitudes lying on a straight line, equal one to another, with their centres of gravity at equal intervals, and even in number; it is therefore clear that the centre of gravity of the magnitude compounded of them all is the point of bisection of the line containing the centres [of gravity] of the middle magnitudes [from Prop. 5, coroll. 2]. And since $\Lambda E = \Gamma\Delta$ and $E\Gamma = \Delta K$, therefore $\Lambda\Gamma = \Gamma K$; so that the centre of gravity of the magnitude compounded of them all is the point Γ. Therefore if A is placed at E and B at Δ, they will balance about Γ.

Proposition 7. *And now, if the magnitudes be incommensurable, they will likewise balance at distances reciprocally proportional to the magnitudes.*

Let $(A + B)$, Γ be incommensurable magnitudes [Figure 2.2] and let ΔE, EZ be distances, and let $(A + B) : \Gamma = E\Delta : EZ$; I say that the centre of gravity of the magnitude composed of both $(A+B)$, Γ is E. For if $(A+B)$ placed at Z do not balance Γ placed at Δ, either $(A + B)$ is too much greater than Γ to balance or less. Let it [first] be too much greater, and let there be subtracted from $(A+B)$ a magnitude less than the excess by which $(A + B)$ is too much greater than Γ to balance, so that the remainder A is commensurate with Γ. Then, since A, Γ are commensurable magnitudes, and

$$A : \Gamma < \Delta E : EZ,$$

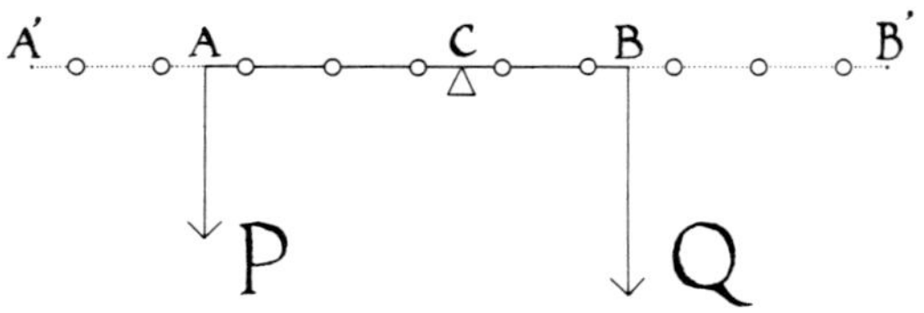

FIGURE 2.3.

A, Γ will not balance at the distances ΔE, EZ, A being placed at Z and Γ at Δ. By the same reasoning, they will not do so if Γ is greater than the magnitude necessary to balance $(A + B)$.[5]

2.2 Interpretations (and Improvements) of Archimedes' Proof

The heart of Archimedes' argument lies in proposition 6, for which one may substitute a shorter, equally expressive form—for example, R. Marcolongo's exposition: let the ratio between the magnitudes of weights P and Q be commensurable; for a clearer demonstration, let (e.g.) $P : Q = 2 : 3$ (Figure 2.3). If fulcrum C is such that $AC : CB = 3 : 2$, then P and Q are in equilibrium. Let us divide AB into five equal parts; AC will contain three and CB will contain two. Let AB be extended to A' and B', so that $AA' = CB$ and $BB' = AC$; C is therefore the middle point of $A'B'$. Let us consider the middle points of the equal parts of AC, etc. For the force P we could substitute $1/4P$ four parallel forces, and for the force Q six $1/6\,Q$ ... applied at these middle points, and symmetrically placed with respect to A and B. Then, by composing two-by-two the forces so symmetrically obtained with respect to C, we shall obtain the resultant applied at C and equal to the sum of P and Q.[6]

E. Mach suggested an even more convincing formulation in the following terms: consider the case of undoubted equilibrium in which two weights of equal magnitude 1 are suspended at equal distances AC, BC from the middle point C (Figure 2.4).

---

[5] *Selections Illustrating the History of Greek Mathematical Works*, Vol. 2, *Aristarcus to Pappus of Alexandria,* ed. and tr. I. B. Thomas (Cambridge, Mass.: 1980), pp. 209–217. Thomas completes Archimedes' proof as follows: "Since $A : \Gamma < \Delta E : EZ$, Δ will be depressed, which is impossible, since there has been taken away from $(A + B)$ a magnitude less than the deduction necessary to produce equilibrium, so that Z remains depressed. Therefore $(A + B)$ is not greater than the magnitude necessary to produce equilibrium; in the same way, it can be proved not to be less; therefore it is equal."

[6] R. Marcolongo, *Meccanica razionale* (Milan, 1917), Vol. 1, pp. 221–222.

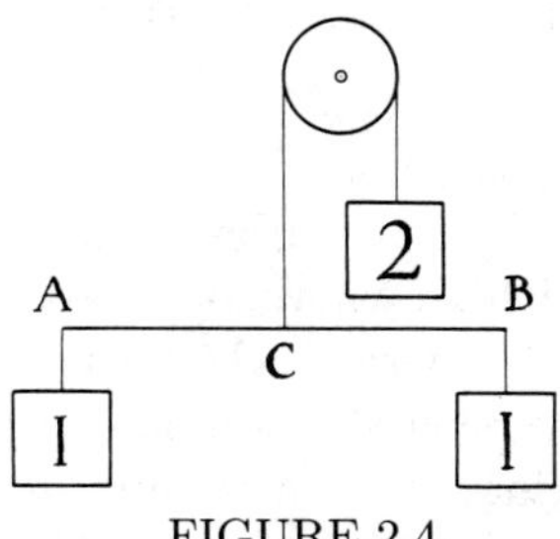

FIGURE 2.4.

If the whole system is suspended on a thread at C, this thread ... will sustain a weight equal to 2 [if we ignore the weight of the rod]. In conclusion, two equal weights applied to the extremities of the arms substitute for the double weight applied to the centre of the rod. Let us now suspend two weights with a ratio of 1 : 2 at the extremities of a lever [Figure 2.5]

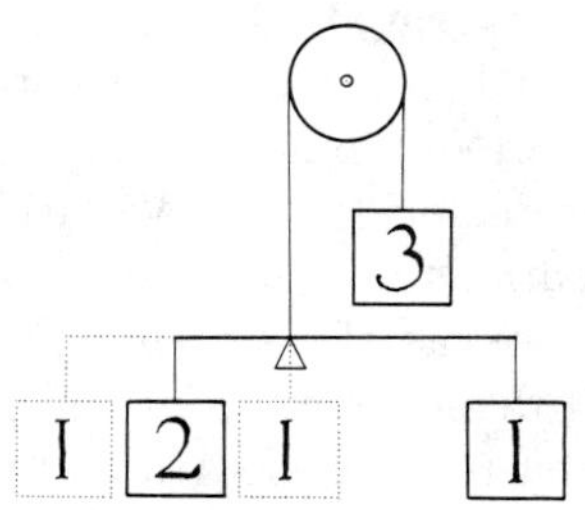

FIGURE 2.5.

whose arms are in the ratio of 2 : 1. Let us imagine that weight 2 could be replaced by two weights 1 applied to either side at a distance 1 from the point of suspension. In this case, as well, we have a perfect symmetry with respect to the suspension point and, therefore, equilibrium. Let us suspend again weights 4 and 3 to lever arms of length 3 and 4 [Figure 2.6]. Let us extend arm 3 by length 4 and arm 4 by length 3, and let us substitute weights 4 and 3 respectively with 4 and 3 pairs of weights symmetrically arranged. Also in this case we have perfect sym-

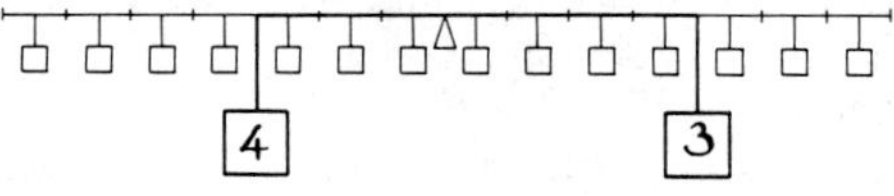

FIGURE 2.6.

> metry. This treatment carried out with particular magnitudes can easily be generalized.[7]

In fact, Mach's purpose was more than a little malicious. This is not a historical account moved by real curiosity about the theme under discussion. Mach already had his own theories about the law of the lever and the "process of ideas" (his words) which suggested Archimedes' treatment. This is why Mach's version seems intended to expose the scantiness of Archimedes' demonstration, making it subordinate to ideal operations which Archimedes' postulates cannot justify. Mach's treatment makes a tacit appeal to the physical intuition behind Archimedes' thesis. Specifically, according to Mach, to admit that moving the application point of half a given weight does not disturb equilibrium if the other half is moved in the opposite direction supposes that equilibrium depends only on the sum of the products of each weight and its distance from the rotation center. We shall return to this by no means trivial objection in due course.

Vailati's interpretation seems much more faithful to Archimedes' original aims. This author—a contemporary of Mach's, and, like him, aware of the need for demonstrative precision—is equally aware of the flaws in *On the Equilibrium of Planes*. He attributes these weaknesses to the fact that Archimedes had provided an exhaustive treatment of a theory of centers of gravity in his text *About Balances*, and that this work justified the deductions in his later demonstration. Vailati intended to reconstruct "the whole set of considerations and arguments which led Archimedes to the conclusions he then took as the starting point from which to proceed towards the classical demonstration of the principle of the lever."[8] He was led to this study by reading an Arabic manuscript in the Leiden library. The manuscript contained one of Hero's "lost" works, *The Elevator* (Βαρουλκός), which was discovered and published by Carra de Vaux in 1893. This text, and gleanings from book 8 of Pappus' *Mathematicae collectiones*, provide a rich sequence of reasoning about the concept of the center of gravity and its properties, probably derived from Archimedes. In proposition 1 of book 8, Pappus defines the center of gravity as a point such that, if a body is suspended by it, the body remains at rest no matter what its initial position. The existence and uniqueness of such a point in any solid was proved (according to Pappus) by Archimedes' and Hero's works. In Hero's *The Elevator*, there is such a demonstration, drawn from Archimedes, but the passage in which it is reported is damaged and mutilated.[9]

---

[7] E. Mach, *Die Mechanik in ihrer Entwickelung historisch-kritisch dargestellt* (Leipzig, 1883; Italian tr. published Turin, 1977) (hereafter cited as Mach, *Die Mechanik*), pp. 44–45.

[8] Vailati, "Del concetto," p. 81.

[9] Pappus of Alexandria, *Collectiones quae supersunt*, ed. Hultsch (Berlin, 1878), pp. 1030–1031; Carra de Vaux, "Les mécaniques ou l'élévateur ... de Héron d'Alexandrie," *Journal Asiatique*, Vol. 2 (1893), pp. 164–165.

Nonetheless Hero's short account and Pappus' fragments gave Vailati the material to outline what he thinks may be Archimedes' original reasoning. In his reconstruction, Vailati resembles the archeologists of his time, who were tempted to create daring—sometimes rash—"restorations" from a few uninformative remains.

According to Vailati, Archimedes based his deductions about the center of gravity on four axioms. The first and fourth exist in Pappus' work; the third corresponds to a passage in Hero, and the second is pure reconstruction. The axioms are as follows:

1. If a weight is suspended by one or two of its points, there exist, in both cases, positions of the body in equilibrium—that is, if the body is placed in such positions, it remains there.

2. If a weight suspended by one of its points is in a position of equilibrium, it also remains in such a position if suspended at any other point of the vertical straight line passing through the point of suspension. The equilibrium still exists if the body is rotated around such a vertical straight line.

3. If a weight which can rotate around a horizonal axis (a balance beam, ξυγόν) is in equilibrium, it will remain in that position if, freed from the axis itself, it is suspended by any two points on the vertical plane passing through it. Furthermore, such a body continues to be in equilibrium if the body is moved parallel to such a vertical plane.

4. The equilibrium is disrupted if the first axis of suspension is replaced by another one, parallel to it and placed on the same horizonal plane.

Several theorems derive from these axioms. If we call the line defined in axiom 2 the "central straight line," and the plane described in axiom 3 the "central plane," it is easily proved that

1. At least one central plane passes through every pair of points.

2. At least one central straight line passes through every point of the body.

3. Every plane passing through a central straight line is a central plane.

On the basis of these axioms, we can establish the properties set forth by Pappus:

4. No parallel central planes can exist.

5. All the central straight lines meet in pairs.

We can also prove Hero's theorem:

6. All the central straight lines pass through one and the same point O.

In fact, "when a straight line meets two intersecting straight lines and does not lie on their plane, it must meet them in their intersecting point."[10]. But since

7. Through every point Ω which is not point O, only one central line passes, and it coincides with the straight line ΩO, it follows that:

8. *If a weight suspended by a given point is to remain in equilibrium, the vertical drawn through such a point must pass through O.*

We find this last proposition in chapter 6 of Archimedes' treatise on the *Squaring of the Parabola.* No other point enjoys this property. Point O is termed the *center of gravity,* according to the description of it given by Pappus, who is also responsible for one further observation:

9. Every central plane goes through O, and every plane going through O is central.[11]

Whether or not it is historically accurate, Vailati's reconstruction is fascinating, since it organizes concepts in ways to which we are no longer accustomed. The concept of the center of gravity is defined by a path independent of the principle of the lever or any other more general proposition about the equilibrium of forces. More modern treatments of rational mechanics prove the concept of the center of gravity by referring to previously established principles of statics. Note that the first four axioms given above are nothing but an accurate description of the physical act of suspending a weight. Their intuitive truth rests on the same basis as the Archimedean concept of weight: the "possibility of being suspended."

According to Vailati, Archimedes' original idea was that he could demonstrate his propositions about the center of gravity starting from this property, suitably supplemented by the observations summarized in the axioms. He could then go on to use them to deduce the law of the lever. But is Vailati's reconstruction correct? From Guidobaldo del Monte and Luca Valerio, we have some evidence that the ancient conception of "the science of weights" did not subordinate the center of gravity to statical principles.[12] Both felt the need to introduce a suitable postulate to prove the existence and uniqueness of such a point. In the following century, J. Wallis claimed to be the first to demonstrate the existence of the center of gravity: "All the writers of mechanics assume that in every heavy body, there is a point called center of gravity, but I do not know whether any one of them has

[10] Carrà de Vaux, *op. cit.*, p. 177

[11] Pappus of Alexandria, *op. cit.*, p. 1032.

[12] Guidobaldo del Monte, *Mechanicorum liber* (Pisa, 1577), p. 2, "*unius corporis unum tantum est centrum gravitatis*"; L. Valerio, *De centro gravitatis solidorum* (Rome, 1604), book 3, p. 7, "*postulatur omnis figurae gravi unum esse centrum gravitatis*".

demonstrated its existence before me."[13] Wallis' approach tends to support Vailati, for Wallis starts from the law of the lever, thereby confirming the change from the old to the new concept.

Returning to Vailati: Archimedes' treatment of the lever is incomplete; it needs the fundamental lemma which Galileo was later to foreshadow in his *Discorsi.* According to Archimedes–Vailati this lemma states that:

1. If we imagine a body to be divided in two, the straight line joining the centers of gravity of each part also contains the center of gravity of the whole body.

2. The position of this center of gravity depends only on the weights of the two parts, and does not change, no matter how their forms change, as long as their respective centers of gravity are not displaced.

The first of these propositions is easily proved, and Archimedes states it explicitly. The second can be deduced from the postulates presented at the beginning of *On the Equilibrium of Planes*, particularly from point 6, if the existence and uniqueness of the center of gravity have been proved and the validity of the "solidification principle" is accepted.

Suspend a body by its center of gravity, C. Divide it into two parts, whose respective centers of gravity lie on a horizontal straight line. Picture this straight line as the only link between the two centers of gravity: a rigid rod, suspended at C. We have made a balance on whose unequal arms the weights of the two bodies are impaled. This balance is at equilibrium; if it were not, we should have to find another point of suspension D at which the system would be at equilibrium. If we rejoined the two parts, the reconstructed body would then have *two* centers of gravity, C and D, which is impossible. Therefore, if the balance suspended at C is in equilibrium, it remains in that state according to postulate 6, even if the forms of the bodies change, as long as their weights and their points of suspension at the ends of the rod (that is, their centers of gravity) remain the same.

This allows us to prove the law of the lever, both in the form given in proposition 6 of book 1 of *On the Equilibrium of Planes* and in the more elegant form presented at the beginning of book 2. In the latter, Archimedes (presumably dissatisfied with his previous deductions) returns to the theme. He considers two parabolic segments, having areas (ab) and (cd) (Figure 2.7). He puts them on a balance with their respective centers of gravity at e and f. He aims to show that the center of gravity of the whole is at h, such that $(ab) : (cd) = fh : he$. To do this, he makes use of a previously demonstrated result about the transformation of a parabolic segment into a rectangle (squaring it).

> Let segments fg and fk be defined equal to eh and segments eg and el similarly be equal to fh. Then lh will be equal to

[13] J. Wallis, *Mechanica sive de motu* (London, 1670).

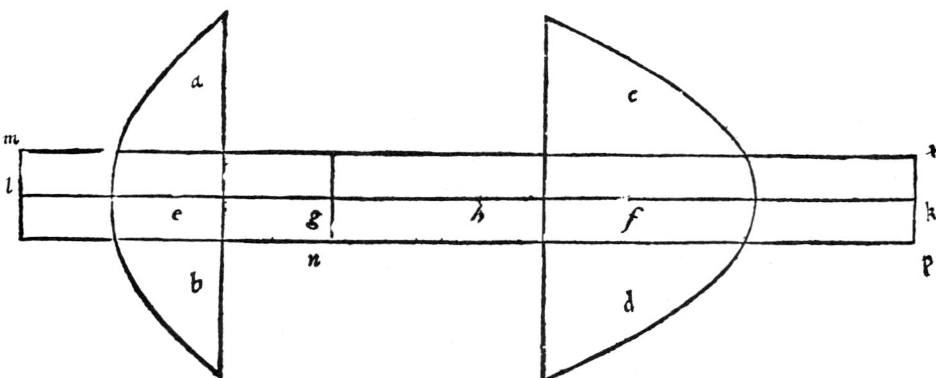

FIGURE 2.7. Archimedes' second demonstration of the law of the lever. From the *Editio Princeps* of Archimedes' Works (Basel, 1544).

> kh. We obtain $lg : gk = (ab) : (cd)$ because lg and gk are both double the length of fh and eh respectively. Let us now consider on [the] line lg a rectangle (mn) equal in area to (ab) whose center of gravity remains at e. Let us now complete rectangle (nx). Then $(mn) : (nx) = lg : gk$, but $(ab) : (cd) = lg : gk$. therefore $(ab) : (cd) = (mn) : (nx)$. Since (ab) is equal to (mn) it follows that (cd) is equal to (nx) and the center of gravity of (nx) remains at f. As lh is equal to hk, the two sides lh and hk divide by half the whole rectangle (mp) whose center of gravity is h. But (mp) is equal to the sum of (mn) and (nx), whereby h is also the center of gravity of the body formed by (ab) and (cd).

This second demonstration of the law of the lever was to be progressively simplified and generalized during the sixteenth and seventeenth centuries, thanks to the increased understanding of the squaring of plane figures. Several authors made notable contributions: Maurolico in his *De momentis aequalibus*, which expands *On the Equilibrium of Planes* into four books; Guidobaldo in his periphrasis of Archimedes' text; G.B. Benedetti in his letter to Vincenzo Mercato; and particularly Luca Valerio in his brief early work of 1582, *Subtilium investigationum liber primus* and later in his major treatise, *De centro gravitatis solidorum libri tres*.[14] An excellent historical essay on Valerio by P.D. Napolitani illustrates the author's intentions

[14] F. Maurolico, *De momentis aequalibus*, in *Admirandi Archimedis Syracusani monumenta omnia mathematica quae extant ...* (Palermo, 1685), pp. 98–99; Guidobaldo del Monte, *Paraphrasis in duos Archimedis de aequiponderantibus* (Pisa, 1588), pp. 60–76; G.B. Benedetti, *Diversarum speculationum mathematicarum et physicarum liber* (Turin: 1585), pp. 380–396; L. Valerio, *Subtilium investigationum liber primus. Seu quadratura circuli* (Rome, 1582), pp. 20–21; L. Valerio, *De centro gravitatis solidorum libri tres* (Rome, 1604).

"in his critical defense of the principles of geometrical statics."[15] While Archimedes had had to confine himself to squared parabolic figures, Valerio believed he could square any plane figure. This is why his demonstration, based (like Archimedes') on the transformation of other forms into rectangles or parallelepipeds, reaches a full generalization, even if the proof of the geometric assumption is flawed.

The same train of thought appears in contributions by Simon Stevin in his *Hypomnemata mathematica* and Galileo in his *Discorsi*. Lagrange, in the first section of his *Méchanique analitique*, refers to them as modern authors "who have rendered Archimedes' demonstration simpler." Lagrange summarizes their arguments as follows: assume that

> the weights applied to the lever are two horizontal parallelepipeds suspended by their midpoints, having equal width and height, but twice the length of the lever arms that inversely correspond to them. The two parallelepipeds are thus inversely suspended from their lever arms, and at the same time they face each other, so as to form a single parallelepiped whose midpoint corresponds exactly with the fulcrum of the lever.[16]

Lagrange's sweeping synthesis was widely admired by Mach and others,[17] but it is perhaps a little *too* sweeping. The law of the lever seems almost to be divorced from physical fact; it becomes simply a way of seeing things, another means of describing reality. One could imagine a homogeneous horizontal prism suspended by its midpoint as a union of two prisms $2m$ and $2n$ long; the distances between the centers of gravity of the two prisms and their common suspension point (that is, n and m) are inversely proportioned to their respective weights. Is the law of the lever only a tautology, then? Does it seem to be deducible, not because it concerns the object, but because it fits a geometric convention?

No; we cannot reduce Archimedes' complex treatment to a circular argument. One cannot divorce the lever from its physical reality; we have too much experience of it as a machine. If Archimedes wished to create a theoretical demonstration according to geometric principles, he also desired to "lift the world" in purely practical terms. We should examine the physical meaning implicit in all these hypotheses and postulates, to reach the reality behind these apparently neutral and self-evident terms.

---

[15] P.D. Napolitani, "Metodo e statica in Valerio," *Boll. di Storia delle Scienze Matematiche*, Vol. 2, No. 1 (1982).

[16] L. Lagrange, *Méchanique analitique* (Paris, 1788), p. 2.

[17] Mach, *Die Mechanik*, p. 46.

2.3 An Alternative Approach: Pseudo-Euclid and Huygens

In 1851, Franz Woepcke discovered a text in an Arabic manuscript in the Bibliothèque Nationale, which he published the same year. The authorship of this text is uncertain. Some historians (Woepcke) attribute it to Euclid himself, others (Curtze, Heiberg) to one of Mûzâ ibn Shâkir's three sons (i.e., to a tenth-century Arabic scholar). In fact, another manuscript indicates that Mûzâ's sons are its authors, but Woepcke himself believed that the latter translated or edited a treatise of Greek origin. Duhem shares this opinion, but refrains from accepting Woepcke's attribution of the original to Euclid.

Whatever its origin, the fragment in question is valuable for its demonstration of the law of the lever—a demonstration which, like Archimedes', derives not from kinematic or dynamic analysis, but from postulates made evident (in Duhem's words) "by their simplicity and by everyday experience."[18] Though the proof is perhaps less effective than Archimedes', it is interesting for two reasons. First, its fundamental propositions coincide surprisingly closely with Huygens'; second, the thesis implicitly depends on the superimposition of equilibria which, as Lagrange was to admit, was the key to the entire question.

In fact, the Arabic manuscript has no trace of the axioms laid down by its author; the text is mutilated. They can, however, be reconstructed *a posteriori* by following the thread of the demonstration. According to Vailati,[19] they are as follows:

1. If we have a rigid horizontal plane capable of revolving around one of its straight lines, and we imagine weights hanging from any of its points so that the plane is in equilibrium, then the equilibrium will still exist if we move the weights to any other position parallel to the fixed straight line.

2. If a horizontal plane from which weights are hanging stays in equilibrium when it is suspended by two of its straight lines which intersect at point O, then the plane will be in equilibrium even if it is suspended from O alone.

Using the principle of superimposed equilibria and these two axioms, we can (somewhat tortuously) derive the following theorem: If we imagine a horizontal straight rod AB (Figure 2.8), suspended by its middle point O,

[18] P. Duhem, *op. cit.*, Vol. 1, p. 69.

[19] G. Vailati, "Di una dimostrazione del principio della leva, attributa ad Euclide," *Boll. di storia e bibliografia matematiche*, supplement to *Giornale di matematiche di Battaglini*, Nov.–Dec. 1897, reprinted in Vailati, *Scritti*, pp. 115–117.

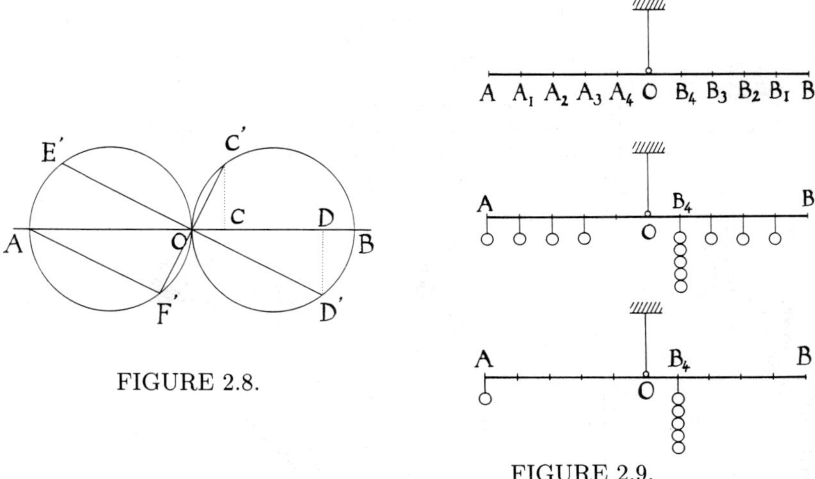

FIGURE 2.8.

FIGURE 2.9.

with three equal weights hung at points A, C, D, then the system will be in equilibrium if $OC = DB$.

To prove this, consider Figure 2.8. The plane is horizontal and passes through AB; on it, we draw two circles with respective diameters AO and OB. If the plane is suspended by the point O, we see by axioms 1 and 2 that three equal weights placed at A, C', D' balance one another, since they can be moved parallel to the suspension axes $E'D'$ and $F'C'$, so as to be symmetrical to the axes themselves. If, then, the plane is suspended by a straight line perpendicular to AB and passing through O, we can move the weights from C' and D' to C and D respectively without disturbing the equilibrium.

Having established this, "pseudo-Euclid" goes on to demonstrate the law of the lever as follows: Consider a horizontal rod AB suspended by its middle point O. Divide the segments AO and OB into (let us say) five equal parts (Figure 2.9). Now apply the theorem just proved by putting three equal weights, first at A, B_4 and B_1, then at A_1, B_4 and B_2, then at A_2, B_4 and B_3, and finally at A_3, B_4 and B_4. If we think of these equilibria existing simultaneously, we have the equilibrium system 1 of Figure 2.9. We can form the equilibrium system 2 of Figure 2.9 by removing from system 1 the pairs of weights placed symmetrically with respect to point O. In fact, system 2 shows that the system is in equilibrium when the ratio between the weights (at A and B_4) is inverse to the ratio between the distances (AO and OB_4) from fulcrum O.

Whatever the validity of this "proof," it does present two useful concepts: first, the recognition that the "efficacy" of a weight with respect to the equilibrium of a plane revolving on an axis depends on the distance of the weight from the axis; and, second, the superimposition of equilibria.

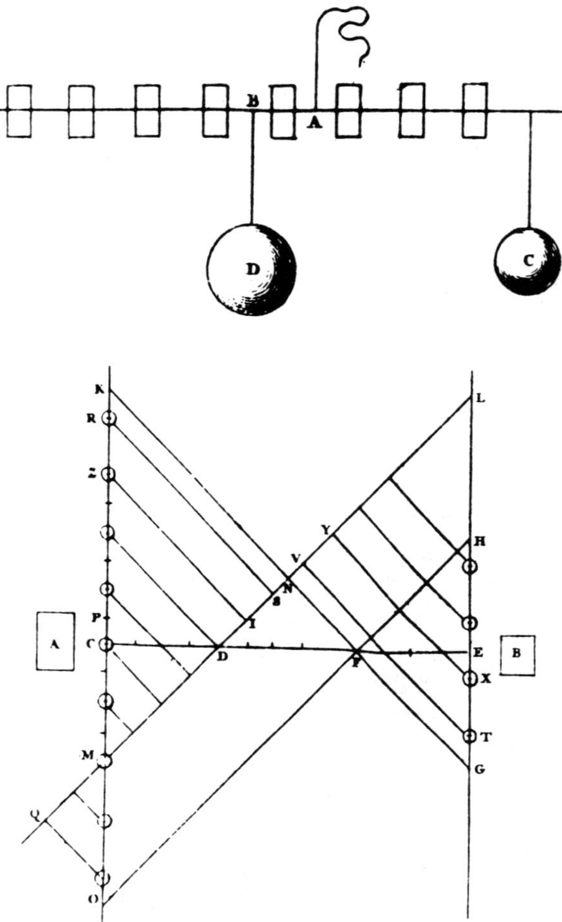

FIGURE 2.10. Huygens' demonstration of the law of the lever. From Huygens' *Opera varia* (Leiden, 1724).

These are surprisingly modern concepts. They constitute the core of Huygens' *Demonstratio equilibrii bilancis*, which was the subject of a letter of his to de la Hire in 1686:

> In Archimedes' demonstration of the fundamental proposition of mechanics, he assumes something we have good reason to doubt. It concerns the following hypothesis: if two equal weights are applied at equal distances on the axis of a balance, whether they are all placed on one side of it with respect to the point of suspension, or whether some of them are transferred onto the opposite side [Figure 2.10] where the point of suspension is A, such weights exert the same deflecting force on the

> balance [*pondera habere eandem vim ad deflectendam libram*] as they would if they were suspended all together at point B, where their common center of gravity is located. Thus, the same counterweight C would balance both of them if they were suspended separately, and if they were all grouped together and suspended at B, or substituted at B by a single weight D equal to their sum.[20]

Huygens put his finger on the sore spot. To suppose that two or more weights applied at equal distances on a horizontal axis can be replaced by the sum of the same weights applied at the central point—this is almost alarming. It seems to be an unlikely consequence of an axiom concerned only with symmetry, but one is loath to add it to the list of other axioms. Not only is it not self-evident, but it risks turning the law of the lever into a tautology. Huygens aims at doing without it up to his penultimate passage. To do this, he starts by introducing postulates 1 and 2 of *On the Equilibrium of Planes*, and then adds a third which allows us to conceive "as inflexible, and deprived of gravity, the lines and planes dealt with by the demonstration."[21] In two further propositions, he rephrases "pseudo-Euclid"'s two axioms:

> *Prop. I.* If on a horizontal plane suspended at a straight line that divides it into two parts, a weight be applied, the more distant such a weight is from the straight line, the stronger is the force [*vis*] it exerts in deflecting the plane. *Prop. II.* If a horizontal plane loaded with several weights is in equilibrium when suspended at one of its straight lines, then the center of gravity of the loaded plane passes through the straight line.[22]

Here, at last, is the demonstration of the law of the lever, set forth in proposition 3: "Two commensurable weights suspended at the extremities of the arms of a lever [*libra*] will be in equilibrium, if the arms are in inverse ratio to the weights." Consider a horizontal plane Π going through the (horizontal) axis of lever AB (Figure 2.11). Suppose that $n+m$ unitary weights are applied at equal distances on segments DE and FG, lying in Π and perpendicular to AB, so that DE contains n and FG includes m. Let C be a point of AB such that $m : n = CB : AC$. We must show that if we put the fulcrum of the loaded lever AB at C, the lever remains at equilibrium. Now let us draw a straight line MCL parallel to line DG, and detach segment $FH(= DE)$ from segment FG. The simple geometry of this figure shows the following:

---

[20] C. Huygens, *Opera varia* (Leiden, 1724), Vol. 1, pp. 282.

[21] *Ibid.*, p. 283.

[22] *Ibid.*, pp. 284–293.

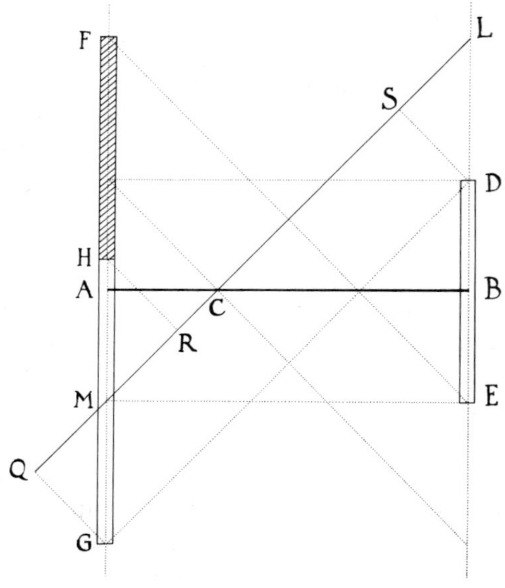

FIGURE 2.11.

1. To every weight on DE corresponds a weight on HF equally distant from ML.

2. To every weight on HM corresponds a weight on MG equally distant from ML.

This ensures, by virtue of the given postulates and propositions 1 and 2, that the plane is in equilibrium if it is suspended by the straight line ML. But thanks to symmetry, the plane is in equilibrium even if it is suspended by the line AB. It follows that C is the center of gravity of the system. The demonstration is completed by substituting the m weights of FG and the n weights of DE by two single weights applied at A and B, respectively equal to m and n.

Huygens' demonstration is almost perfect; the sore spot mentioned by the author in his premiss is not quite gone, but it is reduced to a minimum. The principle of the superimposition of equilibria supports every stage of the argument and, in consequence, we can verify the whole equilibrium by examining one at a time each pair of equidistant weights on both sides of the line ML. This principle has a prominent role in the theory of the lever and, more generally, in mechanics. Lagrange was to admit as much, and singled it out as "as fruitful a principle as that of the superimposition of figures is in Geometry."[23]

---

[23] L. Lagrange, *Mécanique analytique*, 3rd ed. (Paris: 1853), p. 5.

Is the superimposition of equilibria a first principle, to be accepted in and of itself, or does it derive from more evident axioms? This question is worth examination.

2.4 Marchetti's New Approach and Daviet de Foncenex's Improvements

In 1669, seventeen years before Huygens' *Demonstratio*, Alessandro Marchetti, a Tuscan scholar, dedicated an entire treatise entitled *Exercitationes Mechanicae* to the subject of the law of the lever. "I was not intimidated by the theme," he wrote in his introduction,

> though it had been dealt with, first of all by Aristotle, the Peripatetic Prince, then by the divine Archimedes, and by so many other most illustrious men, such as Luca Valerio, the neo-Archimedes of our Age, the most learned and erudite Guldin, and the highly ingenious Galileo—all men to whom not even in thought dare I compare myself, satisfied only to worship their memory. I was not intimidated—I repeat—because, even while admiring their discoveries, I do not lose hope of having reached results which people learned in these matters may judge worthy of some praise. It is not only because I have begun to deal with the concept of moment in a much more diffused and distinct way than anybody else, and perhaps (forgive the presumption) in a clearer and more accurate manner, but also because—unless I am wrong—I have managed to demonstrate that all the hypotheses assumed by Archimedes, and his numerous disciples in mechanics, are all well founded and more evident, so that everybody may perceive and admit them as being truthful.[24]

Marchetti's immediate aim was not so much to establish the conditions for the equilibrium of levers as to shed light on the concept of moment. He took for granted that the moment is the proper object of every proposition connected with equilibrium. Because of this, the postulates put forth at the beginning of his treatment simply state that

1. Equal weights suspended at equal distances have equal moment.
2. Given two equal weights, the weight suspended further from the fulcrum has greater moment.

From these two simple hypotheses or definitions, Marchetti claims to deduce everything. In reality, he is wrong; his postulates are far more complex

[24] A. Marchetti, *Exercitationes mechanicae* (Pisa, 1669), p. ix.

than their simple formulation would suggest. For example, the proof given in proposition 8 (commensurable weights inversely proportionate to their distances have equal moment)[25] shows how the author interprets his second postulate, linking it to an additive property regarding distance. For Marchetti, the second postulate does not simply mean that, if distance c is greater than distance b and the weights P are equal, then the moment $M(P, c)$ is greater than $M(P, b)$, but it also implies that, if $c = b + a$,

$$M(P, b + a) = M(P, b) + M(P, a). \tag{2.1}$$

An analogous property is valid (*patet*; manifest) with respect to two weights P, Q, suspended together at a point situated at distance d from the fulcrum. In this case,

$$M(P + Q, d) = M(P, d) + M(Q, d). \tag{2.2}$$

This is suggested during the proof of proposition 2 (weights of equal moment, suspended at equal distances, are equal). Consider two equal weights E, D, suspended at equal distances and thus having equal moments (postulate 1). Suppose as well that weight $(E + F)$ suspended at the same distance has the same moment. It follows that "the moment of $(E + F)$ is equal to the moment of E, and therefore the whole would be equal to a part thereof, which is absurd."[26] In other words, according to Marchetti, equation (2.2) is related to the metaphysical axiom on the whole and its parts. Given distance, the moment represents a measure of weight, and if such a measure is to be significant, one cannot admit that the whole is equal to one of its parts. Of course (2.2) is rather more exacting; its formulation is presupposed in the demonstration of proposition 4 (among unequal weights suspended at equal distances, the heaviest has the greatest moment) as well as in proposition 8.

At this point, the direction of Marchetti's research seems well defined. He means to show that (2.1) and (2.2) necessarily imply a proportionality between the moment and the product of weight and distance. We know that this can be obtained fairly accurately if we start from the two functional equations and from obvious hypotheses of continuity. We cannot expect a seventeenth-century thinker like Marchetti to produce a demonstration as accurate as Cauchy's, but Marchetti's anticipation is interesting. He starts slowly, first proving that "the moments of commensurate bodies suspended at equal distances are in the same proportion to each other as are their weights" (proposition 14), then establishing that "moments of bodies of equal weights suspended at commensurate distances, are in the same proportion to each other as are their distances" (proposition 17). He extends both results to the case of incommensurate quantities (propositions 15 and

[25] *Ibid.*, pp. 8–10.
[26] *Ibid.*, pp. 2–3.

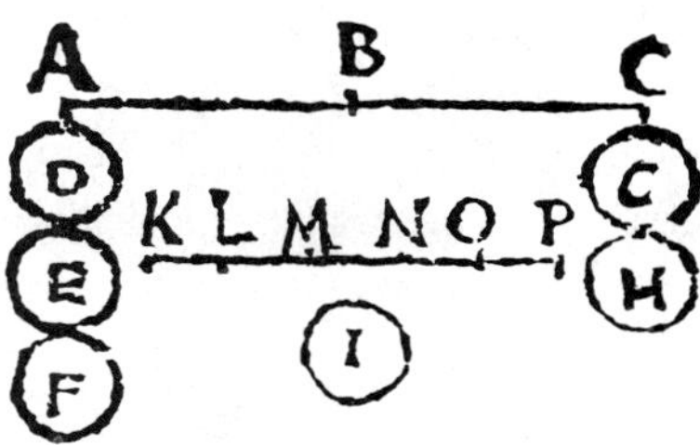

FIGURE 2.12. Marchetti's figure for his fundamental proposition on the concept of moment. From the *Exercitationes Mechanicae* (Pisa, 1669).

18), and finally reaches his goal: "The moments of weights are proportional to the product of weights by distances" (proposition 22).

As an example, we shall mention the demonstration of proposition 14 (Figure 2.12). Let us measure the moments of the equal weights D, E, F, G, H, with equal segments KL, LM, MN, NO, OP. "The sum of moments KL, LM, MN—that is, moment KN—contains the parts NO, OP of moment NP in the same proportion as weight DEF contains the parts of weight GH, whereby KN is to NP, that is, the moment of DEF is to the moment of GH, as the weight DEF is to the weight GH."[27]

We must wait a whole century before the solution of a functional equation expressing Marchetti's postulates becomes instrumental in the demonstration of the law of the lever. The credit for this belongs to a Piedmontese scientist, Daviet de Foncenex in 1760–1761 and to d'Alembert in 1769. We shall examine their contributions in more detail later, but they deserve mention here as well. Daviet de Foncenex tries, in the first instance, to demonstrate Archimedes' rather obscure thesis (already criticized by Huygens)—the possibility of replacing two equal weights suspended at the ends of a horizontal rod by one single body of twice the weight, suspended at the middle of the same rod. Foncenex's demonstration is imperfect (as d'Alembert pointed out) and can therefore be omitted. But the author later analyzes the concept of moment, just as Marchetti had, in order to deduce the law of the lever. The moment (i.e., the "action" or "efficacy") of a weight P placed at a distance x from the fulcrum, Foncenex writes, is a function of P and x in the form $M = P\xi(x)$, where ξ is a function to be determined. Now, by virtue of his previous (flawed) proof, he states that the moment due to P at x is equal to the moment due to the two weights $P/2$ placed at $x-z$ and $x+z$, z being an arbitrary distance. The functional equation follows:

$$P\xi(x) = \frac{P}{2}\xi(x+z) + \frac{P}{2}\xi(x-z). \tag{2.3}$$

[27] *Ibid.*, p. 19.

The solution is immediate. According to Foncenex, if we replace z by the term dx, equation (2.3) becomes $d^2\xi/dx^2 = 0$, whereupon $\xi = Hx + K$ (H, K being constants). He finally observes that if $x = 0$, then $\xi = 0$. It follows that $\xi = Hx$—Marchetti's conclusion of proposition 22.

2.5 De la Hire's Proof, Lagrange's Remarks and Fourier's Contribution

A number of seventeenth-century authors set the law of the lever at the center of their conception of mechanics and struggled to perfect its proof. Among these was Philippe de la Hire, whose *Traité de méchanique* (1695) was later published in the *Mémoires de l'Académie Royale des Sciences.* We shall look at his work more closely later because of his contribution to the statics of arches and vaults; here we shall consider only the general aim of his treatise, which is devoted to the law of the lever. He begins,

> In this work I have tried to demonstrate all the propositions in the ancient geometers' manner, without the assistance of any fundamental axiom or proposition other than the one all writers of mechanics have always assumed. And in order to render it even more evident, I shall demonstrate it with my first proposition by means of another which is more universal, and which provokes no doubts in the field of physics, that is, that in pull of powers, everything being equal on both sides, the pulls are equal [*dans l'effort des puissances, toutes choses étant égales d'un coté et d'autre, les efforts sont égaux*].[28]

The fundamental proposition to which de la Hire alludes is the one we have treated earlier: the obscure point in Archimedes' demonstration of the lever.

> *Proposition I:* Let us consider a straight lever AC at whose extremities two weights or two powers [i.e., forces] are applied, or one power and one weight, equal to each other and with parallel directions, and perpendicular to the lever. At point B that divides the lever into two equal parts there is thus a power H applied there, in a direction parallel to the others, that is, perpendicular to the lever, which is equal to the sum of the other two.[29]

---

[28] P. de la Hire, "Traité de méchanique," *Mémoires de l'Académie Royale des Sciences depuis 1666 jusqu'à 1699*, Vol. 9 (1730), p. aii.

[29] *Ibid.*, pp. 10–11.

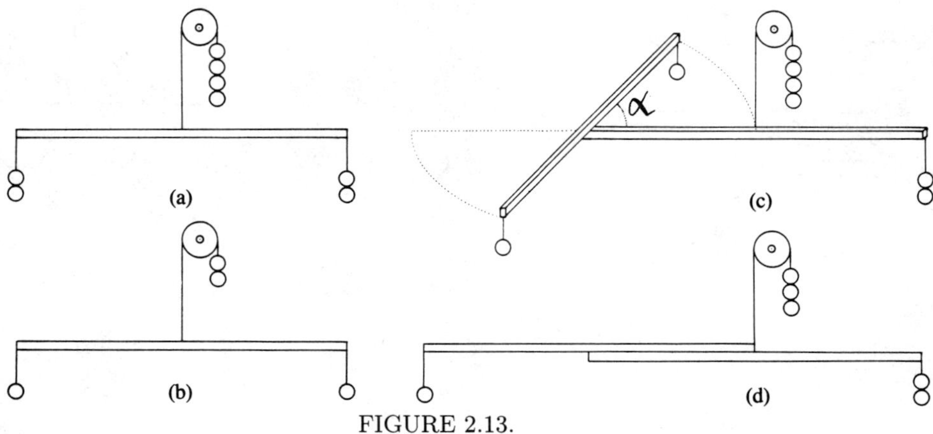

FIGURE 2.13.

To demonstrate this thesis, de la Hire uses the principle of symmetry and three suppositions, set forth in the premiss of the treatise. These suppositions state:

1. The form (shape) of a body is independent of its weight, and can be changed without changing the weight.

2. The direction of the weight is towards the center of the earth. Because of the great distance from the latter, the forces due to weight are parallel to each other.

3. Every force can be transferred along its line of action.[30]

Proposition 1 is proved by transferring the two weights or forces in such a manner as to apply them to the extremes, A and C, of the lever, and by considering the system as a single body whose form can be shrunk into its own center of gravity.

This being so, proposition 3 establishes the law of the lever. Consider Figure 2.13. Let us start with the symmetric situation (Figure 2.13a) and substitute for one of the weights $2P$ the equal arms lever (Figure 2.13b), loaded at its ends with two weights P. The angle α between the two levers (Figure 2.13c) is arbitrary. Therefore, through $\alpha = 0$, we reach the situation shown in Figure 2.13d, which expresses the required thesis.

The merit in de la Hire's proof lies chiefly in the fact that it shows the close relationship between proposition 1 (the load on the fulcrum is the sum of the two weights) and supposition 1 (the weight of the body is independent of its form), the first being deduced from the second. If Vailati's reconstruction is correct, this was implicit in Archimedes, but de la Hire reached the conclusion on his own.

---

[30] *Ibid.*, pp. 8–9.

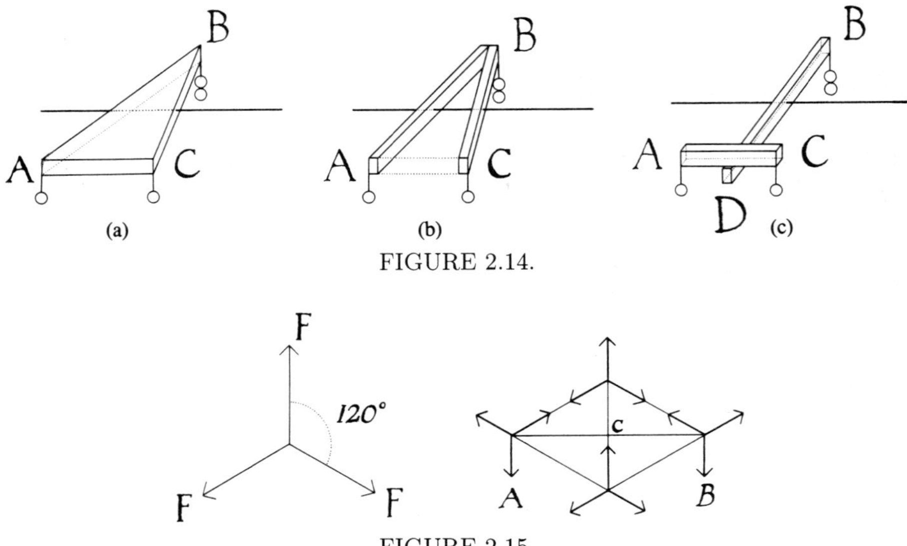

FIGURE 2.14.

FIGURE 2.15.

We may also invert this deduction, getting the thesis of the supposition from that of the proposition. This, in fact, is what d'Alembert does in his treatment "Sur l'équilibre," article 3 in the *Mémoire sur les principes de la méchanique*, published by the Académie Royale des Sciences in 1769. Lagrange later considered d'Alembert's proof to be not entirely satisfactory, since the independence of form and weight can be based on more evident principles.

Consider, Lagrange suggests, a triangle ABC (Figure 2.14a). Load it as shown and suspend it by the axis joining the mean points of the two sides AB and BC. Such a triangle can be interpreted as two levers, AB and BC, each of which stays in equilibrium by symmetry (Figure 2.14b), or as two levers, AC and DB (Figure 2.14c) superimposed in the shape of a capital T. This second interpretation "demonstrates" the thesis of the supposition.[31] Lagrange also cites Fourier's admirable proof of the possibility of substituting a double force, acting on the mean point of the segment joining the two forces, for two parallel equal forces. Fourier's proof is set forth in his work on the principle of virtual velocities (1798), which we will deal with shortly.[32] It is outlined in Figure 2.15.

---

[31] L. Lagrange, *Mécanique analytique*, 2nd ed., Volume 1, (Paris 1811); cf. 3d ed. (Paris, 1853), p. 4.

[32] J.B. Fourier, "Mémoire sur la statique contenant la démonstration du principe des vitesses virtuelles et la théorie des moments," *Journal de l'École Polytechnique, cahier* 5 (1798), reprinted in J.B. Fourier, *Œuvres* (Paris, 1890), Vol. 2, pp. 477–521 (hereafter Fourier, "Mémoire sur la statique").

Consider a system of three concurrent and equal forces at angles of 120°. The equilibrium is assured by symmetry. By arranging four similar systems as in Figure 2.15, we prove the thesis simply by accepting the assumption that the magnitude of two concurrent forces with equal direction is equal to the sum of their magnitudes. In fact, all the non-vertical forces cancel each other out, and only the two vertical forces A and B, balanced by a double force on the mean C, remain.

Fourier uses this result to eliminate all uncertainties from Archimedes' demonstration of the law of the lever, which he puts (as we shall see) at the base of the principle of virtual velocities. But he resorts to an even more basic principle, that of the composition of forces. This becomes in turn the foundation of the law of the lever. The hierarchy of principles established by the Aristotelian author of the *Mechanical Problems*, and confirmed by Archimedes' use of the law of the lever as the basis of everything, is therefore modified. The composition of forces takes on the role which it was to hold in the nineteenth century as the true principle of statics.

2.6 Towards the "Dethronement" of the Law of the Lever: Saccheri and de Maupertuis

In the twenty-fourth letter of the second tome of his *Epistolary* (Paris, 1659) Descartes wrote: "It is a ridiculous thing to want to use the law of the lever for the pulley, as Guidobaldo convinced himself he should do, if I remember correctly." Varignon happened to open his copy of the book at that spot, and quoted Descartes exactly in his *Projet d'une nouvelle méchanique* (Paris, 1687): "*c'est une chose ridicule, que de vouloir employer la raison du Levier dans la Poulie.*" This reflection, Varignon adds, "led me to another one. Is it perhaps more reasonable to resort to the lever for the problem of the inclined plane?" After giving the matter some thought, the author of the new mechanics comes to a conclusion: no simple machine, not even the lever, can embody a law so general that it can be set at the summit of mechanics.[33] What we need is a principle stripped of non-essentials and able to express the deeper truths of equilibrium itself. Descartes and Wallis had already set out along this path, and Varignon took a parallel road with his program for a new mechanics. The search for this principle evokes the great synthesis of Aristotle and his school—the *motus scientia.* According to the fundamental dichotomy *power/act*, there were two possibilities: to focus one's attention on power (potential or virtual motion, based on the principle of virtual velocities), or to concentrate on the actual motion of

[33] R. Descartes, *Epistolae* (Amsterdam, 1668), Vol. 2, p. 107; P. Varignon, *Projet d'une nouvelle méchanique* (Paris, 1687) (hereafter cited as Varignon, *Projet*), preface.

the mobile body, for whose description the principle of the composition of velocities is essential.

From the latter half of the seventeenth century on, the study of mechanics follows two connected but somehow distinct schools of thought. The first descends from Descartes and is more closely related to the medieval tradition, which included the Aristotelian principle of dynamics. At the end of a long and heated debate, this approach evolves into Lagrange's synthesis, which sees the principle of virtual work (or velocity) as the cornerstone of mechanics. The second strategy is that of Varignon and Roberval. It was to dominate statics until the nineteenth-century synthesis of graphic statics, linking up with the instruments of the *Geometrie der Lage*. By incorporating a more or less explicit reference to motion, both schools deviated from Archimedes' assumptions: that statics came second only to geometry, that the axiomatic structure of the former was parallel to and just as accurate as that of the latter, and that statics and geometry were so closely interconnected that they formed a kind of geometric statics or static geometry, as Luca Valerio intended.

The geometrization of statics depends on the rationality of the law of the lever, which can be proved from geometric principles. All other principles seem to require the definition of a relation between velocities and forces – that is, a law of motion. For this reason, a number of late seventeenth-century studies, purposefully or not, questioned the universal validity of the law of the lever. We shall mention only two of them, both by Jesuits who, in those days, were at the center of great scientific debates.

"Give me a fulcrum and I will lift the earth." This proverb, traditionally attributed to Archimedes, was the subject of consideration by some scientists of the sixteenth and seventeenth centuries, including Stevin and Casati. We shall examine the work of Domenico de Riso, a Neapolitan member of the Jesuit College of Nobles.[34] The tone of his dissertation is playfully serious. It aims to oppose Bettino's and Caramuel's opinions about the possibility of constructing a lever which could, in fact, lift the earth by means of a counterweight. In fact, if the center of the earth equals the common center of weights (Figure 2.16, sketch 2), the counterweight at C is ineffective because its displacement would not correspond to a movement approaching the earth. It is hardly worth going through the complicated mechanisms which de Riso suggests for fulfilling Archimedes' promise (see Figure 2.17). But the discussion is amusing and not entirely trivial, because it points out the importance of reconsidering the principles governing the law of the lever when the center of weights is set at a finite distance.

---

[34] See P. Casati, *Terra machinis mota. Dissertationes geometricae, mechanicae, physicae, hydrostaticae, in quibus ... Archimedes Terrae motionem spondens ab arrogantiae suspicione vindicatur* (Rome, 1658); D. de Riso, *Orbis terrarum machinis motus* (Naples, 1682).

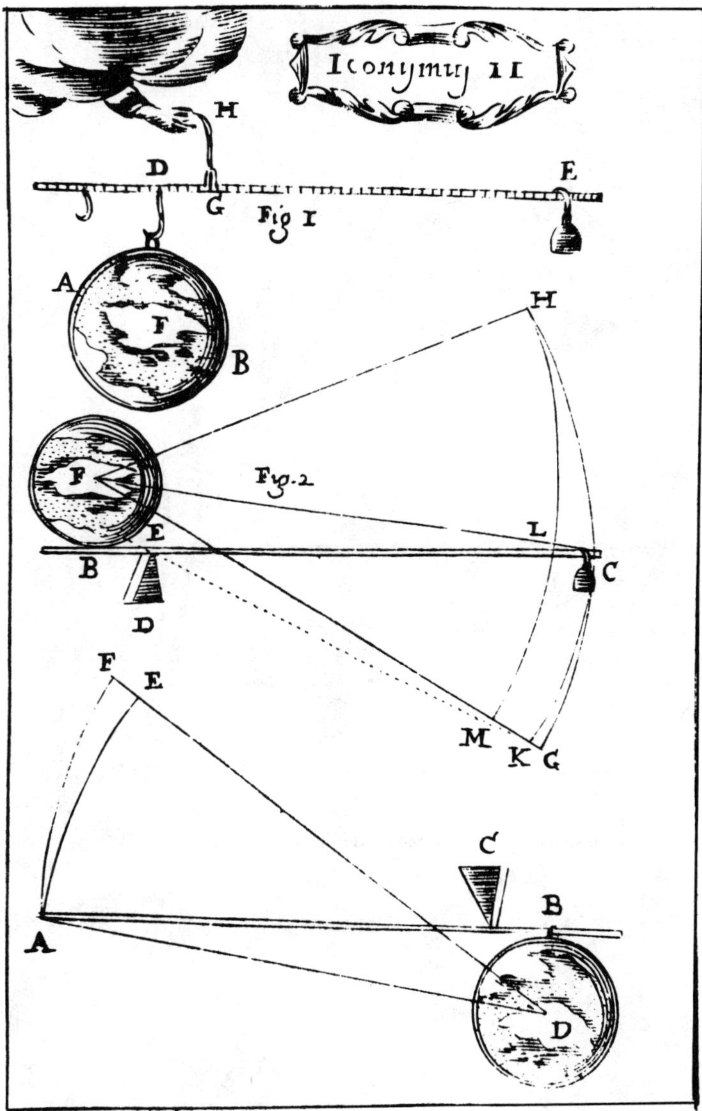

FIGURE 2.16. De Riso's discussion of Archimedes' saying: "Give me a fulcrum and I will lift the earth." From *Orbis terrarum machinis motus* (Naples, 1682).

Such a revision was carried out with great care and seriousness by the Jesuit Girolamo Saccheri. He was born in San Remo in 1667 and taught mathematics at the University of Ticino. With his famous essay *Euclides ab omni naevo vindicatus* (published in Milan in 1733), he began a critical discussion of Euclid's fifth postulate—a discussion which took him to

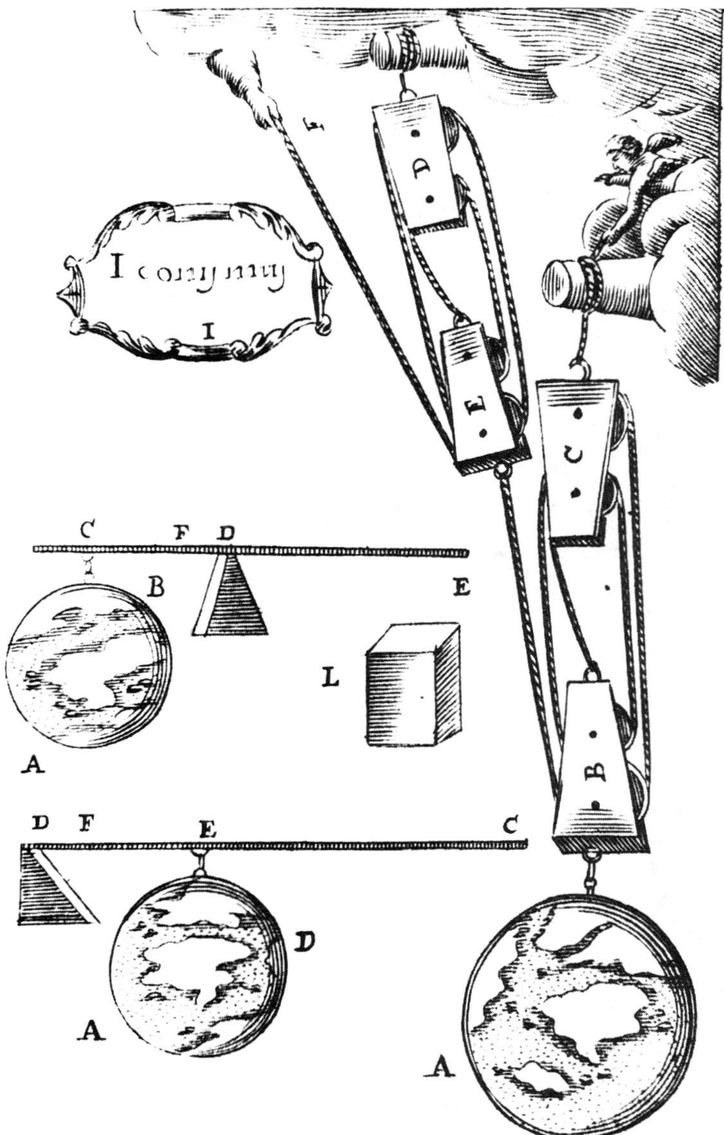

FIGURE 2.17. Machinery suggested by de Riso for fulfilling Archimedes' promise. From *Orbis terrarum machinis motus* (Naples, 1682).

the brink of non-Euclidean geometry. The second book of his treatise *Neo-statica* (Milan, 1708) is entirely dedicated to the problem of the lever. He hypothesizes that the center of weights is *not* situated at an infinite distance, and therefore that gravity cannot be represented by parallel forces. In his opening synopsis, Saccheri explicitly refers to Archimedes' "geometric" demonstration of the law of the lever.

> Archimedes, before all others, and not a few after him, have attempted to demonstrate geometrically that equilibrium subsists if the ratio between the arms is the inverse of the ratio between the weights. But this argument does not appear satisfactory to many geometers. I believe that the reason for dissatisfaction on the part of scholars will emerge from this book. In fact, it is proved here that the law of the lever is valid only if one supposes that the impetus of weights starting from a position of rest is proportional to the distance from the common center of the weights themselves. Moreover, the points of equilibrium are here determined, both in general and in particular, by the common hypothesis, according to which impeta do not vary with the distance.[35]

To understand Saccheri's treatment, we must remember that, to him, "impetus" means the moving force which will give a certain velocity to a body, independent of its weight, and that "momentum" is the force needed to move a given weight at a given velocity. "Weight" is essentially used as it is in common parlance; it is not specifically defined. It is "matter," "body endowed with gravity," "body in itself, devoid of moving force," and so forth. In addition, we should note that in the eighth proposition of his first book, Saccheri demonstrates (in the Aristotelian manner) the law of the composition of impetus—that is, the composition of velocities—and it is precisely from this exposition that the development of the second book derives.

Saccheri introduces two axioms and one postulate about the problem of the lever. The axioms state that (Figure 2.18a):

1. If an inflexible horizontal straight line CR, fixed at N, is acted on by an impetus in the direction of ND, the line does not move.

2. If the direction of impetus does not pass through the fixed point N, the line CR rotates around N.

The postulate states that on every horizontal straight line fixed at N and loaded with a weight at R, it is always possible to find a point C where a weight may be placed so as to effect equilibrium (Figure 2.18b).

Having established this, the author goes on to state a geometrical lemma (proposition 1): given a parallelogram $ABCD$ (Figure 2.19), any straight line CR intersects the diagonal BD at a point N such that $CN : NR = DA : DR$. Consider the similar triangles BNC and RND. Suppose now (proposition 2) that two "moving forces" act on a weight D in the directions CD and RD; the forces have magnitude F_{CD} and F_{RD} such that F_{CD} :

[35] G. Saccheri, *Neostatica* (Milan, 1708), p. 54.

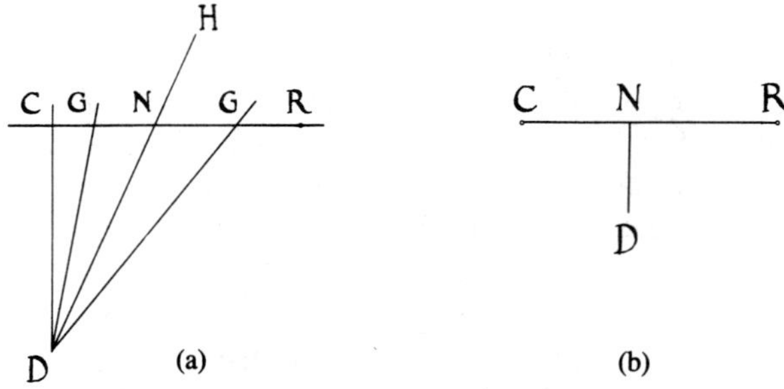

FIGURE 2.18.

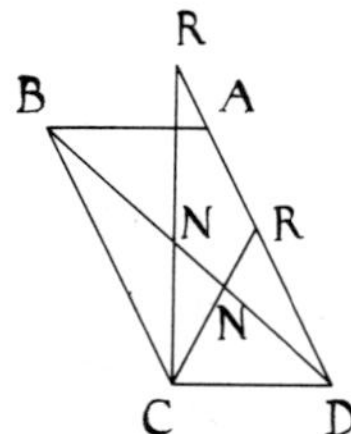

FIGURE 2.19.

$F_{RD} = CD : AD$. Then the weight D will be forced to describe a straight line DB intersecting CR at point N so that

$$CN : NR = (CD : RD) : (F_{CD} : F_{RD}). \tag{2.4}$$

This can be directly attributed to proposition 1. As a result of the first axiom, if the straight line CR is fixed at N and stressed at C and R by F_{CD} and F_{RD} respectively (Figure 2.20), then it stays in equilibrium (propositions 4 and 5).

Unfortunately, the idiomatic confusion caused by failure to define terms like "weight," "impetus" and "momentum" makes the reading of subsequent propositions uncertain. It is not, however, at all difficult to understand Saccheri's aim: the straight line CR represents a lever with its fulcrum at N, subject to "moving forces" at C and R directed towards the center of weights D. If the weights (we would call them "masses") at C and R are

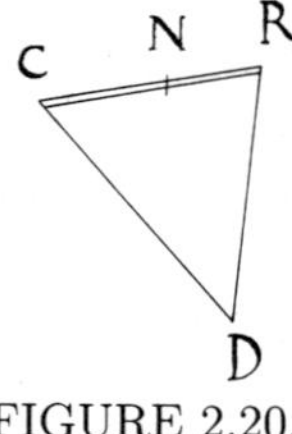

FIGURE 2.20.

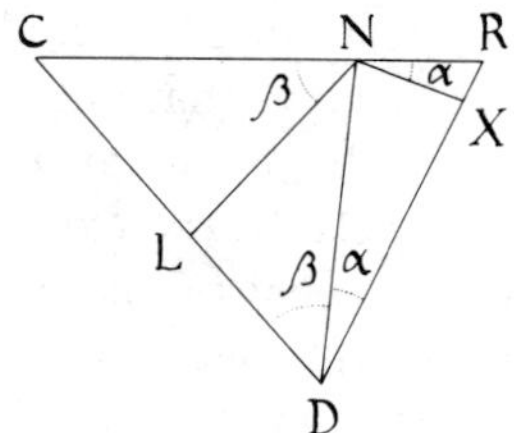

FIGURE 2.21.

equal, then the moving forces are identifiable with the "impeta" (proposition 6). If, on the contrary, the masses are unequal, the moving forces may be identified with the "momenta" (proposition 7). In this case, because "momentum" is defined as the product of weight and impetus, equation (2.1) can be rewritten as follows:

$$CN : NR = (CD : RD) : [(p_C.I_{CD}) : (p_R.I_{RD})] \qquad (2.5)$$

where p_R and p_C are the weights at R and C respectively, and I_{RD}, I_{CD} are the respective impeta.

Proposition 8 states this outright: "The ratio of CN to NR is in direct proportion to the ratio of CD to RD and in inverse proportion to both weight to weight and the impetus in the direction RD or DR to the impetus in the direction CD or DC."[36] In more modern terms, still in keeping with Saccheri's intention, we would substitute masses for weights and gravitational accelerations for impetus.

At this point, the author can formulate his main thesis: if—and only if—the impetus I_{CD}, I_{RD} are proportional to the distances from the common center of the weights,[37] equation (2.5) becomes:

$$CN : NR = p_R : p_C. \qquad (2.6)$$

That is, the principle of the lever (proposition 9ff): the equilibrium point N divides the lever in reciprocal ratio to the weights. If, on the contrary, "the natural impetus of the weights towards the common center are supposed to be constantly equal to any distance," then equation (2.5) establishes a "new" law of the lever, in which LN substitutes for the distance CN and NX for NR (Figure 2.21). NL is the segment obtained by intersecting CD with a straight line passing through N, such that $\widehat{CNL} = \widehat{CDN}$;

[36] *Ibid.*, p. 61.

[37] This hypothesis caused a famous dispute, started in 1635 by Jean de Beaugrand, who claimed the hypothesis as his own discovery. In 1636 he published his "result" under the title *Géostatique.* In the same year, Fermat intervened on his behalf, upon which a number of others weighed in, including Mersenne, Étienne Pascal, Roberval, and Descartes.

NX is similarly derived (proposition 14). The book concludes with other propositions, in which Saccheri shows that only in his own hypothesis (that is, the proportion of the impetus to the distance from the common center of weights) is it possible to demonstrate the existence and uniqueness of the center of gravity. Of course, the usual results of "common" statics coincide with those of neostatics when the distance from the common center of weights is considered infinite (proposition 21).

Saccheri's ingenious text is confirmed by de Maupertuis' famous *Mémoire*, presented in Paris in 1740.[38] De Maupertuis has little faith in rational proofs of the principles of mechanics and statics: "Never has anyone given a *rigorous* general demonstration of these principles."[39] But his point of view differs from Saccheri's. In his opinion, the demonstration of principles is a matter for "some superior science" which would embrace all the laws of nature, but to pave the way for such a synthesis, partial and easily verified principles may be useful. The principle of living forces (very fashionable, in those days) is of the latter sort, as is the law of rest which de Maupertuis now proposes:

> Let a system of heavy bodies be given, or bodies attracted towards some centers by forces, with each force acting on each body as a power n of its distance from its center. In order that all these bodies remain at rest, it is required that the sum of the products of each mass by the "magnitude of the force," as well as by the power $n+1$ of the distance from the center of the force (definable as the *sum of the rest forces*) be a *maximum* or *minimum*.[40]

The demonstration of this is fairly simple, and rests on the decomposition of forces and the principle of virtual velocities. In the interests of brevity, we shall omit it, but its consequences are interesting. First, Torricelli's principle reappears: if gravity is the sole force involved and if the center of weights is at an infinite distance, then equilibrium requires that the center of gravity of the system be as low or as high as possible.[41] With reference to the law of the lever, de Maupertuis reaches results which are completely analogous to Saccheri's. Consider the lever ACB (Figure 2.22) with fulcrum C. Let the masses A, B be attracted to the common center of weights F

---

[38] P. Moreau de Maupertuis, "Loi du repos des corps," *Mémoires de l'Académie Royale des Sciences* (1740), pp. 170–176.

[39] *Ibid.*, p. 170.

[40] *Ibid.*, p. 171; "magnitude of the force" means the central force acting on the mass unit at unit distance.

[41] *Ibid.*, p. 173.

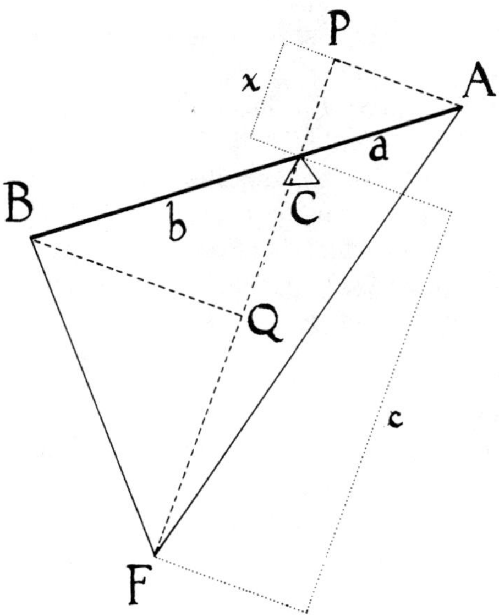

FIGURE 2.22.

by a force with magnitude p, and proportional to power n of the distance from F. If we call $CA = a$, $CB = b$, $CF = c$, and $CP = x$, the result is:

$$FA = \sqrt{c^2 + a^2 + 2cx}, \qquad FB = \sqrt{c^2 + b^2 - \frac{2bcx}{a}}.$$

The rest law therefore requires that

$$pA(c^2 + a^2 + 2cx)^{(n+1)/2} + pB\left(c^2 + b^2 - \frac{2bc}{a}x\right)^{(n+1)/2} = \text{extr.}$$

whence we obtain

$$Aa(c^2 + a^2 + 2cx)^{(n-1)/2} = Bb\left(c^2 + b^2 - \frac{2bc}{a}x\right)^{(n-1)/2}.$$

This last equation, as de Maupertuis observes,

> shows that if the center of the force is at an infinite distance, as is supposed for heavy bodies studied in ordinary mechanics, it is clear that, whatever the power of the distance according to which such force acts, the terms a^2, b^2, and those containing x are negligible compared to c^2. It is sufficient, in order to have equilibrium, that $Aa = Bb$, that is that the masses of

> the two bodies be in inverse proportion to the lever arms; the equilibrium will exist in all the lever positions independently from x. If $n = 1$, that is, if the force acts in direct proportion to the distance from center F, we again have, for a condition of equilibrium, $Aa = Bb$. From this it is evident that in this hypothesis there is still a point C, around which the system of the two bodies is always in equilibrium if it is in equilibrium in one position. This means that, in these two hypotheses, there is a center of gravity which is always the same in every position. But beyond these two hypotheses, it is evident that for the law of the rest, the existence of such a center is impossible.[42]

These are the same conclusions as Saccheri's. The once-sovereign law of the lever, the foundation and pillar of the old mechanics, has today lost most of its fascination. It applies to particular and limited cases, and is subordinate to a specific hypothesis about the force of gravity. Other principles have taken over—those very principles which provided ammunition for the coup. They are the principles of the composition of "impetus," of "velocities" and of "forces," as Saccheri called them, and the principle of virtual velocities which underlies de Maupertuis' law of rest. To this last, we should now turn our attention.

---

[42] *Ibid.*, p. 175.

3

The Principle of Virtual Velocities

3.1 Medieval Roots

Of the two propositions—the equation of virtual velocities and the rule of the composition of forces—the former is clearly richer in tradition and well-known references. It seems to promise more to the advancement of knowledge and to the solution of the eternal "why?"

As we have shown, the origins of the concept of virtual velocities go back as far as Aristotle; he and the anonymous author of the *Mechanical Problems* used it to approach the enigma of movement, and to explain the seeming disproportion between cause and effect created by such tools as the lever. As they showed, the principle of virtual velocities established a balance or a proportion of sorts; it ensures that something remains unchanged. We shall follow the evolution of this concept, giving special attention to the theoretical problem of the nature of its equation. Is it a principle or a theorem? If the former, what is its "metaphysical" basis? Does it in fact have one? If it is a theorem, what is the proposition from which it derives?

We will not discuss the Greek origins of the equation, but we should note that this prehistory includes fragments attributed to Euclid,[1] Hero's *The Elevator* (in which the discussion of virtual velocities is interwoven with the definition of the law of motion),[2] and medieval texts by Jordanus de Nemore and his school.[3] The Jordanic demonstrations exhibit a principle which Descartes would accept as the only basis for statics, and which,

---

[1] Paris, Bibliothèque Nationale, "Liber Euclidis de gravi et levi et de comparatione corporum ad invicem," fonds latin, ms 10260.

[2] G. Vailati, "Il principio dei lavori virtuali da Aristotele a Erone d'Alessandria," *Atti R. Accad. Sci. Torino*, Vol. 32 (1897) (hereafter cited as Vailati, *Principio*), reprinted in Vailati, *Scritti*, pp. 91–106.

[3] Paris, Bibliothèque Nationale, *Elementa Jordani super demonstrationem ponderis*, fonds latin, ms 10252; *see also* the *"Commentaire peripatéticien" of Elementa Jordani*, fonds latin, ms 7378a, and *Liber Jordanis de ratione ponderis*, fonds latin, ms 8680a; cf. P. Duhem, *op. cit.*, Vol. 1, pp. 112–147. The major publication about Jordanus was given, in recent years, by J. Høyrup, "Jordanus de Nemore, 13th century mathematical innovator: an essay on intellectual context, achievement, and failure," *Archive for History of Exact Sciences,* Vol. 38 (1988), pp. 307–363.

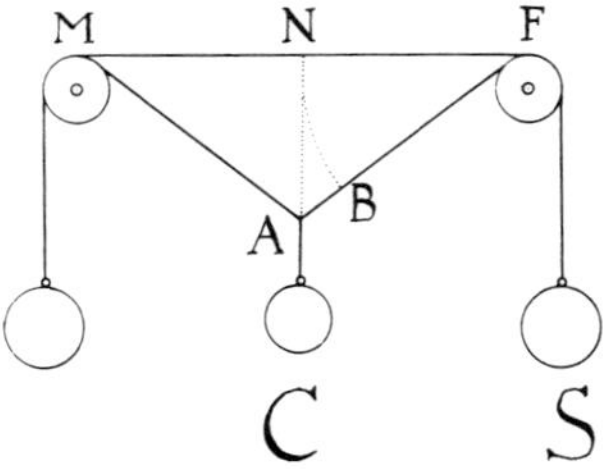

FIGURE 3.1.

thanks to Bernoulli, would become the principle of virtual work: if force x can lift weight y to height z, it can also lift weight ky to height z/k.[4]

The medieval tradition of Jordanus' school continued into the Renaissance and was taught as a matter of course. But modern authors have incorrectly attributed the concept to the post-medieval new formulation of static principles. For example, G.B. Venturi, referring to Leonardo da Vinci's manuscripts, attributes the discovery of the principle of virtual work to him. Leonardo's notes contain the following statement:

> 1 August 1499. Let cord FM [Figure 3.1] be stretched by two equal forces F, M. Apply in the middle of the cord, at N, a small weight C; it will make point N descend as far as A, causing weight S to rise at the same time. With radius FN, trace arc NB, [and] the movement of weight S will be BA. Point N will ascend until there is the proportion $C : S = BA : NA$, that is, until the respective movements of the two weights are between them in inverse ratio of the weights themselves.[5]

Nicolò Tartaglia's *Quesiti et inventioni diverse* (Venice, 1546) is presented as an original work, with no reference to the past. The principle of virtual work becomes the pivot around which the most important propositions revolve. For example, in the twenty-ninth *Quesito, propositione seconda*, Tartaglia proves "*per absurdum*" that since "the ratio of the powers of heavy bodies of a same kind equals the one of their velocities—in descending," equilibrium requires that "the ratio of their opposite movements—that is, of their ascent—is the same, but transmutatively."[6] Consider a lever; at the end of one arm there is a body *ab* subject to power *de*, and at the end of the other arm there is a body *c* subject to power *f*. "Therefore

[4] P. Duhem, *op. cit.*, Vol. 1, p. 121.

[5] G.B. Venturi, *Essai sur les ouvrages physico-mathématiques de Léonard de Vinci ...* (Paris, 1797), p. 17. Venturi refers to Vol. N, p. 103 of Leonardo's manuscripts kept in the Bibliothèque Nationale of Paris.

[6] N. Tartaglia, *Quesiti et inventioni diverse...* (Venice, 1554), pp. 87r–87v.

the velocity of body c, in the opposite motion, will be to the velocity of body ab, as the power de will be to power f."[7]

Part of Tartaglia's book concerns mechanics. It is in the form of a dialogue with Don Diego Hurtado de Mendoza, imperial ambassador to Venice. It contains many propositions of the medieval science of weight, translated almost word-for-word from Jordanus de Nemore and his school. This is evidence not that Tartaglia was a plagiarist, but that the medieval tradition was accepted as common knowledge, not requiring attribution. According to Vailati,[8] Jordanus and his follower ("Leonardo's forerunner," according to Duhem)[9] may have set forth in their works previously known concepts and results.

We have Tartaglia to thank for his edition of the *Jordani opusculum de ponderositate*, which Duhem discovered in manuscript under the title *Liber Jordani de ratione ponderis* and attributed to "Leonardo's forerunner." In 1565, after Tartaglia's death, the famous Venetian editor Curtio Trojano published this work. In it, the medieval author tends to base his arguments on the analysis of displacements and virtual velocities, excluding all other principles however evident and intuitive they may be. To mention only one of Vailati's significant quotations: in question 4, it is demonstrated that, if two bodies hanging from a balance are in equilibrium, the equilibrium is not disturbed if we vary the length of the wires by which the bodies are suspended. This is drawn from the fact that the virtual displacements of the weights do not depend on the length of the wires.

> In order to find another work in which such an absolute centralization and, I would almost say, such a despotic subjugation of the whole of statics to the principle of virtual momenta [i.e., virtual velocities] is pursued, ... and in which any intervention and initiative is so rigorously denied to direct intuition (which is so widely made use of in Archimedes' treatment), we must wait for Descartes, whose little work entitled *Explicatio machinarum atque instrumentorum quorum ope gravissima quaeque pondera sublevantur*, constitutes the first attempt of those made later to construct the entire edifice of statics according to a plan that was then brought to an end with Lagrange's *Mécanique analytique*.[10]

Descartes' and Lagrange's works will form an important part of our discussion; we must examine the latter to understand the meaning of the equation of virtual velocities as a *reductio ad unum* of the principles and methods—even the very concepts—of mechanics. Before we turn to these

---

[7] *Ibid.*, p. 88r.

[8] G. Vailati, "Principio," p. 100.

[9] P. Duhem, *op. cit.*, Vol. 1, pp. 134ff.

[10] Vailati, "Principio," p. 103.

two authors, we should give a short account of those to whom Lagrange himself attributes the origin of such a principle: Guidobaldo del Monte and Galileo.

3.2 Guidobaldo del Monte, Galileo, and the Principle of Virtual Velocities

While Lagrange is wrong in attributing the principle of virtual velocities to Guidobaldo del Monte,[11] there is no doubt that the latter's *Mechanicorum liber* refers often to the calculation of "spaces" which are described by the "moving power" and the "moved weight."[12] This calculation allows Guidobaldo to explain the properties of the lever. For example, proposition 4 states "If the power moves the weight suspended by the lever, the space of the moving power is to the space of the moved weight as the distance between the support and the power is to the distance between the same support and the hanging weight." From this, the author deduces that "The space of the power that moves has a greater proportion [with respect] to the space of the moved weight than the weight to the same power."[13]

We should, above all, mention the very accurate analysis which Guidobaldo makes of the tackle, where the equation of virtual (or effective) work is particularly in evidence. To begin with, in the first to ninth propositions, he examines various tackles from the simplest to the most complex, determining the tensions of the threads and the powers needed to equilibrate the weights. For example, the second proposition deals with the case of the pulley system shown in Figure 3.2. "The power which is in G will be half of weight A."[14] In the case of Figure 3.3, proposition 5 proves that "the power supporting the weight will be a third of weight A."[15] And so on, until we reach proposition 9 (Figure 3.4), in which "the power will be a fifth of the weight."[16] In propositions 10 through 28 (the last), Guidobaldo develops the kinematic analysis, calculating geometrically the ratio between the displacement of the weight and the corresponding displacement of the power for each type of pulley. Proposition 11 (Figure 3.5) asserts that "the space described by the power that moves the weight is twice the space of the same moved weight."[17] His treatment is often intricate because of the variety of examples given, and because the pulley systems which he invents to

---

[11] L. Lagrange, *Mécanique analytique*, 3rd ed. (Paris, 1853), p. 18.
[12] Guidobaldo del Monte, *Mechanicorum liber* (Pesaro, 1577).
[13] G. del Monte, *Le mechaniche*, tr. F. Pigafetta (Venice, 1581), pp. 38v–39r.
[14] *Ibid.*, p. 59r.
[15] *Ibid.*, p. 63r.
[16] *Ibid.*, p. 68r.
[17] *Ibid.*, p. 73v.

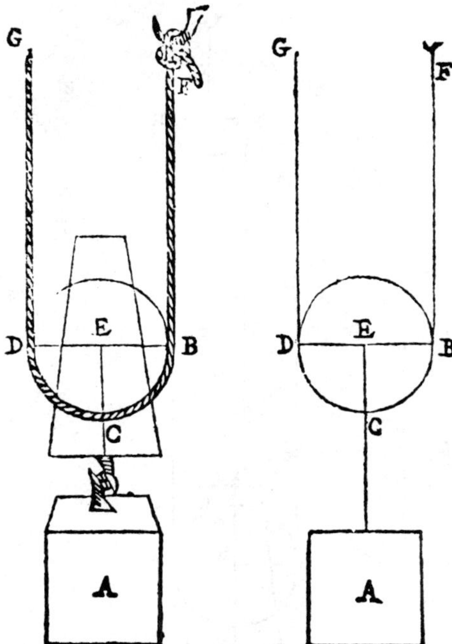

FIGURE 3.2. "The power which is in G will be half of weight A." From *Le mechaniche* (Venice, 1581).

obtain specific ratios between weight and counterweight are often complex. He always reaches the same result, given in the corollary to proposition 26 (Figure 3.6): "The space of the power that moves always has a greater proportion towards the space of the moved weight than the weight has towards the same power."[18] Therefore, equilibrium requires that "the space of the power that moves is to the space of the weight as the weight is to the power that supports it."[19]

It is difficult to understand why Mach, in his *History of Mechanics*, refused to recognize the remarkable value of Guidobaldo's considerations. Mach claimed that Guidobaldo "lacks a comprehension of the principle which eliminates the prodigious character of the action of machines," and attributes the analysis of pulleys given above to Simon Stevin. Stevin's *Hypomnemata mathematica* contains the proposition "as the space of the actor is to the space of the sufferer, so is the power of the sufferer to the power of the actor" ("*ut spatium agentis, ad spatium patientis, sic potentia patientis, ad potentiam agentis*")[20]—a principle already stated in Guidobaldo. This

---

[18] *Ibid.*, p. 99v.

[19] *Ibid.*, p. 97r.

[20] S. Stevin, *Hypomnemata mathematica* (Leiden, 1608), Vol. 4, Book 3, p. 172

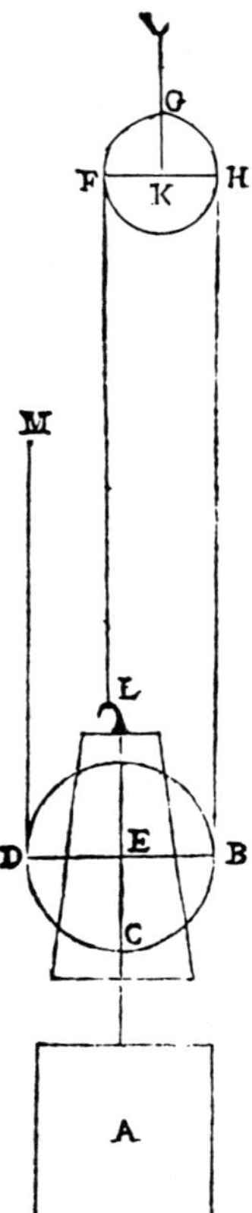

FIGURE 3.3. "The power supporting the weight will be a third of weight A." From *Le mechaniche* (Venice, 1581).

is even more curious since Stevin (as we know) disliked the principle of virtual work. In fact, in his *Hypomnemata* (or, more precisely in the original edition of his *Beghinselen der Weeghconst*) Stevin set himself resolutely against the "Aristotelian" method, which based the equilibrium of the lever on the consideration of the circular arches virtually described by the application points of weights. Stevin is almost sarcastic: "What is immobile does not describe circles, but two weights in equilibrium are immobile; thus two weights in equilibrium do not describe circles." With this curious syllogism, bombastically produced in the medieval *EAE* form, Stevin established the party of those who thought little of the principle of virtual velocities, seeing it as a criterion, not an explanation, of equilibrium.[21]

Galileo's position was different. The treatise on mechanics which Lagrange cites in his attribution is inserted at the end of the first volume of

[21] *See* P. Varignon, *Projet.*

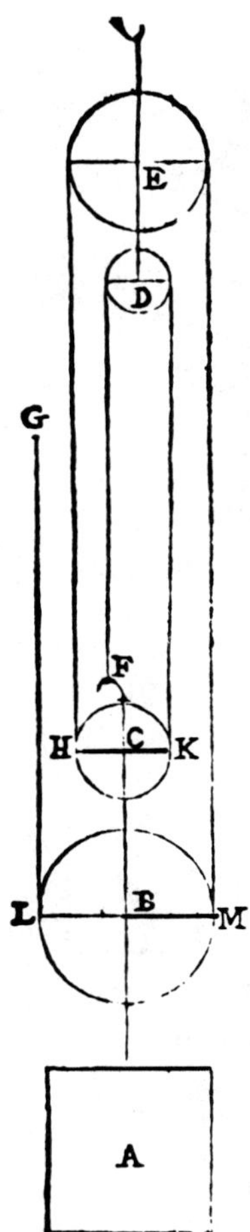

FIGURE 3.4. "The power will be a fifth of the weight." From *Le mechaniche* (Venice, 1581).

the edition of Galileo's works published at Bologna in 1665 with a different sequence of pagination. On page 30 of this treatise (which dates from between 1593 and 1599) we find the following remarkable explanation of equilibrium on an inclined plane (Figure 3.7):

> One must observe that although mobile E will have run over the whole line AC at the same time as the other body F will have descended by an equal distance, body E will not be distanced from the common center of heavy things farther than perpendicular CB, just as body F, descending perpendicularly, will have covered a distance equal to the whole line AC. And as heavy bodies do not resist transversal movements, unless they are distanced from the center of the earth, and as mobile E in

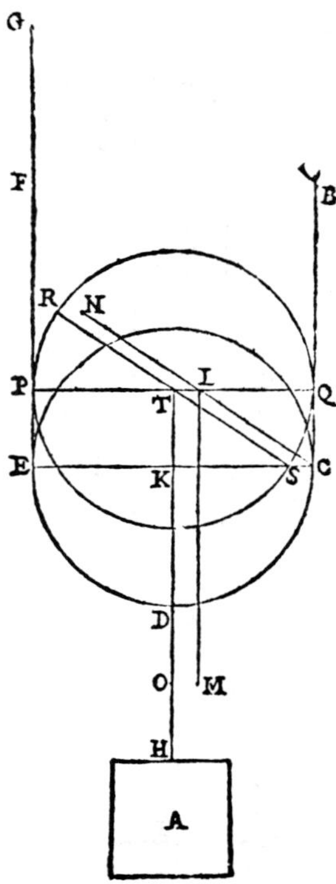

FIGURE 3.5. The space described by the powers... is twice the space of the ... weight." From *Le mechaniche* (Venice, 1581).

> the whole movement AC has not ascended higher than line CB, but the other body F descended perpendicularly to the same extent as line AC, we can thus legitimately say that the path [viaggio] of force E maintains the same proportion as line AC to CB, that is, weight E to weight F.

Not only does Galileo enunciate and apply the equation of virtual velocities, but he also gives a brilliant justification of it. In fact, such an equation can be drawn from the evident principle that the equilibrium of the system is preserved if the barycenter of the system itself does not descend. Finally, Torricelli, in his *De motu gravium naturaliter descendentium et projectorum*,[22] explicitly formulates this principle, the ancestor of the *extremum* principles in statics.

---

[22] *See* E. Torricelli, *Opera geometrica Evangelistae Torricelli* (Florence, 1644), p. 99. *See* also *Opere di Evangelista Torricelli*, ed. G. Loria e G. Vassura, Vol. 2 (Faenza, 1919), p. 105.

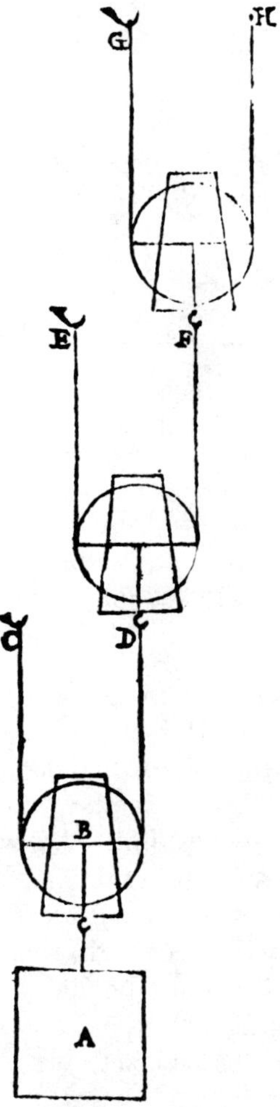

FIGURE 3.6. Guidobaldo's figure for supporting his formulation of the principle of virtual velocities. From *Le mechaniche* (Venice, 1581).

3.3 Descartes: "Explicatio Machinarum Unico Tantum Principio"

Descartes' treatise on mechanics seems expressly designed to fit the doctrine that "a long book is a great evil" (τὸ μέγα βιβλίον ἴσον μεγάλῳ κακῷ), a saying quoted by N. Poisson in his preface to the 1704 edition

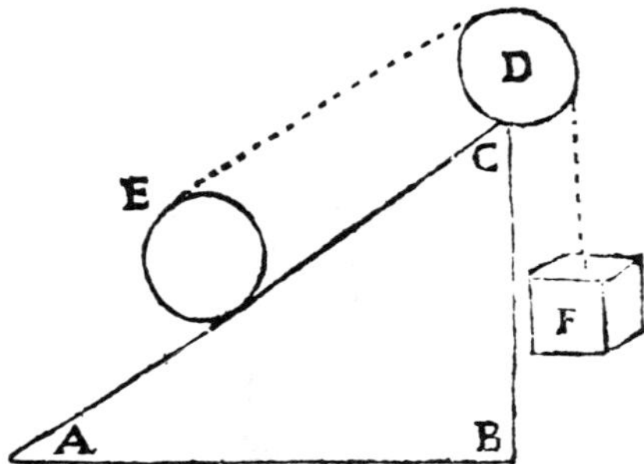

FIGURE 3.7. Galileo's solution of the problem of the inclined plane by means of the equation of virtual velocities. From the *Opere* (Bologna, 1665).

of Descartes' *Opuscula posthuma.*[23] Descartes' "book," entitled *Explicatio machinarum atque instrumentorum, quorum ope gravissima quaeque pondera parvis viribus levari possunt*, is only nine pages long. Its beginning summarizes the whole:

> The invention of all these machines is founded on only one principle which affirms: with the same forces with which we can lift a weight of (e.g.) 100 lbs to the height of two feet, we can also lift a weight of 200 lbs to the height of one foot. And this principle must necessarily be admitted, if we consider that there must always and necessarily be a proportion between action and the effect produced by it: so that if a certain action is able to lift a weight of 100 lbs two feet high, then the weight that the same action lifts one foot must be 200 lbs. In fact, lifting 100 lbs one foot and yet again 100 lbs one foot is the same as lifting 200 lbs one foot or 100 lbs two feet.[24]

This passage—which is almost exactly what Descartes wrote to Constantijn Huygens in a letter dated October 5, 1637—is particularly interesting. Not only does it state the only principle to which all machines are later reduced, but it does so in terms of the concept of action.

Aristotle recognized in *act* the aspect which makes movement intelligible; it is evidence of a constant and necessary proportion between "before" and "after". The Greek term ἐνέργεια (*energeia*) corresponds even better than "act" to the Cartesian "action." There is, however, a difference. Aristotle sought a constant proportion among the three elements which determine a local motion: the body (i.e., weight), the distance moved, and the time

[23] R. Descartes, *Opuscula posthuma physica et mathematica* (Amsterdam, 1704), preface by N. Poisson, p. 11.

[24] R. Descartes, "De mechanica tractatus una cum elucidationibus N. Poissonii," in *Opuscula posthuma physica et mathematica* (Amsterdam, 1704), p. 13

needed for movement. But Descartes limited himself to the first two, the weight and the distance covered. In a letter to Boswell in 1646, he consciously expresses this choice:

> I do not deny the objective truth of what mechanicians are used to saying, and that is: the greater the velocity of the long arm end of a lever compared with the other end, the less force it needs to move. But I deny that the velocity or slowness may be the cause of this effect.[25]

To discover this cause is Descartes' aim. His principle seems to answer this purpose perfectly. If we ignore time, to say that the act of lifting 100 lbs by two feet once is the same as lifting 100 lbs by one foot twice is so obvious that it is almost a tautology. If, on the other hand, we take time into account, correlating weight with velocity, the question becomes more complicated. In a letter to Father Mersenne (September 12, 1638) Descartes observes: "It is impossible to say something correct and definite [*bene et solide*] about velocity without having thoroughly clarified what heaviness is, and thus the whole system of the world."[26] For this reason, he wants to see in his own principle an explanation which, he thinks, Galileo's treatment of the balance and lever lacks. Galileo explains what happens, Descartes writes to Mersenne on November 15, 1638, but not *why* it happens, as Descartes himself does. Perhaps he believes that his principle explains, simply and exactly, the "efficacy" of force, distinguishing it from force itself.

Actually, this is the distinction between force and momentum which we have discussed above. *Explicatio machinarum* attributes the power of lifting a weight to a certain height to the efficacy, or action, of a force. While the weight is characterized by only one "dimension" (its weight), the action or efficacy is defined by two, weight and distance. In his letter of September 12th to Mersenne, Descartes states this clearly: "The force of which I have spoken always has two dimensions and is not the force necessary to [assure the equilibrium of a weight on its support]: this latter has only one dimension."[27] This corresponds to the notion of momentum as composed of two elements, t and n, where t is a measure of force and n is a geometric quantity. The two-dimensional force (in Descartes' terms, which in this case are anything but "clear and distinct") is exactly analogous to momentum. Velocity has to be excluded; if we introduced space and time, "we should have to attribute three dimensions to force."[28]

All of Descartes' statics then proceeds from the consideration of a two-dimensional quantity tn, where t is the force in itself or the weight, and n

---

[25] R. Descartes, *Epistolae*, (Amsterdam, 1668) (hereafter Descartes, *Epistolae*), part 2, letter 23, p. 102.

[26] *Ibid.*, part 1, letter 74, p. 256.

[27] *Ibid.*, p. 254.

[28] *Ibid.*, p. 256.

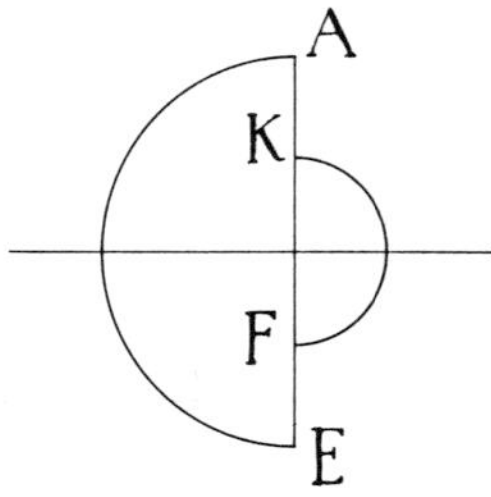

FIGURE 3.8.

is the distance (the height) to which the weight is lifted. The equilibrium rule based on tn governs all simple machines. For example, the general principle permits us to deduce the law of the lever by computing the ratio between the diameters AE and KF, representing the heights to which the weight and the counterweight can be lifted (Figure 3.8). A few passages in the letters imply that Descartes noticed the "infinitesimal character" of his principle of virtual work—this, at least, is René Dugas' opinion.[29] Moreover, in his analysis of collisions, Descartes introduces a law of conservation which is related to the quantity of motion (i.e., to the product of mass and velocity). As the discussions of live forces make clear, this implies the consideration of yet another "bidimensional" quantity tn for the measure of the action of force; in this quantity, n is a temporal interval. The distinction between the two two-dimensional quantities lies at the heart of the great debate on the true measure of forces—a debate in which disagreements on the nature of live forces were lively indeed.

3.4 Bernoulli and Varignon

We can finally turn to the most important contribution to the development of the principle of virtual work, Johann Bernoulli's letter to Varignon. The breadth of the formulation which Bernoulli proposed changed the nature of the principle itself, which became more than an explanation of simple machines (like Descartes'). Bernoulli's concept is a fundamental, all-embracing criterion for defining equilibrium in any problem of statics.

When Varignon received Bernoulli's letter, the completed manuscript of his *Nouvelle mécanique* was on his desk. Thirty years earlier, he had promised that it would be his crowning achievement as a scientist. It identified the composition of movements and forces as the supreme principle of mechanics. This choice was, he thought, justified by the great generality and simplicity of the parallelogram rule, which could justify other basic propositions such as the law of the lever. Varignon welcomed Bernoulli's

[29] R. Dugas, *Histoire de la mécanique* (Neuchatel, 1950), p. 150.

proposal, but not without uneasiness. The new proposition could hold its own in the field; it was beautifully general and simple, and equally (if not more) efficacious in the solution to problems.

What was he to do? There was only one way in which he could defend his own choice (and incidentally confirm his pre-eminence), and that was to prove that the principle of virtual velocities as Bernoulli presented it was a corollary, however noble and important, of the principle of the composition of forces.

And this is what Varignon did. Under the title *Corollaire général de la théorie précédente*,[30] he treated Bernoulli's thesis and general demonstration, with all its manifold applications. As Varignon writes,

> In a letter from Basle dated January 26, 1717, Mr. [Johann] Bernoulli, after having defined there what he meant by the word *energy*, as will be seen in the following definition, declared to me that *in every equilibrium of any forces whatsoever, in whatever way they are applied to one another, either indirectly or directly, the sum of positive energies will be equal to the sum of negative energies, taken positively.* This proposition seemed so general and beautiful to me that, seeing that I could easily deduce it from the preceding theory, I asked him for permission, which he gave me, to add it here with the demonstration obtained from this theory, but which he did not send me.[31]

Definition 32 is taken verbatim from Bernoulli's letter and introduces the concepts of virtual velocities and energy:

> Given several forces acting in various directions that hold in equilibrium a point, a line, a surface or a body, let us imagine applying to the whole system of these forces a slight movement, either parallel to itself in any one direction or around any one fixed point. It is easy to see that, because of this movement, each of the forces will go forward or backward in its own direction, unless the direction of the slight movement is perpendicular to one of the forces.... [These] advances and withdrawals are what I call *virtual velocity*. ...[32]

Consider Figure 3.9. If P is a point where force F is applied, and Pp is the slight movement of P which moves PF into pf, "we thus draw PC perpendicular to fp; we have Cp for the virtual velocity of force F, and

---

30 P. Varignon, *Nouvelle mécanique*, (Paris, 1725) Vol. 2, Section 9, pp. 174–223)

31 *Ibid.*, p. 174.

32 *Ibid.*, p. 175.

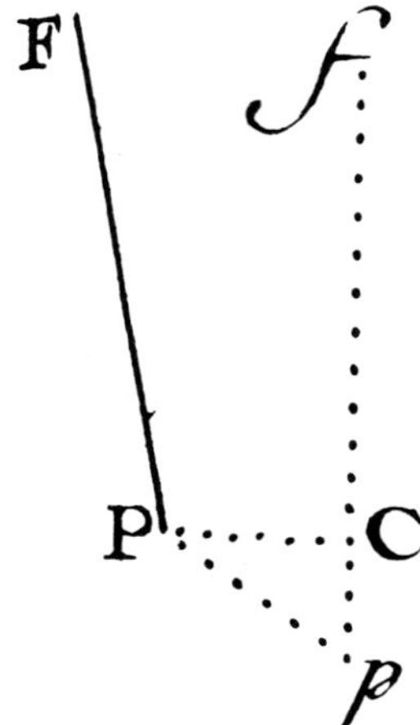

FIGURE 3.9. Bernoulli's definition of virtual velocity and energy. From the *Nouvelle mécanique* (Paris, 1725).

the product $F \times Cp$ is what we call *energy*."[33] Thus the energy is either negative or positive depending on whether angle $\widehat{FPp}$ is obtuse or acute.

Having stated this, Varignon then reformulates Bernoulli's general proposition as theorem 40, and gives a detailed demonstration of it in several parts in order to include every case (including impact). Figure 3.10 shows one such case. Point A is subjected to weight K and forces P, Q, R, S, T, whose intensity is measured by segments AB, AC, AE, AF and AM respectively. If we assign a small segment Aa on AK, and draw segments ap, aq, ar, as, at, all perpendicular to the directions of the respective forces, the law of composition of forces lets us affirm (for example) that projection b of force P on the vertical AK is given by $b = P.Ap/Aa$. By analogy, we have

$$c = Q.Aq/Aa, \quad e = R.Ar/Aa, \quad f = S.As/Aa, \quad m = T.At/Aa.$$

But the equilibrium according to direction AK requires that

$$c + e + f - b - m = K.$$

Therefore we must establish that

$$Q.Aq + R.Ar + S.As - P.Ap - T.At = K.Aa$$

or, rearranging terms and generalizing, that

$$K.Aa + P.Ap + T.At + \text{etc.} = Q.Aq + R.Ar + S.As + \text{etc.}$$

This equation corresponds exactly to Bernoulli's.

[33] *Ibid.*, p. 176.

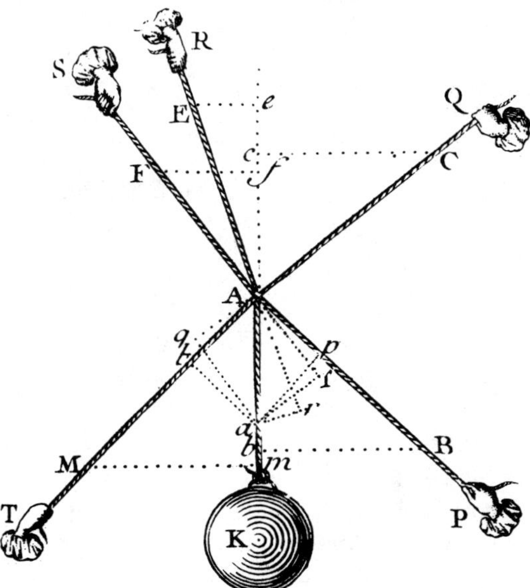

FIGURE 3.10. Varignon's demonstration of the principle of virtual velocities. From the *Nouvelle mécanique* (Paris, 1725).

With Bernoulli and Varignon, the principle of virtual velocities becomes sufficiently general; it can be seen as the very basis of statics. At the same time, in contrast with this generality, Varignon's statement slightly devalues the principle by presenting it as a corollary of a stronger, more general theory.

3.5 Riccati's "Universal Principle of Statics"

Later authors, in discussing the principle of virtual velocities, usually followed Varignon's lead. On one hand, the principle clearly had great value; it gave unity and coherence to all of statics. On the other hand, many thinkers were convinced that the principle had to be demonstrated, not in terms of its adherence to physical reality, but in terms of its coherence with other principles which they believed to be more evident—for example, the composition of forces and the law of the lever.

One may see Vincenzo Riccati's essay *De'principj della meccanica* (Venice, 1772) as something of a deviation from this trend. It sets out concepts in an order and synthesis which the author had long worked upon. Riccati studied the "universal principle of statics" and was thoroughly conscious of the problems involved, at the core of which was the uncertainty inherent in the concept of "power"—that is, of force.

> What are these powers and what is their nature? I and all tne philosophers who want to be sincere, are compelled to make a frank confession that we do not know. As certain as their existence may be, their nature and essence is just as obscure; neither is there any gleam of hope that we may know sooner or later. Since physics has been engaged in seeking the essence of powers, it has not put one foot forward, neither has it discovered a truth; and it has done nothing else but produce unsubstantial hypotheses and enrich the history of opinions.[34]

If nothing else can be said about the essence of force or power, we can at least establish criteria which allow us to measure and compare the "proportion between the values of some powers." This necessarily leads to the consideration of equilibrium, where it is obvious that the effects or actions of various powers are equal. "But no matter how such equality may be obtained, and whatever quantities constitute it, it is still very uncertain; neither is there any hope of knowing about it, if a safe and certain law of equilibrium is not first established."[35]

Riccati proceeds with determination and with the strictest methodological exactitude. He proposes a crucial experiment. We shall construct two bodies which are exactly the same in matter, form and volume. We shall set them in the same region, at the same distance from the center of the earth, so that the power (the force of gravity) cannot distinguish them from each other (a "principle of indifference"). We have, therefore, two powers which are exactly equal. We can now establish that the power of one of these bodies is in proportion to the power of *both* bodies as 1 is to 2. Generalizing from this, we obtain a set of powers "the proportion of whose values is known."[36]

Now consider a lever. We can set it in equilibrium using different powers provided that the ratio of the distances from the fulcrum is inverse to the ratio between those powers. Experimentally, therefore, the powers do not have to be equal in order to establish an equilibrium; rather their *actions* must be equal. This is the concept of action of power; in this way, equilibrium implies the equality of actions and *vice versa*.[37]

Riccati now tries to find the physical basis for the concept of action, which his experiment has distinguished from power. He gives the example of a body suspended by a thread. As long as it is hanging, the power of gravity is applied to it, but has no effect (no action). If we cut the thread, the same power "successively and continuously repeats its impulses" on the body, causing it to fall. This is precisely the action "which is in proportion

---

[34] V. Riccati, *De'principj della meccanica* (Venice, 1772), p. 8.
[35] *Ibid.*, p. 10.
[36] *Ibid.*, p. 11
[37] *Ibid.*, p. 12.

to both the power and the number of its impulses."[38] All that remains is to discover the meaning of the term "number of impulses." "What is the number to which the power is proportional and according to what element does it repeat its impulses?" To this question, Riccati replies with an interesting argument.

> Power exercises its action in time and through space. It may equally happen that it repeats its impulses in proportion to time or in proportion to space. In the former case, it would be an action in proportion to power and time, and in the latter case in proportion to power and space.[39]

This statement echoes tradition, as well as more recent discussion on living forces. The repetition of impulses is basically an image helping to establish that the force or power is realized in space *and* time. The action of power, on the other hand, is two-dimensional (to use Descartes' expression); it can operate either in space or in time, but not in both simultaneously. We have to choose which dimension will, together with power, characterize the action. Riccati continues:

> The matter is awkward, and because of the different opinions —and, even more, because of the obstinate aims of the writers— entangled and confused. Therefore, we must be careful in order to advance with an accurate and precise method. I establish the following principle that stands up to every refutation: That measure of action must be rejected at once, if it does not maintain the necessary equality of actions that are contrary in equilibrium; and only the one that invariably conserves the above-mentioned equality must be accepted. With such a principle at hand, let us again take up the only equilibrium that we know of, that is, the one of the lever, in which the powers that are normal to it and that are supposed to be unequal are in inverse proportion to the distances. And let us examine which of the two measures conserves in it the equality of actions.[40]

Of course, the examination is quickly over. It shows that

> the measure of the action is to be taken not in terms of the time in which it is exercised, nor of the space that is described by the point to which the power is applied, but of the space through which the approach to, or the withdrawal from, the center of the power is made.[41]

---

[38] *Ibid.*, p. 13.
[39] *Ibid.*, p. 16
[40] *Ibid.*, pp. 16–17.
[41] *Ibid.*, p. 20.

Riccati means that the displacement by which we multiply the power is the component of the effective displacement in the direction of the power. He concludes: "If you describe what happens, you find the general principle of statics. If, conceiving a minimal motion, the spontaneous actions of the powers are equal to the induced actions of the other powers, there will definitely be equilibrium."[42]

This tortuous but coherent path leads Riccati to a proposition which is exactly analogous to the principle of virtual velocities. In fact, it is impossible to tell them apart. Riccati himself admits this, but claims that his "general principle of statics" has a demonstrative foundation which the principle of virtual velocities lacks, at least in its usual presentation.

> As they [the "writers of statics"] have never explained what those virtual velocities were, which they introduced into mechanics, and have shown but few examples, they have made them proportional to spaces taken in the directions of the powers. The principle [of virtual velocities] was then supplied by no other evidence than that which could give it a valid induction that revealed it as always leading to already-known truths. But the progress we have made has dissipated all the fog and has shown that the virtual velocities, or, rather, the access and recess spaces, are those quantities by which an action is measured, and to which the number of impulses of the power is proportional. The principle is, then, nothing but the equality of spontaneous and induced actions required by the nature of equilibrium.[43]

Riccati's treatment deserves appreciation for its methodology. He appeals to experience not to derive a general law by induction, but to disprove one of two alternative hypotheses. He uses the lever, first to establish that equilibrium does *not* depend on the "equality" of powers in themselves, and second, to disprove the hypothesis that the action of power is proportional to time. The general principle seems to be subordinate to the law of the lever, but this is not true; that law is only an instrument, useful in research. Although it had always been presented as a "safe and certain" equilibrium, the law of the lever lacks intrinsic value as a foundation because it is only a consequence of the general principle.

---

[42] *Ibid.*, pp. 20–21

[43] *Ibid.*, pp. 21–22.

3.6 Lagrange's First Demonstration

Like Riccati, Lagrange sought to give an autonomous, unified foundation to statics. His *Mécanique analytique*, perhaps more than any other treatment, carries out the *reductio ad unum* of the principles and methods of the subject, relating everything to the principle of virtual velocities. The first edition was published in 1788; it presents the principle as an established axiom. Lagrange gives no specific demonstration other than the one represented by the treatise as a whole, but the splendor of the synthesis which the principle allows and the extraordinary efficacy of the analytic methods which Lagrange succeeds in drawing from it are proofs of a sort.

Between the first edition (1788) and the second (1811), a number of scientists set about to prove the principle and were for the most part successful. The technique they used was, of course, to deduce the principle of virtual velocities from the other two, especially from the principle of the composition of forces. In the second edition, Lagrange himself addresses the question of the proof of the principle, recognizing "that it is not sufficiently evident by itself to be established as a basic principle."[44] But in contrast to his predecessors, he tries to prove the equation of virtual velocities independently of both the law of the lever and the composition of forces. Instead, he refers to another "general principle," which we may call the principle of pulleys. It is a strange principle, whose intuitive truth is connected with those particular systems of pulleys or *poliplastes* which had been considered by Guidobaldo del Monte. It states: "the power [of a pulley] is in proportion to the weight carried by the mobile tackle just as the unity is in proportion to the number of cords that meet at this tackle, assuming they are all parallel, and ignoring friction and the stiffness of the cord."[45]

The physical properties of the *poliplaste* are not important in this discussion. Lagrange, in effect, frees this object from its material existence and turns it into a pure instrument of thought. He uses it to create a geometric and kinematic image of force or power. If we assume that the tension of the cord is uniform throughout its length, and idealize it as a perfectly flexible thread wound over pulleys of infinitesimal diameter, we can think of the weight as being supported by a number of equal "powers," one for each of the cords supporting the pulley. The number of cords thus becomes a direct measure of the weight. The concept of power can be eliminated; for it, we substitute "a cord uniformly taut," which, as it winds two or three times around the mobile pulley, doubles or trebles its magnitude. If we blot out any notions of "force," "effort" and "tension," and consider

[44] L. Lagrange, *Mécanique analytique*, 2nd ed. (Paris, 1811), 1st Part, 1st Section, No. 18, *see* 3rd ed. (Paris, 1853), p. 21. The demonstration reported here had already been published in the *Journal de l'École Polytechnique*, Vol. 2, (1798).

[45] *Ibid.*

an ideal *poliplaste* (one with infinitesimal pulleys) from a purely geometric and kinematic point of view, we can then translate all the static concepts into terms of cords and numbers of cords. For instance, a "power" (that is, a counterbalance that supports a given weight by means of a *poliplaste*) is nothing but the free end of a taut cord. In contrast, a weight is simply the number of cords that support a mobile pulley. We may represent any force applied in a certain direction at a certain point by imagining that we attach a mobile pulley at that point with the cords oriented in the given direction. In this way, we can transform any system which is subject to forces and weights into a compound system of *poliplastes*, all wound around by the same cord, which ends in a single counterbalance.

> In this way, a single weight shall produce, by means of the cord that winds round all the pulleys, different forces acting on the different points of the system—following the directions of the cords which arrive at the pulleys attached at these points—and these different forces will be in the same proportion to the weight as the number of cords will be to unity. Thus the forces themselves will be represented by the number of cords that act together in order to create them by means of their tension.
>
> Now, in order that the system subject to these various forces be in equilibrium, it is necessary that the counterbalance cannot descend as a result of any infinitesimal displacement of the points of the system.[46]

We have come to the heart of Lagrange's argument. Having eliminated force, by replacing it with the concept of "the number of cords that bind a mobile pulley to a fixed one," Lagrange has now changed the problem of equilibrium into a strictly geometric/kinematic one. Remembering Galileo and Torricelli, he redefines equilibrium as a situation in which a counterbalance cannot descend. (For want of space, we will not discuss the objections raised to this definition,[47] and shall simply follow the line of Lagrange's reasoning.) The analysis is immediate. We apply an infinitesimal displacement to the system, and then, as the mobile pulley moves towards or away from the fixed one, we can reckon how much the cord winds or unwinds from the *poliplaste*.

Let α, β, γ, ... be these infinitesimal displacements relative to the *poliplastes* supplied with P, Q, R, ... number of cords. If we assume that the whole cord cannot be extended, the counterbalance moves over the distance $P\alpha + Q\beta + R\gamma + \cdots$. Thus, if we confine our attention—as Lagrange does—to reversible displacements, "it will be necessary for the equilibrium of the forces represented by the numbers P, Q, R, etc., that we have the

---

[46] *Ibid.*, p. 22. Cf. *Œuvres de Lagrange*, Vol. 11, pp. 23–24.

[47] *See ibid.*, J. Bertrand's note.

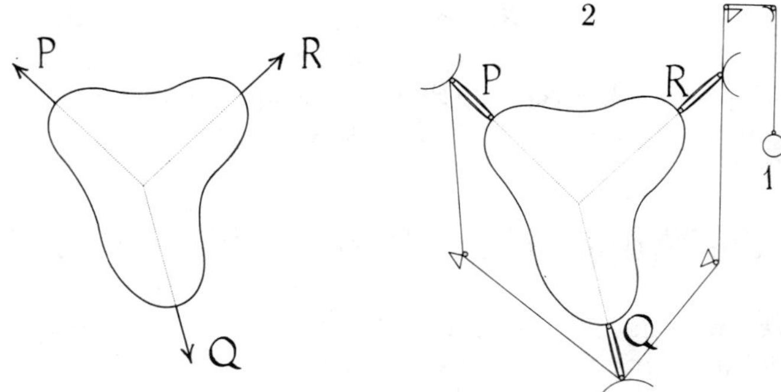

FIGURE 3.11.

equation $P\alpha + Q\beta + R\gamma + \cdots = 0$, which is the analytical expression of the general principle of virtual velocities."[48]

Lagrange's demonstration is surprising. So many hypotheses have to be introduced into the system (inelastic cord, perfect flexibility, uniform tension, absence of friction, etc.) that we can echo Mariotte's complaint about Archimedes' demonstration of the law of the lever: it tries to prove a proposition by means of an even more obscure proposition.[49]

On the other hand, Lagrange's argument is so effective that it cannot be dismissed. Mach himself, despite his usual disdain for proofs of the principles of statics, frankly said as much.[50] Lagrange's pulleys are not, in fact, a physical system to be interpreted by mechanical concepts and laws. Rather they are a geometric metaphor which allows us to eliminate the concept of force; Lagrange reduces force to such a level that it is effective but not necessary, either to the definition or to the description of equilibrium. If a force (or power) is nothing but a number of cords, then the principle of virtual velocities becomes an imaginative way of re-stating a geometric/kinematic proposition, which states that the height of the end of a curve, subject to complicated conditions, does not vary if the curve's length does not change and the other end is fixed.

The transition from system 1 in Figure 3.11, in which the powers are acting, to system 2 in the same figure, in which the forces are replaced by pulleys, is therefore significant. It does not offer an explanation, but it does propose a reduction, in the spirit of epistemological reductionism. Lagrange's explanation is simultaneously embarrassing and convincing; it does not prove the principle of virtual velocities, but rather the coherence of a project of reduction which eliminates the enigma of force. In fact,

---

[48] *Ibid.*, p. 22.

[49] E. Mariotte, "Essai de logique," in *Œuvres*, rev. ed., (The Hague, 1740) Vol. 2, p. 696.

[50] Mach, *Die Mechanik*, p. 96.

Lagrange makes a sort of "exchange of parts." He tries to demonstrate the principle of virtual velocities by trading cords in a pulley for forces. But in fact the principle itself supplies the means for making the reduction, allowing us to see the numbers of cords as forces.

3.7 The Approaches of Fossombroni and Fourier

We have dwelt at some length on Riccati's and Lagrange's "demonstrations" because both try to prove the principle of virtual velocities through an autonomous argument—that is, independently of other principles of statics. Now, however, we should deal with other authors who offered proofs which were more or less general and elegant but substantially correct, starting from the law of the lever and the principle of composition of forces. The matter is easily resolved when the constraints affecting the system can be expressed by equations, allowing us to assume that the virtual displacements are "invertible"—that is, that any displacement may occur in the opposite direction. (If not, the equation of virtual velocities turns into an inequality and requires hypotheses about constraints.) In fact, the *nexus inter se* of the principles of statics can be decided by strictly formal processes. For this reason, we may keep ourselves to a succinct account.

Most accounts of the history of mechanics ascribe the first general demonstration of the principle of virtual velocities to Fourier. His "*Mémoire sur la statique*" appeared in 1798, but two years earlier, as Fourier himself points out,[51] Vittorio Fossombroni gave substantial contributions to this subject in a long essay entitled *Memoria sul principio delle velocità virtuali*.[52] Fossombroni's work proves the principle in successive passages from the particular to the general, for both rigid and deformable systems. The essay is, perhaps, rather wordy, but the proof is meticulous. "In the present work," Fossombroni writes, "I made use of the composition and the decomposition of forces, a doctrine about which there can be no doubts."[53]

In the case of a point subjected to forces P_1, P_2, P_3, ... in directions p_1, p_2, p_3, ... respectively, the deduction is immediate: if α_i, β_i, γ_i, are the angles formed by the line p_i with the orthogonal axes x, y, z, the equilibrium of the point is described by means of the law of composition and decomposition of forces. The equations are:

$$\sum_i P_i \cos\alpha_i = 0, \quad \sum_i P_i \cos\beta_i = 0, \quad \sum_i P_i \cos\gamma_i = 0. \tag{3.1}$$

[51] J.B. Fourier, "Mémoire sur la statique," p. 518.

[52] V. Fossombroni, *Memoria sul principio delle velocità virtuali* (Florence, 1796).

[53] *Ibid.*, p. 26

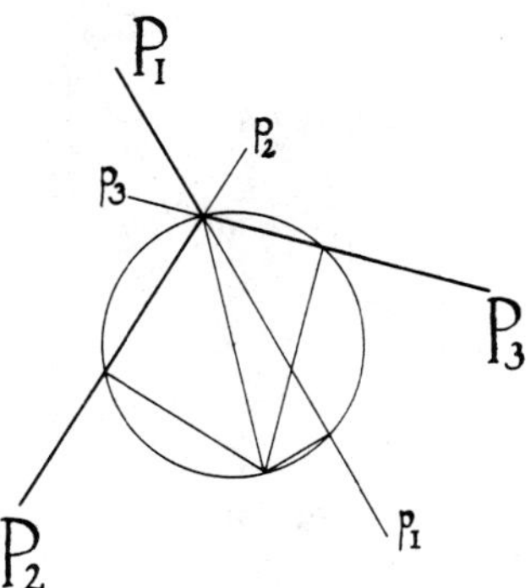

FIGURE 3.12.

If we impose on the point a displacement Δr, whose components are Δx, Δy, Δz, its projection on line p_i is given by

$$\Delta p_i = \cos\alpha_i \Delta x + \cos\beta_i \Delta y + \cos\gamma_i \Delta z. \tag{3.2}$$

Multiplying the three equations of equilibrium by Δx, Δy and Δz, and summing them yields

$$\sum_i P_i(\cos\alpha_i \Delta x + \cos\beta_i \Delta y + \cos\gamma_i \Delta z) = 0$$

and therefore, the following "equation of moments" may be deduced:

$$\sum_i P_i \Delta p_i = 0 \tag{3.3}$$

where Δp_i is assumed to be infinitesimal.[54] Fossombroni discusses at length the possibility of Δp_i being finite. He aims to distinguish the "equation of moments" (3.3), in which Δp_i is assumed to be infinitesimal and which represents the principle of virtual velocities in Lagrange's form, from the "equation of forces," in which Δp_i is finite. The latter, "while not being an indubitable criterion of equilibrium," makes it possible to describe noteworthy properties of equilibrium itself "if this is demonstrated." For example, equation (3.3) tells us that

> if, from the point to which the forces are applied I make a line of any dimension and direction, and on that diameter construct a sphere whose surface intersects all the p_i lines [Figure 3.12], whenever the sum of the products of these intersections multiplied by the respective forces is equal to zero, the given point will be in equilibrium.[55]

---

[54] *Ibid.*, p. 37.
[55] *Ibid.*, p. 40.

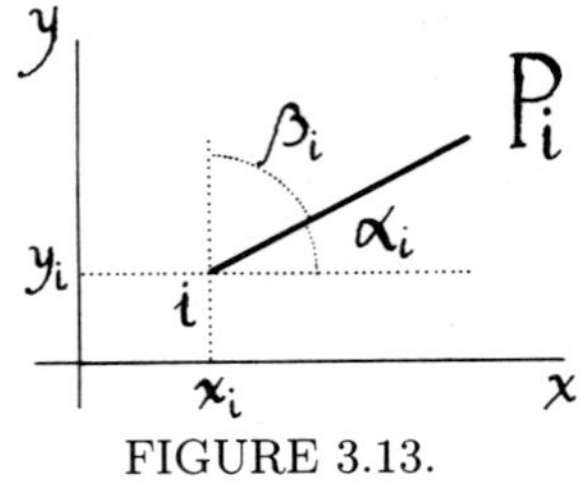

FIGURE 3.13.

In the case of rigid systems, consisting of more than one point, Fossombroni's demonstration proceeds to formulate both the cardinal equations of equilibrium and the expression of the most general rigid displacement of the system. The two groups of equations together yield in every circumstance "the equation of moments deduced from the principle of virtual velocities." For example, consider a system composed of three points subject to forces in the same plane. The equations of equilibrium are (*see* Figure 3.13):

$$\begin{aligned} P_1 \cos\alpha_1 + P_2 \cos\alpha_2 + P_3 \cos\alpha_3 &= 0, \quad (3.4a)\\ P_1 \cos\beta_1 + P_2 \cos\beta_2 + P_3 \cos\beta_3 &= 0, \quad (3.4b)\\ P_1(y_1 \cos\alpha_1 - x_1 \cos\beta_1) + P_2(y_2 \cos\alpha_2 - x_2 \cos\beta_2) & \\ + P_3(y_3 \cos\alpha_3 - x_2 \cos\beta_3) &= 0. \quad (3.4c) \end{aligned}$$

There are five geometrical/kinematic conditions expressing the "invariability of the distances between points and between each one of them from the center of the axes identified with the center of rotation. Fossombroni deduces these conditions in different ways, comparing "Lagrange's method" with others of his own invention. Finally he gives them the form

$$\begin{array}{ll} x_1\delta x_1 + y_1\delta y_1 = 0, & x_1\delta y_2 - x_2\delta y_1 = 0, \\ x_1\delta y_3 - x_3\delta y_1 = 0, & y_1\delta x_2 - y_2\delta x_1 = 0, \\ \text{and} \quad y_1\delta x_3 - y_3\delta x_1 = 0. & \end{array} \quad (3.5)$$

By means of equations (3.5), equation (3.4c) can be rewritten as follows:

$$\begin{aligned} P_1(\delta x_1 \cos\alpha_1 + \delta y_1 \cos\beta_1) + P_2(\delta x_2 \cos\alpha_2 + \delta y_2 \cos\beta_2) & \\ + P_3(\delta x_3 \cos\alpha_3 + \delta y_3 \cos\beta_3) = 0. & \quad (3.6) \end{aligned}$$

We multiply equation (3.4a) by Δx and equation (3.4b) by Δy, and then add the results to equation (3.6). Furthermore, as the most general displacement of the system we take

$$dx_i = \Delta x + \delta x_i \quad \text{and} \quad dy_i = \Delta y + \delta y_i,$$

where δx_i, δy_i obey equations (3.5). The outcome is

$$\begin{aligned} P_1(dx_1 \cos\alpha_1 + dy_1 \cos\beta_1) + P_2(dx_2 \cos\alpha_2 + dy_2 \cos\beta_2) & \\ + P_3(dx_3 \cos\alpha_3 + dy_3 \cos\beta_3) = 0, & \end{aligned}$$

that is, the "equation of moments."[56]

The analysis of "systems in any way variable in the distances"—that is, of deformable systems—is carried out in the second part of the work. Fossombroni modifies the conditions of equations (3.5) by introducing "arbitrary functions with respect to the arbitrary quality of the system, but determined by the nature of the different systems."[57] In the case of the previous plane system of three points, he substitutes the following for conditions (3.5):

$$x_1\delta x_1 + y_1\delta y_1 = 0, \quad x_1\delta y_2 - x_2\delta y_1 = m'\delta y_2 + n'\delta y_1,$$
$$x_1\delta y_3 - x_3\delta y_1 = m''\delta y_3 + n''\delta y_1, \quad y_1\delta x_2 - y_2\delta x_1 = r'\delta x_2 + s'\delta x_1$$
$$\text{and} \quad y_1\delta x_3 - y_3\delta x_1 = r''\delta x_3 + s''\delta x_1,$$

where m', n', ... r'' and s'' are, in fact, those "arbitrary functions."

Fossombroni's approach to the description of deformable systems, while perfectly legitimate, is not a happy one because of the complicated formulae it leads to, and because the functions it leads to have scant meaning. For this reason, we will abandon his *Memoria*, interesting as it is. Taking its cue from the demonstration of the principle of virtual velocities, and following the lines laid down by Lagrange, it opened up the way for promising results in the analysis of deformation. It inspired later research on a grand scale in Italy during the first decades of the nineteenth century—for example, the work of Gabrio Piola in this field.

Fourier's "Mémoire sur la statique," cited above, stands out for various reasons: its breadth, the detail of its treatment, the extraordinary perspicacity with which it examines some themes in depth (e.g. the theory of the stability of equilibrium by means of an extremum principle), and the richness of its demonstrative arguments. All these virtues make the work extraordinarily difficult to summarize, and summarize we must, considering only the demonstration of the principle of virtual velocities. Fourier starts from other principles of statics, assuming these to be known and certain: "we suppose to be known the principle of the lever, as it is demonstrated in the books of Archimedes or, what yields the same issue, the theorem of the composition of forces given by Stevin."[58] In defining the moment of a force as the product of that force and the virtual velocity of the point at which it is applied (changing the sign of the latter), Fourier immediately recognizes, as Fossombroni had before him, that in the case of a single point subjected to more than one force, the law of composition of forces means that the "moment of the resultant" equals the sum of the "moments" of the elements. At equilibrium, this sum must be nil.[59]

[56] *Ibid.*, pp. 117–135.
[57] *Ibid.*, p. 136.
[58] Fourier, "Mémoire sur la statique," p. 479.
[59] *Ibid.*, p. 481.

The author then considers two points r, s, whose distance is l_{rs}. The points attract or repel each other with forces T_{rs}. The "total moment"—that is, the sum of the "moments" of the two forces—equals $\mp T_{rs} dl_{rs}$, depending on whether the forces repel or attract. Therefore, the "moment" is proportional "to the complete differential of the distance between the two points." It follows that, if the distance between r and s is invariable, the "moment" is nil.[60] By analogy, the "total moment" of the mutual reaction between two smooth, inflexible surfaces which come into contact[61] is also nil, as is the "moment" of the tensions acting on a non-extensible line joining two solid bodies.[62] Setting aside the assumption that the line is not extensible, Fourier maintains that the reactive forces act according to the variation of l_{rs}. Because of the convention which he adopted for the sign of "moment," this implies that the "moment" of such reactive forces is essentially negative.

In reality, this definition of the sign of reactive "moment" brings a new postulate into place. This makes it possible to express the principle of virtual velocities in the form of an inequality involving only the forces applied. For example, in the case of the unilateral contact of a system with fixed supports, it is clear, Fourier states,

> that there are probably displacements which do not satisfy the equation of condition, and it is also clear that for these displacements, the moment of the resultant is necessarily positive, since the direction of these forces must be perpendicular to the resistant surfaces. For this reason, the sum of the moments of the applied forces is positive for all displacements of this kind, but it is impossible to attribute to a rigid body in equilibrium a virtual displacement such that the total moment of the applied forces is negative. On the other hand, if the reactions are considered as forces—which clears the way, as is known, to estimate these reactions—the body can be considered free, and the sum of the moments is nil for all the possible displacements.[63]

A similar result holds for an "indefinite mass of hard bodies" which rest on each other. If the virtual displacement maintains the contact between them, while perhaps changing the *point* of the contact, the "moment" of the reactions is nil. If the bodies tend to separate, it is negative.

> If all the forces acting on all the bodies be taken into account, it is certain that the sum of their moments must be nil for all the conceivable displacements, including those that are

---

[60] *Ibid.*, p. 482.
[61] *Ibid.*, p. 484.
[62] *Ibid.*, p. 489.
[63] *Ibid.*, p. 488.

> forbidden by the impenetrability of the solids. Now, for the displacements which are compatible with these latter conditions, the moment of all the pressures is nil or negative. Therefore, the sum of the moments of the applied forces alone is nil or positive for all the possible displacements.[64]

We shall pass over Fourier's final case, that of an incompressible fluid, and go on to the second part of the *Mémoire*. So far, Fourier has only illustrated "that property of moments according to which the value of the moment of the applied forces is the same as the moment corresponding to the resultant"[65] for every system. The reference to the principle of virtual velocities is quite clear. If the equilibrium of a system can be reduced, by means of the composition of forces, to the direct opposition of two equal resultants, it follows that the sum of the moments of the applied forces is nil. We cannot make this reduction "without demonstrating, at the same time, the truth of the principle of virtual velocities."[66] Fourier adds, a little smugly, this straightforward observation sums up "all the special demonstrations given by Varignon." But this is not enough; the demonstration must be completed by showing the truth of the *inverse proposition*: if the principle of virtual velocities is valid, then equilibrium is ensured for any system.

The second part of the *Mémoire* is, in fact, devoted to this objective. Fourier aims to reduce the principle of virtual velocities to the law of the lever. To do this, he implicitly establishes a kind of "identity criterion" for mechanical systems which we can express in this way. They must be *identical point systems*, subject to the same external forces but with different constraints. Mechanically, two such systems are indistinguishable if their constraints permit the same elementary mobility of their points. Thanks to this implicit criterion, Fourier can replace a given system with "a body which is simpler but susceptible to being displaced in the same way." In this way "the conditions of equilibrium of the system depend on the properties of equilibrium of the system that replaces it."[67]

For example, consider a system in which the two points P and Q are constrained in such a way that the displacement dp of P in direction p' corresponds to the displacement dq of Q in direction q' (Figure 3.14). Fourier imagines another system in which a suitable knee-joint lever has P and Q as its endpoints and point R as its fulcrum. Point R lies on the line of intersection r between planes α and β, perpendicular to dp and dq respectively. This lever $hh'h''$ is obtained by dropping in α the perpendicular h from P to r until it meets R, then by drawing line h' in β through R, perpendicular

---

[64] *Ibid.*, pp. 490–491.
[65] *Ibid.*, p. 488.
[66] *Ibid.*
[67] *Ibid.*, pp. 494–495.

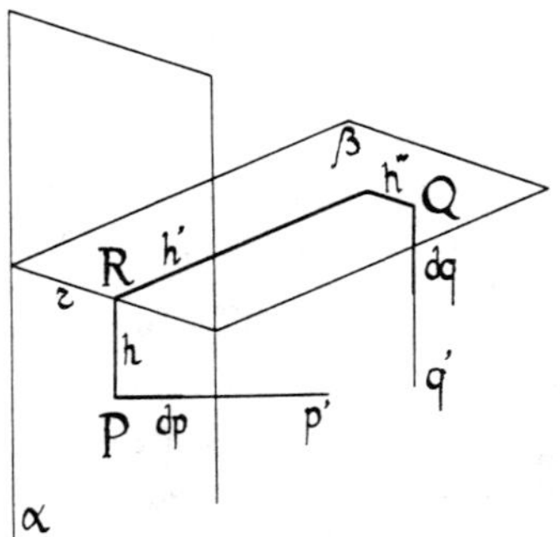

FIGURE 3.14.

to r, and finally by tracing perpendicular h'' from Q to h' in β. Under these conditions, dp corresponds to dq. We could visualize the same construction for any other pair of points, finally substituting the primary system by "assembling the levers," which effects the same virtual displacement system. Now, the law of the lever ensures that the system is in equilibrium when the sum of the moments of the forces acting on the points of the lever system is zero. Therefore, because of the identity criterion noted above, the primary system will also be in equilibrium. Fourier arrives at this conclusion by way of a *per absurdum* reasoning, but that reasoning is only an oblique reformulation of the identity criterion.[68] He goes on to discuss the case of unilateral constraints in an auxiliary system, also using the law of the lever (§§20, 21 and 22). Fourier then proposes a different ideal model of a lever, this one with pulleys, which allows him to connect the equation of virtual velocities to the Galileo–Torricelli principle.

In part 3 of his *Mémoire* Fourier deals with the stability of equilibrium, while part 4 presents the improvement, already mentioned in part, in the proof of the law of the lever. Part 5 is devoted to the "comparison of the different propositions demonstrated in this work, to make their mutual dependence clear." One passage in particular delineates the concept of statics which most attracted Fourier, and to which he made such a notable contribution:

> The properties of equilibrium depend ... entirely on the consideration of the moments; one deduces these with unexceptionable exactitude from the single principle of the lever. The equilibrium of the lever depends on that of the balance and we have applied this latter to the case of three opposed forces which clearly cancel each other out.[69]

[68] *Ibid.*, pp. 495–497.
[69] *Ibid.*, p. 513.

This is curiously like the passage at the beginning of the *Mechanical Problems*, in which the Aristotelian author declares that the objective of mechanics is to relate all machines to the lever, the lever to the balance, and the balance to the circle. The initial and final terms are different, but the structure of the argument is the same. In place of machines, we have moments, in place of the circle, the law of the composition of forces. The principle of virtual velocities rests on the concept of "moment," while basic questions about the nature of the laws of statics are concentrated on the composition of forces. Fourier emphasizes that the concept of "moment" is not "a simple combination of abstract ideas," but rather is the more meaningful "exponent" of the physical reality studied by statics.

Not the least of Fourier's merits is that of considering irreversible displacements. Other writers of the period dealt with them, but limited their analyses to cases of constraints expressed by equations. Only after about 30 years were Fourier's ideas taken up by Augustin Cournot[70] and later by Gauss (1829–1837) and Ostrogradski (1838).

3.8 The Principle of Virtual Velocities and Constraints: Poinsot's and Ampère's Contributions and Lagrange's Second Proof

Poinsot's and Ampère's demonstrations, as well as Lagrange's second proof, differ from Fourier's in two ways. First, all three see the principle of virtual work (or velocities) as a simple interpretation of formulae which can be derived analytically from the equations of constraint. Second, none of them takes into account the possibility of irreversible displacements.

The second note in the appendix of Louis Poinsot's article "Théorie générale de l'équilibre et des mouvements des systèmes"[71] demonstrates the principle of virtual velocities, equating this principle with the general theorem which is the subject of the paper. This general theorem concerns a mechanical system with h points, whose coordinates are subject to an arbitrary number of constraints, and which is governed by three axioms, introduced at the beginning of the article:

> (i) If certain forces are together in equilibrium on any variable system, the equilibrium does not cease, assuming that the system suddenly becomes invariable, that is, assuming that it

[70] Augustin Cournot, "Extension du principe des vitesses virtuelles au cas où les conditions des liaisons du système sont exprimées par inégalités," *Bull. de Férussac*, Vol. 8 (1827).

[71] Louis Poinsot, "Théorie générale de l'équilibre et des mouvements des systèmes," *Journal de l'École Polytechnique*, Vol. 6, 13ème cahier (1806), pp. 206–241.

> is, so to speak, solidified. ... (ii) The forces applied to different points of the system can be decomposed along the directions of the straight lines that join pairwise those points into equal and opposed pairs of forces. ... (iii) Two equal and opposed forces are, of necessity, in equilibrium.[72]

From these axioms, Poinsot took a very different path from that of the *Mécanique analytique*, which had concentrated on the *principle* of virtual velocities. Nonetheless, he arrives at the aforementioned general theorem, whose tone is distinctly Lagrangian:

> whichever the equations may be that govern the coordinates of the points, each of these equations requires, for equilibrium, the application, at these points and in the direction of the coordinate axes, of forces proportional to the first derivatives of this equation with respect to the relative coordinates.[73]

"Thus," he continues,

> by representing by $L = 0$, $M = 0$, etc., any equations between the coordinates x, y, z, x', y', z', etc., of the different points, and by representing by λ, μ, etc., the indeterminate coefficients, the total forces X, Y, Z, X', Y', Z', etc., which must be applied to these points according to the direction of the axes, will be expressed as follows:
>
> $$\begin{aligned} X &= \lambda\left(\frac{dL}{dx}\right) + \mu\left(\frac{dM}{dx}\right) &&+ \text{etc.} \\ Y &= \lambda\left(\frac{dL}{dy}\right) + \mu\left(\frac{dM}{dy}\right) &&+ \text{etc.} \\ Z &= \lambda\left(\frac{dL}{dz}\right) + \mu\left(\frac{dM}{dz}\right) &&+ \text{etc.} \\ X' &= \lambda\left(\frac{dL}{dx'}\right) + \mu\left(\frac{dM}{dx'}\right) &&+ \text{etc.} \\ Y' &= \lambda\left(\frac{dL}{dy'}\right) + \mu\left(\frac{dM}{dy'}\right) &&+ \text{etc.} \\ Z' &= \lambda\left(\frac{dL}{dz'}\right) + \mu\left(\frac{dM}{dz'}\right) &&+ \text{etc.}, \end{aligned} \tag{3.7}$$
>
> and so forth. Eliminating the indeterminates λ, μ, etc., from those equations, we are left with the conditions of equilibrium in the literal sense, that is, the relations that must exist between

[72] *Ibid.*, p. 207.
[73] *Ibid.*, p. 219.

> the single applied forces and the coordinates of their points of application for the equilibrium of the system.[74]

Naturally the number of equations (3.7) and the coordinates of the points of the system are equal, while the indeterminates λ, μ, etc., are equal in number to the constraints. When the system is statically determinate, the constraints may not number more than $3h - 6$.

Poinsot is justly proud (as we see in his conclusion) of this theorem of his. It enables him to establish directly the general equations of equilibrium and movement.

> From the law of equilibrium laid down here we could have easily deduced that if one causes an infinitesimal disturbance to the equilibrium of any system of bodies without the constraints being violated, the sum of the moments of all the forces applied—i.e., according to Galileo's meaning, the sum of their products times the relative paths taken by their points of application in the direction of these forces—must always be zero; and this constitutes the principle of virtual work. But this principle is still nothing but a simple corollary of the preceding theory.[75]

Note 2 makes this more explicit. Given the conditions

$$f(x, y, z; x', y', z'; \ldots) = 0, \qquad \varphi(x, y, z; x', y', z'; \ldots) = 0, \quad \text{etc.,} \tag{3.8}$$

if velocities $\dot{x}$, $\dot{y}$, ... are given to the points of the system, in order that equations (3.8) not be violated, it is necessary that

$$f'_x \dot{x} + f'_y \dot{y} + \cdots = 0, \qquad \varphi'_x \dot{x} + \varphi'_y \dot{y} + \cdots = 0, \qquad \text{etc.,} \tag{3.9}$$

where the prime and the subscripts indicate the partial derivative of the function, and a superscript dot denotes differentiation with respect to time. Multiplying the first equation of (3.9) by a parameter λ, the second one by μ, and so on, one obtains:

$$[\lambda f'_x + \mu\varphi'_x + \cdots]\dot{x} + [\lambda f'_y + \mu\varphi'_y + \cdots]\dot{y} + \cdots = 0. \tag{3.10}$$

But from the general theorem, the quantities in brackets in equation (3.10) may be interpreted as the forces which keep the system in equilibrium. Thus the principle of virtual velocities is demonstrated. Moreover, this argument allows us to understand the real meaning that must be assigned to this principle:

> the general problem of statics consists not only in looking for the relations between the forces which actually balance each

---

[74] *Ibid.*, p. 220.
[75] *Ibid.*, p. 233.

> other in the system, but also the general expression of the forces which can balance each other in all the configurations that the system may take on by virtue of the equations of condition. The general equation given by the principle of virtual velocities is not, then, if it is possible to say so in these terms, the relation of a moment. It must not simply take into consideration the equilibrium of the system in the configuration it is in; but it must also consider all the series of configurations in which it can be, since it is this series or succession of configurations which characterizes it and constitutes its definition.[76]

This observation is really perceptive. Most demonstrations of the principle of virtual velocities do indeed tend to take the principle back to other principles, and to the laws of some simple machine such as the lever. But these, Poinsot says, "seem to us to be rather proofs than exact demonstrations" since they do not manage to explain why we should consider a disturbance of equilibrium. What substantiates the principle of virtual velocities is instead a suitable definition of "system" which includes not only the existing configuration, but the set of all possible configurations. Once this is admitted—as Poinsot has proposed in his demonstrative argument—the equation of virtual velocities becomes no more than a corollary (*une suite*) of the general definition.

We should briefly mention A.M. Ampère's work on the demonstration of the principle of virtual work. It was published in the same *Cahier* of the *Journal de l'Ecole Polytechnique* in which Poinsot's work appeared.[77] Ampère writes:

> Among the mathematical achievements that marked the end of the last century, one of the most remarkable is, without doubt, that of having expressed all the cases of equilibrium and movement in one single equation. This equation results from the combination of the principle of virtual velocities and the other principle by which d'Alembert has reduced dynamics to statics.[78]

Ampère begins by recognizing the pre-eminent role that the principle under discussion plays at the vertex of the science of equilibrium and movement. He calls it "a branch of mathematical analysis" and reduces every one of its results to an "application of one formula." The reference is, as always,

[76] *Ibid.*, p. 240.

[77] A.M. Ampère, "Démonstration générale du principe des vitesses virtuelles, dégagée de la considération des infiniment petits," *Journal de l'École Polytechnique*, Vol. 6, 13ème cahier (1806), pp. 247–269.

[78] *Ibid.*, p. 247.

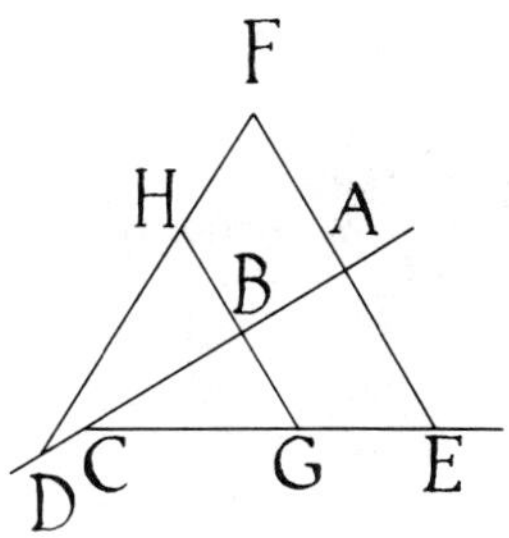

FIGURE 3.15.

to Lagrange's *Mécanique analytique.* Among others who have presented demonstrations of the principle, he cites Carnot,[79] who reduced it to the law of the lever, and Laplace, who deduced it "from more general considerations," namely the principle of the composition of forces together with a principle of superimposition of equilibria.[80]

We have not yet mentioned the two authors cited by Ampère because their propositions reduce, in the end, to methods already used by others. Laplace arrived at the equation of virtual work by considering separately the effects of internal forces acting on a system of points, assuming that all the points except two are fixed, and finally adding their partial contributions.[81] Ampère puts forward a serious objection to precisely this simplification, and advances a counter-example.

> Let A, B, C, D, be four unfixed points on line AD free in space [Figure 3.15]; supposing that points A, B, sliding along this line, take with them two other straight lines EF, GH, perpendicular to AB, whose parts AE, AF, BG, BH have a given length. Since points C and D must always find themselves at the intersection of lines EGC, FHD with line AD, it is clear that the ratios between the three distances AB, AC, AD, are determined and that by virtue of those conditions one cannot assume that two of these points are fixed without the other two becoming motionless, although the conditions stipulated make it possible, on the other hand, for the four points to move all together in every direction and for the system that they create to be transported arbitrarily in space.[82]

---

[79] L.N.M. Carnot, *Essai sur les machines en général* (Paris, 1783); L.N.M. Carnot, *Principes fondamentaux de l'équilibre et du mouvement* (Paris, 1803) (hereafter cited as Carnot, *Principes*).

[80] P.S. Laplace, *Traité de mécanique céleste* (1799), 2d ed. (Paris, 1829), pp. 36–46.

[81] *Ibid.*, pp. 38–41; *see* L.A. Ampère, *op. cit.*, p. 248.

[82] *Ibid.*, pp. 248–249.

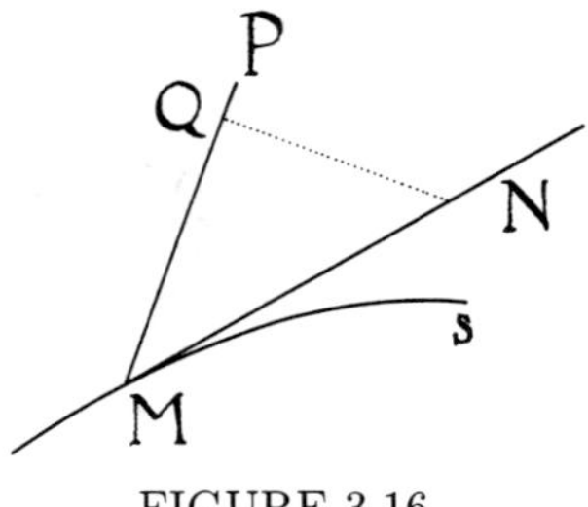

FIGURE 3.16.

Ampère is equally critical of the other hypothesis usually adopted for considering several constraints one at a time:

> A little reflection reveals that it would be necessary to show *a priori* that the effects produced by the coalescing of several conditions result from the effects of each single condition, without their being modified by their union; this truth should be a consequence of the equations of equilibrium, rather than a means of obtaining them.[83]

To deal with these criticisms, Ampère resolves to consider all the constraints together, starting from the case in which they are one fewer than the coordinates of the points of the system. In this case, the displacements of the points can be expressed in terms of one variable u, and every point must follow a determinate curve. Consider the point M to which force P is applied. Let s be the aforementioned curve (Figure 3.16). If $M = (x, y, z)$, a point N on the tangent of s at M has coordinates $(x+x'i, y+y'i, z+z'i)$, where a prime denotes differentiation with respect to u and i is a parameter. The projection MQ of MN on the line of force P is expressed by

$$MQ = \frac{(x-a)x'i + (y-b)y'i + (z-c)z'i}{\sqrt{(x-a)^2 + (y-b)^2 + (z-c)^2}} \tag{3.11}$$

where a, b, c are the coordinates of any point different from M on line MP.

The "moment" of P, in the terminology of Ampère and his contemporaries, is the product $P.MQ$, which is positive or negative depending on whether the angle $\widehat{PMN}$ is acute or obtuse. With an eye to "avoiding any consideration of infinitely small quantities," Ampère recommends the cancellation of the common factor i from equation (3.11). He demonstrates the principle of virtual velocities from the law of composition and decomposition of forces by the following observation: "where the points are forced to move on determinate curves, if the force applied is decomposed at an unfixed point, according to the perpendicular and the tangent to the curve,

[83] *Ibid.*, p. 248.

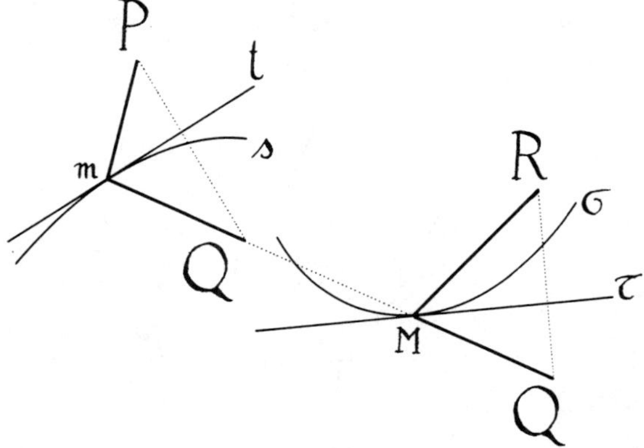

FIGURE 3.17.

only the latter is effective for equilibrium."[84] It follows that if two forces have the same component with respect to the tangent of the curve, "one can be substituted for the other without any change in the state of equilibrium and motion of the system."[85] Thus, for example (Figure 3.17), force P and force Q applied to m are interchangeable when $P \cos \widehat{Pmt} = Q \cos \widehat{Qmt}$, t being the tangent to the curve relative to m. Moreover, if we translate force Q along its straight line of action until it meets another point M of the system, we can substitute a force R of any direction for force Q on M, provided that $R \cos \widehat{RM\tau} = Q \cos \widehat{QM\tau}$, where τ is the tangent to the curve corresponding to M. By eliminating Q from these two equations, we make the formulae explicit, letting $m = (x, y, z)$, $M = (X, Y, Z)$, and arbitrarily fixing two points (a, b, c) and (A, B, C) of the straight lines mP and MR respectively. It follows that

$$P\frac{(x-a)x' + (y-b)y' + (z-c)z'}{\sqrt{(x-a)^2 + (y-b)^2 + (z-c)^2}} = R\frac{(X-A)X' + (Y-B)Y' + (Z-C)Z'}{\sqrt{(X-A)^2 + (Y-B)^2 + (Z-C)^2}}. \quad (3.12)$$

This formula tells us what the relation should be between a force P applied to m and a force R applied to M, so that one could exchange places with the other without disturbing the equilibrium. But, significantly, the formula also expresses the equality of the "moment" of P with the "moment" of Q. The two forces can therefore be exchanged, if their moments are equal.[86]

This puts us within reach of Ampère's objective. We connect an arbitrary point m of the system with an auxiliary point M, and connect M with a

---

[84] *Ibid.*, p. 254.
[85] *Ibid.*, p. 255.
[86] *Ibid.*, pp. 256–257.

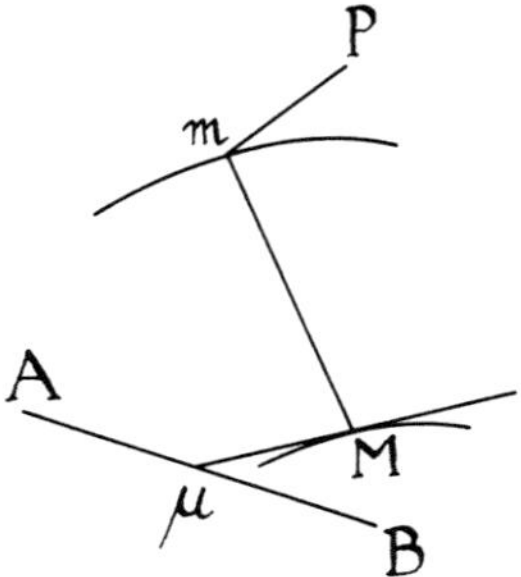

FIGURE 3.18.

point μ which is forced to move along a straight line AB by means of two rigid, non-extensible lines mM, $M\mu$, united at M so as to rotate freely around each other (Figure 3.18). Simple kinematic considerations enable us to determine a curve for M such that, when m undergoes the displacement imposed by the system by a given variation of the parameter u, μ undergoes a specific displacement on straight line AB. Now expand the given system by these two lines mM, $M\mu$, repeating the operation for all points m_i, so that when the points m_i are displaced in line with the system, the related points μ_i are displaced to the same extent on the straight line AB.

Finally, let us replace the forces P_i acting on m_i by an equal number of forces S_i directed along AB. By virtue of equation (3.12) and, for simplicity, making the line AB coincide with axis x, we obtain

$$P_i = \frac{(x_i - a_i)x_i' + (y_i - b_i)y_i' + (z_i - c_i)z_i'}{\sqrt{(x_i - a_i)^2 + (y_i - b_i)^2 + (z_i - c_i)^2}} = S_i \xi_i',$$

where ξ_i is the abscissa of point μ_i.

> One can *add up* all these equations and, observing that the distances between the unfixed points on AB are constant, so that ξ' has the same value for all these points, one will have
>
> $$\sum_i P_i \frac{(x_i - a_i)x_i' + (y_i - b_i)y_i' + (z_i - c_i)z_i'}{\sqrt{(x_i - a_i)^2 + (y_i - b_i)^2 + (z_i - c_i)^2}} = \left(\sum_i S_i\right)\xi'.$$
>
> But, it is necessary for equilibrium that $\sum_i S_i = 0$ and, reciprocally, when this condition is satisfied, the system of forces S (and consequently the system of forces P) is necessarily in equilibrium; thus, thanks to the last equation, it follows that the only condition necessary for the equilibrium of the given system is that the sum of the moments of forces P_i, i.e., the first member of this equation, equals zero.[87]

[87] *Ibid.*, p. 260.

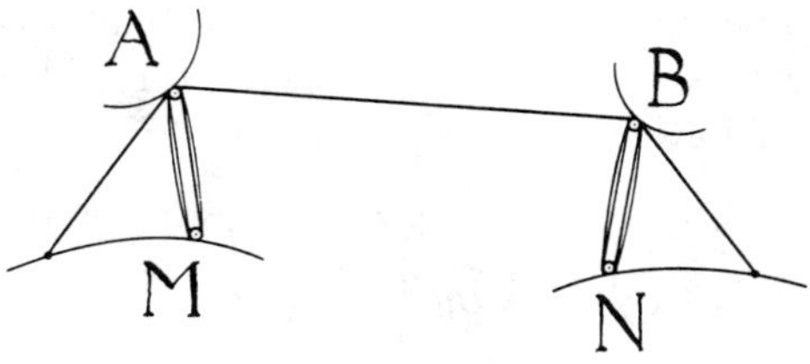

FIGURE 3.19.

Ampère's essay is a stimulating variation of the demonstrations discussed above. It deserves attention for its objective—that is, of not resorting to the hypothesis of superimposition of equilibria. But does Ampère actually avoid this hypothesis? Does it not impinge on the composition and decomposition of forces, which is fundamental to Ampère's argument? This is less certain, and must be left for later.

We should end this review of the demonstrations of the principle of virtual velocities by making only a passing reference to Lagrange's second proof. In his second edition of the *Théorie des fonctions analytiques*,[88] Lagrange returns to his "principle of the pulley" in order to give a final demonstration of it. As we know, this principle makes it possible to reduce the whole question to the kinematic/geometric determination of the displacement undergone by the end of the cord connecting the *poliplastes*. It presupposes, however, that one can always represent the mutual action of two bodies in equilibrium by means of an ideal system of pulleys connected by a non-extensible cord. In other words, it assumes that *any* constraint can be represented by an equation of the type

$$\begin{aligned} m\sqrt{(x-a)^2+(y-b)^2+(z-c)^2} \\ +n\sqrt{(\xi-\alpha)^2+(\eta-\beta)^2+(\zeta-\gamma)^2}-d=0. \end{aligned} \tag{3.13}$$

This formula expresses the inextensibility of a cord of length d (Figure 3.19) which connects two bodies, M with coordinates x, y, z, and N with coordinates ξ, η, ζ. The cord to which M is attached passes through a fixed pulley A with coordinates a, b, c; then it passes through M again and goes through pulley A a second time, and so on for m times. After leaving the fixed pulley a, the cord passes through a second fixed pulley B with coordinates α, β, γ; from here it goes to body N, then back to B, then to N again, and so on for n times, before finally being firmly attached to N.

Now, let the equation

$$F(x,y,z;\xi,\eta,\zeta)=0$$

[88] L. Lagrange, *Théorie des fonctions analytiques* (2d ed., 1813), Part 3, Chapter 5, sections 27, 28, 29, 30.

be an arbitrary equation of constraint between M and N, and denote the left-hand side of equation (3.13) by $f(x, y, z; \xi, \eta, \zeta)$. The function f must obey, first, the equations

$$\left(\frac{\partial f}{\partial x}\right)^2 + \left(\frac{\partial f}{\partial y}\right)^2 + \left(\frac{\partial f}{\partial z}\right)^2 = 1$$

and

$$\left(\frac{\partial f}{\partial \xi}\right)^2 + \left(\frac{\partial f}{\partial \eta}\right)^2 + \left(\frac{\partial f}{\partial \zeta}\right)^2 = 1.$$

Moreover, it must satisfy the following "contact conditions":

$$\frac{\partial f}{\partial x} = \frac{\partial F}{\partial x}, \qquad \frac{\partial f}{\partial y} = \frac{\partial F}{\partial y}, \qquad \frac{\partial f}{\partial z} = \frac{\partial F}{\partial z},$$
$$\frac{\partial f}{\partial \xi} = \frac{\partial F}{\partial \xi}, \qquad \frac{\partial f}{\partial \eta} = \frac{\partial F}{\partial \eta}, \qquad \frac{\partial f}{\partial \zeta} = \frac{\partial F}{\partial \zeta}.$$

In general, there are eight conditions to which the seven arbitrary constants a, b, c, α, β, γ, d correspond, plus the two indeterminate coefficients m, n of equation (3.13): therefore, it can be said that any constraint $F(x, y, z; \xi, \eta, \zeta) = 0$ produces on the bodies M and N the same forces that are engendered by the cord.

> The conclusion is that in a system with two bodies whose constraint depends on the equation $F(x, y, z; \xi, \eta, \zeta) = 0$, their mutual action produces the forces $\Pi\frac{\partial F}{\partial x}$, $\Pi\frac{\partial F}{\partial y}$, $\Pi\frac{\partial F}{\partial z}$ on the first of the bodies, according to the three coordinates x, y, z, and the forces $\Pi\frac{\partial F}{\partial \xi}$, $\Pi\frac{\partial F}{\partial \eta}$, $\Pi\frac{\partial F}{\partial \zeta}$ on the second, according to the rectangular coordinates ξ, η, ζ, Π being an undeterminate coefficient.[89]

The generalization is clear, and the argument which then leads to the equation of virtual velocities is equally obvious. "There is nothing surprising," Lagrange concludes

> in recognizing that the principle of virtual velocities becomes a natural consequence of the formulae which express the forces by means of the constraints, since the consideration of a cord which, because of its uniform tension, acts on all the bodies and produces assigned forces in them, is sufficient to lead on to a direct and general demonstration of this principle, as I have shown in the second edition of the work referred to [i.e., *Mécanique analytique*].[90]

---

[89] L. Lagrange, *Théorie des fonctions analytiques*, 2d ed. (Paris, 1813), p. 355.
[90] *Ibid.*, p. 357.

In reality, Lagrange's second demonstration is more useful as a "reductive" interpretation of forces in terms of equations of constraint, explicitly using the principle of the composition of forces.

Here everything converges: from Fossombroni to Fourier, from Poinsot to Ampère, and finally to Lagrange—all of them arrive at the same goal. Even later, when the argument was resumed towards the end of the nineteenth century by scientists like C. Neumann and L. Boltzmann,[91] the composition of forces was an essential point of reference for the thesis. The rule of the composition of forces—that is, in the end, the parallelogram rule—turns out to be *le vrai principe*, as Sturm said. On it are centered the most difficult, and most stimulating, questions of statics.

---

[91] C. Neumann, "Ueber eine einfache Methode zur Begründung des Princips der virtuellen Verrückungen," *Mathematische Annalen*, Vol. 27 (1886), pp. 502–505; L. Boltzmann, *Vorlesungen ueber die Principe der Mechanik* (Leipzig, 1897), Vol. 1, pp. 115–155.

4

The Parallelogram of Forces

4.1 Daniel Bernoulli's Claim

In his splendid (if disputable) introduction to *Principes fondamentaux de l'équilibre et du mouvement*, Carnot states that "there are two ways to envisage mechanics in its principles. The first views it as a *theory of forces*, that is, of causes that provoke movements. The second considers it as a *theory of movements* in themselves."[1] We know that Carnot unquestionably preferred the second, because the first "has the disadvantage of being based on a metaphysical and obscure notion, that is, *force*" —a cause which cannot be measured except by its effects. If the first way can be differentiated from the second, the distinction must be an essential one, for the first way is alien to the language of mathematics. "Frankly," Carnot adds,

> all the demonstrations in which the word *force* is employed bring about an absolutely inevitable character of obscurity: that is why, in this sense, in my opinion, there cannot be any exact demonstration of the parallelogram of forces; the existence alone of the word *force* in the enunciation of the proposition makes this demonstration impossible because of the nature of things itself.[2]

Carnot's position is extreme. His adherence to the "second way" of approaching mechanics is so zealous and intransigent that it rejects every basic idea (*idée primitive*) which does not involve matter, space and time, or rest and movement. A body is "a determinate part of matter"; mass is "the actual space occupied by a body" insofar as it differs from volume, that is, from the "apparent space" that the same body occupies. Accelerating or delaying force is "a velocity divided by time"; motive power is "the product of a mass multiplied by an accelerating or delaying force." And finally, "force" or "power" generally means either a motive power or a quantity of motion defined as "the product of mass and velocity."[3] Mechanics is strictly restricted to kinematic and geometric analysis. The concept of force disappears, except as a term for certain compound quantities involving space and time. The parallelogram of forces is deduced from the composition and decomposition of motion: "the accelerating, delaying and

---

[1] Carnot, *Principes*, p. xj.
[2] *Ibid.*, p. xiij.
[3] *Ibid.*, p. 12.

motive forces are represented, as are velocities and quantities of motion, by segments of straight lines, proportional to these forces and having the same direction, and subject to the same decomposition."[4]

Even if we deny Carnot's reduction, we must recognize that the elucidation of the parallelogram of forces in modern mechanics started with the composition and decomposition of movements (see Section 1.8, above). Of the three basic principles of statics, the parallelogram rule certainly derives from Carnot's "second way," whereas the historical origins of the other two principles are more complex. The equation of virtual velocities provided a tool for understanding the concept of force, which was somewhere between the first and second ways. The law of the lever gave rise to the first way, following Archimedes' demonstration, presenting statics as independent of the analysis of movement.

On the other hand, we have seen that the three statical principles are reciprocally involved. The composition of forces stands out as the first principle and the other two depend on it. This primacy of the parallelogram rule forces us to reconsider the true structure of mechanics. There are two possible hypotheses. According to the first, the parallelogram rule cannot be demonstrated without referring to the law of motion, and force is defined only in terms of its relation with "real" motion. This means that the law of motion is nearly a definition (as Carnot claimed). Furthermore, this implies that any demonstration of the other principles of statics is fictitious—a *Scheinbeweis*, as Mach was to claim. Therefore statics loses all autonomy and becomes a concealed description of geometric-kinematic relations. According to the second hypothesis, on the other hand, the parallelogram rule can be demonstrated in strictly geometric terms, without any reference to the composition of movements. In this case, a clean separation is established between statics and *mechanica sive motus scientia*. Statics becomes a sort of meta-theory with respect to mechanics. To give statics this theoretical basis was Daniel Bernoulli's declared aim.

In February, 1726, Daniel Bernoulli put the question of demonstrating the parallelogram rule to the Academy of Sciences of St. Petersburg.[5] He was the first to approach the problem. The nature of mechanical principles was the topic of the day, dividing the most eminent scientists into warring metaphysical and scientific factions.

The first section of Bernoulli's essay presents a twofold conception of mechanics: on one side, statics; on the other, dynamics. The former depends

---

[4] *Ibid.*, p. 31.

[5] D. Bernoulli, "Examen principiorum mechanicae et demonstrationes geometricae de compositione et resolutione virium," *Commentarii Academiae Scientiarum Imperialis Petropolitanae*, Vol. 1 (1726); 2d ed. published in Bologna (1740); also 3d ed., *Die Werke von Daniel Bernoulli*, Vol. 3, Basel-Boston-Stuttgart (1987). Quotations are drawn from the Bologna edition; *see* Vol. 1 (1740), pp. 122–137.

on necessary geometrical truths, deducible *a priori*, while the latter is based on experience and governed by "contingent" laws.

> That part of mechanics which deals with the equilibrium of *potentiae* can be entirely deduced from the composition and resolution of forces, as Pierre Varignon well demonstrated; so, if to this principle we add the other one, according to which increases of velocity are proportional to forces with respect to time elements, we obtain the second part of mechanics, the one that deals with the motion of bodies.[6]

This latter principle, "which we owe to Galileo," is established by experience ("if we judge correctly," the author adds, in light of the controversy over the nature of mechanical principles). Therefore we should include it with contingent truths, not necessary ones.

> As a matter of fact, nature might have determined that increments of velocity in moving bodies should be proportional to elements of times multiplied by any function of the pressures, so that if t denotes time, p the pressure, and v the velocity, we should no longer have $dv = p\,dt$, but, for example, $dv = p^2\,dt$ or $dv = p^3\,dt$, etc., from which different laws of motion would derive.

But the composition of forces is another matter: "Of this, I found a strictly geometric demonstration, because of which I came to verify that the theorems of statics are no less true by necessity than are geometric ones [*theoremata statica non minus necessario vera esse, quam sunt geometrica.*]"[7] The second section of the essay is devoted to this demonstration. A number of later scientists have called it prolix or imperfect. But it laid out the main route for demonstration of the parallelogram rule; subsequent work was by way of refinement, not substantial improvement. Rather than delving into Bernoulli's formal passages, we shall only recall the definitions and hypotheses which govern his demonstration.

First, the definitions. Bernoulli's definition of force (power, in his terms) goes as follows: "By *power AB* I mean the power expressed by AB: powers are equivalent when they draw a point with the same force in the same direction. [*Per potentiam AB intelligo potentiam expressam per AB: potentias sibi aequivalere dico, quando eadem vi, per eandem directionem punctum trahunt.*]" The first of these propositions introduces a symbol that not only denotes force, but also provides a full connotation of it. The chosen geometric entity is the oriented segment AB. This relation between the meaning of force and its geometrical representation is so firmly established

[6] *Ibid.*, p. 122.
[7] *Ibid.*, p. 123.

that force nearly takes on the properties of its "bearer of meaning;"[8] force becomes anything to which the characteristics of a line segment (length and direction) may apply.

By establishing an equivalence among forces, the second proposition clearly differentiates the two attributes which force adopts from the line. Bernoulli takes direction as a basic notion, obvious from the linear representation of force. We need no physical hypothesis to distinguish forces with different directions; this can be demonstrated by the appropriate geometric rules.

The magnitude of a force, identified with the length of segment AB, is a different matter. Bernoulli's term (*eadem vi*) recalls the obscure term *vis* which the Cartesians and Leibnizians disputed for many years. The debate on the principle of the conservation of "live forces" (*principium conservationis virium vivarum*) led to an energetic formulation of mechanics. In the first section of his essay, Bernoulli turned his attention to reducing that principle to a consequence of the law of motion. By affirming that two forces have equal magnitude if "they draw a point with the same force," he subordinates the measure of magnitude to the adoption of a law of motion—in other words, to the contingent truth which is the basis of all mechanical theory.

4.2 Daniel Bernoulli's First Geometrical Demonstration

The hypotheses or axioms which Bernoulli introduces at the beginning of his demonstration belong to the unquestionable sphere of metaphysical truth. The first hypothesis postulates simply that the force can be decomposed into its equivalents ("*Potentiis quibuscunque possunt substitui earundem aequivalentes*"). The second defines the sum of two "conspiring" forces (i.e., concurrent forces of equal or opposite direction) applied to the same point. It assumes that this sum has the same direction and magnitude as the sum or difference of the individual magnitudes. According to Bernoulli, this "necessary truth" follows the metaphysical principle that the whole equals the sum of its parts. Finally, the third hypothesis proposes that the resultant of two coplanar forces of equal magnitude bisects the angle formed by its components, by the metaphysical principle of symmetry or of "sufficient reason."

Bernoulli's demonstration progresses from the particular to the general. First, he ascertains the parallelogram rule for two equal orthogonal forces (*propositio* 2), then he calculates the magnitude of the resultant for two unequal orthogonal forces (*propositio* 3) and he observes that, if we knew

[8] T.E. Hill, *The Concept of Meaning* (London, 1974), chapter 1.

the direction of the resultant force, the problem could be solved in general terms (*corollary* 2). He then discusses the rhombus in depth, in order to determine the magnitude of the resultant of two equal and non-orthogonal forces (*propositiones* 4, 5, 7, and 8), pointing out at the same time (*propositio* 6) that the solution of this fully defines the resultant of two orthogonal forces partly discussed in *propositio* 3. At this point, by virtue of *propositio* 3 and *corollary* 2, the parallelogram rule is finally ascertained.

A look at the first passages of Bernoulli's complex argument may help to convey some of the power, as well as the limitations, of his theory. He states a general lemma (*propositio* 1) rather succinctly. It regards the possibility of stating that three concurrent forces are in equilibrium if their magnitude is a multiple or submultiple of the magnitude of three equilibrated forces arranged in a similar way among themselves. In fact, this lemma is much more challenging than Bernoulli suspects, and his formulation does not properly treat all its consequences. For example, he retains the implicit assumption that no direction is distinguished in space—that is, if we rotate the original force pair, the resultant is rotated in the same way. If we agree that the resultant is unique, we realize immediately that this rotational automorphy implies that the resultant lies in the same plane and on the bisector of the angle between its component forces (i.e., hypothesis 3). But the inverse of this statement cannot be established without further hypotheses. Furthermore, for the following propositions to be valid, we have to assume that equilibrium remains even if the relation between the magnitude of one of the forces of the new system and a corresponding one in the preceding system cannot be expressed by a rational number. (A. Cauchy was later to resolve this point.)[9]

The demonstration of the parallelogram rule follows immediately from the preceding lemma. *Propositio* 2 considers the case of two orthogonal forces DA, DC, acting on the same point with the same magnitude a. From hypothesis 3, we know the direction of the resultant DB; we only have to demonstrate that its magnitude x equals $a\sqrt{2}$. Consider the line DB and its perpendicular EG (Figure 4.1), with segments DE, DG, DH, with length y, so that

$$x : a = a : y. \tag{4.1}$$

With this, by virtue of the previous lemma, the force of DA can be interpreted as the resultant of DE and DH, and force DC as the resultant of DG and DH. But DE and DC cancel out, because they are equal and opposite (hypothesis 2). The system is therefore reduced to the two conspiring

[9] A. Cauchy, "Sur la résultante et les projections de plusieurs forces appliquées à un seul point," *Exercices de mathématique* (Paris, 1826), Vol. 1, pp. 29–43 (hereafter Cauchy, "Sur la résultante"). *See* 2d ed., *Œuvres Complètes d'Augustin Cauchy*, Series 2, Vol. 6, pp. 43–61.

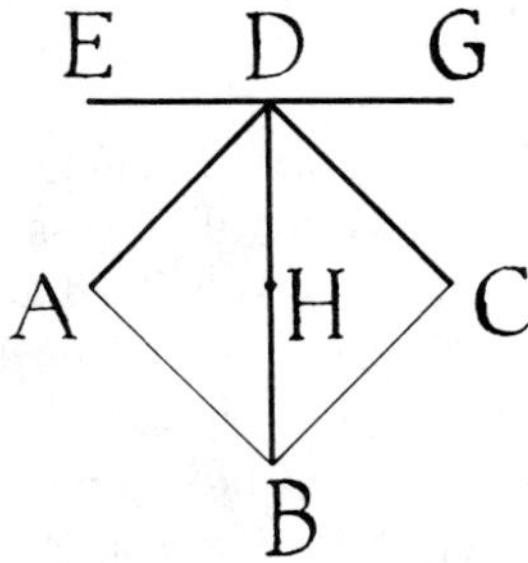

FIGURE 4.1.

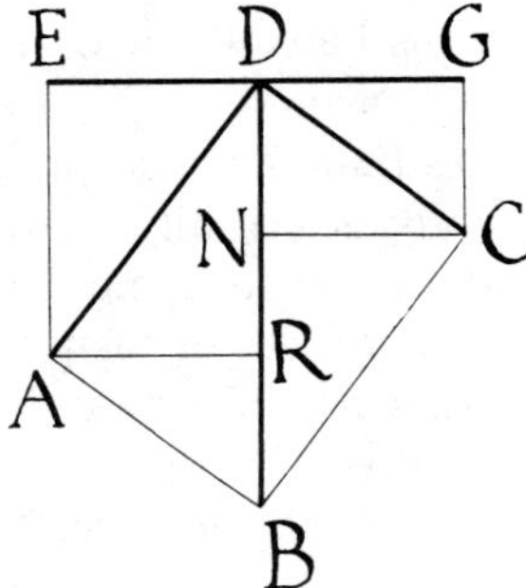

FIGURE 4.2.

forces DH, whose resultant DB has magnitude $2y$; that is, $x = 2y$. From equation (4.1), it follows that $y = a^2/x$, so that $x = 2a^2/x$, or $x = a\sqrt{2}$.

By a completely analogous procedure we can demonstrate (propositio 3) that magnitude x of resultant DB of the two orthogonal forces DA, DC, with magnitudes a and b respectively, is given by $x = \sqrt{a^2 + b^2}$ (Figure 4.2). In this case, we interpret DA as the resultant of DE and DR such that $DE = ab/x$, $DR = a^2/x$, and DC as the resultant of DG and DN such that $DG = ab/x$, $DN = b^2/x$. The initial lemma confirms this interpretation. Once again, DE and DG are equal and opposite and cancel out, so that the system is reduced to the sum of the conspiring forces DR and DN. This leads us to the thesis $x = a^2/x + b^2/x$.

Up to now, as well as later, Bernoulli has constantly directed his attention to determining the *magnitude* of the resultant. The only *a priori* assumptions he makes concern the direction of the resultant; they are taken from hypothesis 3 and are based on the principle of symmetry. This approach is particularly evident in his long and complex examination of the case in which the parallelogram is reduced to a rhombus (*propositiones* 4ff). This characterization of the magnitude of the resultant for every pair of equal forces allows Bernoulli to predict the direction of the resultant when component forces are general.

4.3 Bülffinger's Paradox

The same volume of the *Commentarii Academiae Scientiarum Imperialis Petropolitanae* which published Bernoulli's essay also includes two rather dense and intricate articles, one by Jakob Hermann and the other by Georg B. Bülffinger. Both concern the principles of mechanics and the measure of "live" and "dead" forces.[10] Their main points are irrelevant to this discussion. Both deal with the disagreement between Cartesians and Leibnizians, generally siding with the latter. This dissension was engendered by the ambiguity and misleading evidence surrounding the word "force." The term was too closely bound to ordinary language, and the disputants were unable to look for conceptual differences which were latent in the term.

But this does not concern us here. We will limit our attention to those remarks which have a direct bearing on the parallelogram rule. From the start, Bülffinger demonstrates that the concepts of "live force" ("the principle by which a natural body is actually moved") and of "dead force" ("what tends to produce motion but does not produce it because of some hindrance") are homogeneous. He sees in them two aspects of a single entity, force, which has magnitude and direction. But this prejudice compels him to face irresolvable difficulties; his arguments are ingenious but ineffectual in overcoming them, although they lead in directions which were to prove fruitful.

Bülffinger starts by accepting Leibniz's view of live forces, and offering a curious geometrical demonstration of it.[11] He sees the rectangle as a paradigm for the composition of forces. Because "live" forces are proportional to the square of the velocity, it is evident that "in every rectangular parallelogram, action and live force according to diagonal AD are equal to the sum of actions and living forces according to the sides AB and AC." [section I, theorem 3, p. 55]

Of course, if the parallelogram is not rectangular, we cannot use the Pythagorean theorem. Furthermore, in Bülffinger's opinion, we find a paradox: if the usual rule of the composition of forces "that Stevin had [established] as a supreme principle of statics"[12] is to hold, the action of the resultant is *less* than the lateral actions if the component forces form an obtuse angle, and *greater* if the angle is acute. Does this negate the metaphysical relation between cause and effect? Bülffinger's response is summarized by Vincenzo Riccati:

---

[10] J. Hermann, "De mensura virium corporum," *Commentarii Academia Scientiarum Imperialis Petropolitanae*, Vol 1 (1726); (Bologna, 1740), pp. 3–43; G.B. Bülffinger, "De viribus corpori moto insitis et illarum mensura," *ibid.*, pp. 43–117.

[11] G.B. Bülffinger, *op. cit.*, pp. 50–52.

[12] *Ibid.*, p. 94

> If the directions of the lateral forces form a right angle, the action of one is neither an advantage nor a disadvantage for the action of the other: if they are connected together, they act in the same way as if they were separate. But if the angle is obtuse, because of the opposition of the forces, the action of one is weakened by the action of the other; besides, if the angle is acute, the two actions help each other.[13]

As a response, this is almost as vague and nonsensical as the question it purports to answer. Nonetheless, Bülffinger's treatment deserves attention. The misleading "homogeneity" which he postulates between living and dead forces leads him to interpolate a common element, the concept of "action." "The action of the dead force," he affirms, "is measured by the product of magnitude and the distance travelled [*ex facto intensitatis in viam*]."[14] This allows him to represent the action of a force AB by means of the square of segment AB, giving the dead forces the same argument—and the same problems—as the live ones.

4.4 Riccati's Solution

Bülffinger's paradox (a good example of the linguistic and conceptual mess which preoccupied the scientific community in the first decades of the eighteenth century) was finally clarified by Vincenzo Riccati. He returns many times to the problem of live and dead forces in the course of his work. He first addressed the topic December 1744 during his public dissertation at Bologna's Luigi Gonzaga college; Fr. Giovanni Bernardo was his thesis supervisor and presented him for the examination. He took it up again in his notable *Dialogo ... dove ... delle forze vive e dell'azioni delle forze morte si tien discorso* ("Dialogue ... about live forces and the actions of dead forces"), and yet again in his letters to Fr. Virgilio Cavina on the principles of mechanics (published in Venice in 1772).[15]

Riccati was disquieted by the apparent disproportion between cause and effect which follows the composition or decomposition of forces. How can it be, he asks, that "that growth comes into being, such that the equipollent power is greater than the action of the lateral powers?"[16] The nub of the question, Riccati thinks, is to be found in the correct definition of force and in the proper choice of criteria to include the definition of the equipollence

---

[13] V. Riccati, *Dialogo dove ne' congressi di più giornate delle forze vive e dell'azioni delle forze morte si tien discorso* (Bologna, 1749) (hereafter cited as *Dialogo*), p. 213.

[14] G.B. Bülffinger, *op. cit.*, p. 95

[15] V. Riccati, *De'principj della meccanica* (Venice, 1772).

[16] Riccati, *Dialogo*, p. 214.

of two forces. Unlike Daniel Bernoulli, he makes no claim to offer a general geometrical model for force (or power, *potenza*). Instead, he limits himself to a highly expressive physical model. He starts with an elastic cord which is already taut, and on which an infinitesimal elongation is imposed. In Riccati's view, this physical model suggests the introduction of the concept of *action exercised by a force* as a quantity proportional both to the pre-existing tension and to the additional stretching.[17] Moreover, he assumes the following:

1. The tension of the cord remains constant with respect to the infinitesimal variation of the superimposed deformation (first hypothesis).

2. "The elastic cords which are already taut restore and shorten themselves, but resist a new elongation" (third hypothesis).

3. The "element of a live force engendered in a body to which one or more elastic cords are applied is equal to the sum of all the actions, if all the cords become shorter, and to the difference, if some shorten and others are stretched" (fourth hypothesis).

Finally, he gives two definitions:

1. "We will call a power equipollent to one, two or more powers whose action is equal to the composite action of all the others while the body travels the same distance."

2. "We will say that whatever body to which one or two elastic cords are applied, provided it is free to follow any direction, is stimulated in a free direction. If, however, it is compelled to take a determined path—e.g., if it were placed in a canal which it could not leave—we will say that it is stimulated in a necessary direction."[18]

The demonstration of the parallelogram rule and the resolution of Bülffinger's paradox follow immediately from these hypotheses and definitions. We need only remember the first of the theorems which Riccati proposes, which allows the calculation of the component according to AD of a force directed according to AS, of magnitude $\overline{AB}$. This component is given by $\overline{AH}$. In fact, assuming that direction AD is "necessary" (Figure 4.3), and considering an infinitesimal displacement Aa, the action of AB is measured by $\overline{AB} \cdot \overline{Ap}$; the action of the component X with reference to AD, $X \cdot \overline{Aa}$, must attain the same value. This leads to

$$X = \frac{\overline{AB} \cdot \overline{Ap}}{\overline{Aa}} \tag{4.2}$$

---

[17] *Ibid.*, p. 215, hypothesis 2

[18] *Ibid.*, p. 216

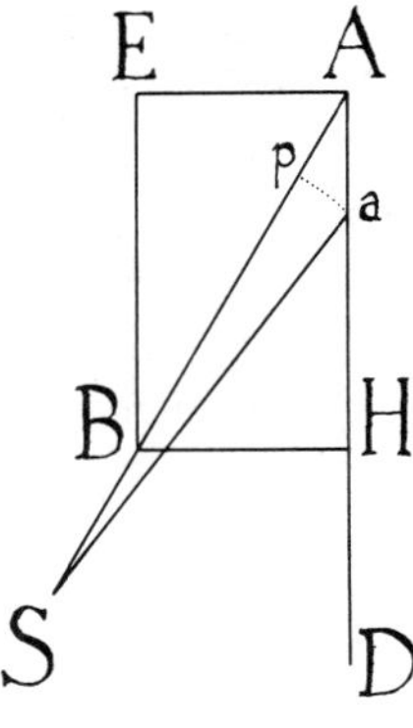

FIGURE 4.3.

and therefore $X = \overline{AH}$.

Riccati's results are not surprising. It seems obvious that, if we accept the principle of virtual work, with its implicit definition of the component of a force according to the direction of an arbitrary displacement, the parallelogram rule is a natural consequence.[19] Similarly, it seems plausible to deduce the law of composition of forces for forces which act concurrently on one point, if we allow the other fundamental law of statics, the law of the lever. This second demonstration was thoroughly developed by Philippe de la Hire in his *Traité de mécanique*, published in 1695.[20] The same line of reasoning appeared in the early nineteenth century in works by Luigi Marini and Louis Poinsot. It was the subject of a treatment by Gaspard Monge and was later included in common didactic works on elementary mechanics, with slight variations and simplifications which were not always real improvements.[21]

We shall not examine these demonstrations in any detail, but it is useful to note that the development of the principles of statics from virtual work, from the law of the lever, and from the principle of the composition of forces raises a subtle question which has never been fully clarified. The middle term of the syllogisms generally used consists of a number of geometric propositions, and thus rely on Euclidean geometry. But the role of geometry is different in the definition of statical principles. For example, the composition of concurrent forces apparently does not depend on Euclid's fifth postulate, while the law of the lever and the composition of parallel forces take on different expressions depending on whose geometry

[19] Cf. Mach, *Die Mechanik*, p. 75.

[20] P. de la Hire, *Traité de méchanique* (Paris, 1695).

[21] L. Marini, *Tentamen de motu composito* (Rome, 1804); L. Poinsot, *Élements de statique* (Paris, 1803); G. Monge, *Traité élémentaire de statique* (Paris, 1810).

we refer to—Euclid's, Lobachevski-Bolyai's or Riemann's.[22] Moreover, the geometry implied by the set of static principles affects the meaning of the word "force," giving it different connotations.

A. Genocchi[23] presents a critical analysis of the relationship between statics and geometry, and explores the possibility of statically deducing the fundamental formulae of trigonometry. In a sense, therefore, he reverses the demonstrative procedures followed by Daniel Bernoulli and others; instead of basing the principles of statics on the "necessary truths" of geometry, he uses statics as an accurate and effective means to express the empirical relevance of geometrical axioms.

4.5 Foncenex's Memoir and Lagrange's Criticism

The turning point in the history of the parallelogram rule came around 1760. The Piedmontese-Savoyard mathematician François Daviet de Foncenex, whom we have already met, published an important essay, "Sur les principes fondamentaux de la mécanique." A good part of this work is dedicated to the subject of the composition of forces and to the law of the lever. Its author was born in Thonon in 1734 and died in Casale in 1799. He was a brilliant young man, a friend and companion to another young Turinese, Luigi Lagrange. Foncenex wrote two extraordinary essays almost simultaneously. One was the essay on the principles of mechanics which we have just mentioned, and another concerned the logarithms of negative quantities. Both were published in the *Mélanges de philosophie et de mathématique de la société royale de Turin pour les années 1760–1761*. And that was the whole of his output. There was nothing more. The king of Sardinia appointed the promising scientist governor of Sassari and commander of his navy.

The historian of mathematics Montucla was astonished by the brevity of Foncenex's career. But Delambre, in his biography of Lagrange, has a more malicious suggestion. These works—especially the one on the principles of mechanics—he says, are the fruit of a collaboration in which Lagrange "provided Foncenex with the analytical part of his memoirs."[24] This hypothesis is interesting, but (at least for the composition of forces) it is

---

[22] J. Andrade, *Leçons de mécanique physique* (Paris, 1898), remark II; R. Bonola, *La geometria non euclidea* (Bologna, 1906).

[23] A. Genocchi, "Sur un mémoire de Daviet de Foncenex et sur les géometries non-euclidiennes," *Atti Accad. Reale Sci. Torino*, Vol. 29, No. 2 (1877), pp. 366–371; A. Genocchi, "Dei primi principi della meccanica e della geometria in relazione al postulato di Euclide," *Atti Accad. Reale Sci. Torino*, Vol. 2, No. 3 (1869), pp. 153–189.

[24] "Notice sur la vie et les ouvrages de M. le Comte Lagrange, par M. le Ch.er Delambre," *Mémoires de l'Institut de France, classe mathématique et physique*, Vol. 13 (1812), pp. xxxiv–lxxx.

probably wrong. Lagrange had his own opinions on the subject and they clearly contradicted Foncenex's. In his *Théorie des fonctions analytiques* Lagrange wanted to reduce the problem of the composition of forces to a trivial application of the normal analytical procedure, so as to transform the parametrical representation of a curve into its Cartesian equation. In his terms, the parallelogram rule is only a simple geometric interpretation of the result of eliminating the temporal parameter—and this interpretation holds both in the case of a linear dependence (the composition of velocities) and in the case of a quadratic dependence (the composition of forces).[25]

Lagrange's thesis seems inclined rather to eliminate the question than to try to clarify it. Even if it is true that the various demonstrations of the parallelogram rule disguise a "composition of spaces," it is also true that the hypotheses needed to deduce the parallelogram rule can be less restrictive than Lagrange's.

4.6 Foncenex's Fundamental Lemma

Foncenex's aims are very different. His essay goes to the heart of the dispute which divided eighteenth-century science, the "grand metaphysical problem" (as d'Alembert called it) posed by the Prussian Academy of Sciences: are the laws of statics and mechanics necessary or contingent truths? Bernoulli had proposed a solution of sorts; the law of motion is contingent, and the law of statics is necessary. Euler, in his *Mechanica sive motus scientia analytice exposita* (1736), came down on the side of the rationalists, adding the law of motion to the list of necessary truths. But it is d'Alembert to whom Foncenex explicitly refers.

The second edition of d'Alembert's *Traité de dynamique* was hot off the press in 1758. In it, the author pointed the way towards a rational synthesis of mechanics which depended on only three principles: the law of inertia, the principle of composition of forces, and the principle of equilibrium. This reduction, d'Alembert claimed, would permit a full clarification of the matter.

[25] L. Lagrange, *Théorie des fonctions analytiques* (Paris, 1797), part 3, ch. 2, p. 324. "This way of considering the composition of velocities and of forces, seems the most natural to me; and it has the advantage of demonstrating clearly *why* the composition of forces necessarily follows the same laws as the one of velocity. Since we can consider forces independently from their movement, we are tempted to derive their composition from purely geometrical and analytical principles; but it would not be impossible to prove that all the demonstrations furnished for the composition of forces are only a composition of spaces in disguise."

> We believe we have demonstrated in this work that a body which is left alone must persist in its state of rest or of uniform movement; we believe we have also demonstrated that, if it tends to move at once along two sides of a parallelogram, the diagonal is the direction spontaneously assumed. ... In conclusion, we have demonstrated that every law concerning the communication of motion between bodies is reduced to the laws of equilibrium, and that these are reduced in turn to those of the equilibrium of two equal bodies, stimulated in opposite directions by equal virtual velocities. ... From all these considerations, it follows that the laws of statics and of mechanics set forth in this book are those which result only from the existence of matter and of movement. Now, experience proves that these laws, in reality, are to be observed in the bodies that surround us. Thus the laws of equilibrium and of movement, which we know from observation, are necessary truths.[26]

(The *Traité de dynamique* really deserves a separate essay; while its language is outdated, its extraordinary epistemological suggestions are worth attention.[27].)

We return to Foncenex's memoir: he shares d'Alembert's aims to a large extent, and accepts the same synthesis of mechanical principles. Three of his essay's sections discuss the law of inertia, the composition of forces and equilibrium. The fourth section, on the law of the lever, may be the most important because of its mathematical inventiveness. Foncenex departs from d'Alembert on the subject of inertia, believing that a uniform rectilinear movement described by free bodies is not a description of a fact, but a choice of a common measure for every other type of motion. Therefore "true or imaginary causes" are introduced when deviations or changes appear.[28]

While Foncenex sets the law of inertia at one remove from the principles of mechanics, he gives an accurate mathematical demonstration of the composition of forces. The second section of his essay is devoted to this subject. Here is the fundamental lemma:

> If two equal forces, whose magnitude and direction are expressed by segments CA and CB, act on any body C, it is obvious that the body will not be able to follow both of these

[26] d'Alembert, *Traité de dynamique*, pp. xxvij–xxix

[27] *See* E. Benvenuto, "Se le leggi della statica e della meccanica siano di verità necessaria o contingente," *Atti Istituto di discipline scientifiche* (1978), University of Genoa.

[28] F. Daviet de Foncenex, "Sur les principes fondamentaux de la méchanique," *Mélanges de philosophie et de mathématique de la Société Royale de Turin*, Vol. 2 (1760–1761), p. 302.

forces at the same time, because it cannot move both according to CA and to CB. It will therefore take a direction CM different from CA and CB, and line CM must necessarily bisect the angle $A\widehat{C}B$, since hypothetically CA and CB are equal, and all that tends to distinguish CM from CA tends equally to distinguish CM from CB. Given this, it is still evident that we can imagine a third force CM which, by itself, gives the same effect of CA and CB acting on body C. Yet the magnitude of force CM can only depend on the magnitude of CA and CB, and on the value of their angle $A\widehat{C}B$. Consequently, if we set

$$CA = CB = a, \quad CM = z, \quad A\widehat{C}B = \varphi,$$

we obtain

$$z = \text{funct.}(a, \varphi).$$

Now, because force CM is of the same nature as CA, it is necessary that both contain the same number of dimensions; and this gives us

$$z = CM = \text{funct.}(a, \varphi) = a\,\text{funct.}(\varphi)$$

because the dimension of φ is zero.[29]

The importance of this lemma lies not in its demonstrative strength—which, as Foncenex recognizes, is not great enough for his purposes—but rather in his decision to express the magnitude of the resultant as a function of only two quantities, relative to the component forces. He says nothing *a priori* about this function, although it is implicitly continuous and differentiable as many times as necessary. But from the outset he uses a principle of reduction according to which the resultant must be defined. That is, the magnitude z of the resultant CM is integrally determined by the magnitude and angle of the components. No other factor can affect the determination of z—not its position with respect to a reference frame, nor the effects of the motion, nor anything else. Foncenex's vocabulary is consistent, and he introduces new hypotheses only if they help to define these terms. His procedure is analogous to the axiomatic method; in both cases, the method first defines an axiomatic "alphabet."

The last part of the lemma—that is, inferring from $z = \text{funct.}(a, \varphi)$ to $z = a \cdot \text{funct.}(\varphi)$—is rather weak. Therefore the author invokes an *ad hoc* principle: z must be proportional to a if φ is to be constant.[30] D'Alembert had followed this principle, and successive authors were to do so: Poisson,

[29] *Ibid.*, pp. 305–306.
[30] *Ibid.*, p. 306

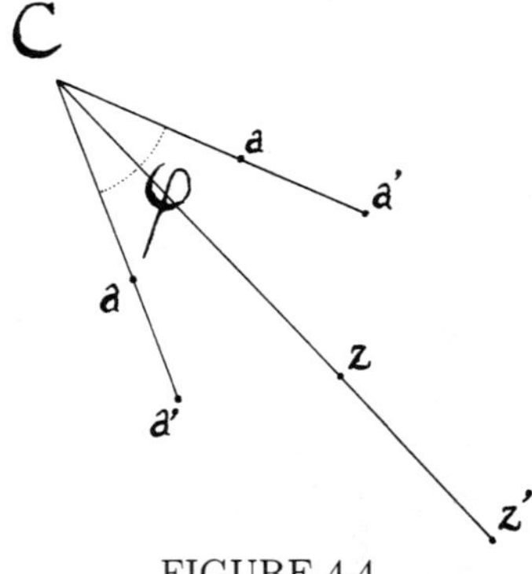

FIGURE 4.4.

Ferrari and Siacci among them.[31] Cauchy used a different demonstrative technique, and De Sinno (a nineteenth-century Neapolitan mathematician) argued as follows: Fixing φ, we can set $z = F(a)$. Now consider two other forces a', disposed like a (Figure 4.4). Their resultant has a magnitude $F(a')$ and a direction like that of z. Therefore, if the addition principle for forces of equal direction holds, than the total resultant is $F(a) + F(a')$ and at the same time also $F(a + a')$. From this we obtain

$$F(a) + F(a') = F(a + a').$$

This is Cauchy's functional equation the most general continuous solution of which is of the form $F(a) = fa$, with $f = \text{constant}$.[32] For each φ, then, we can conclude that the expression of z is

$$z = af(\varphi).$$

4.7 Foncenex's and D'Alembert's Functional Equation

At this point, we have to return to the determination of $f = f(\varphi)$. Evidently this cannot be obtained by considering a single system of two forces a with

---

[31] J. d'Alembert, "Mémoire sur les principes de la mécanique," *Mémoires de l'Académie Royale des Sciences* (1769) (hereafter cited as d'Alembert, "Mémoire sur les principes"), pp. 278–286; S.D. Poisson, *Traité de mécanique*, 2d ed. (Paris, 1833), section 26; M. Ferrari, *Lezioni di meccanica razionale* (Naples, 1868), p. 4; F. Siacci, "Sulla composizione delle forze nella statica," notes 1, 2, 3, *Rendiconti della regia accademia delle scienze fisiche e matematiche di Napoli* (1899). A. Cauchy, "Sur la résultante"; G.B. De Sinno, *Elementi di meccanica generale* (Naples, 1853), p. 6.

[32] A.L. Cauchy, *Cours d'analyse de l'École Polytechnique*, Vol. 1 (Paris, 1821), reprinted in *Œuvres* (Paris, 1897), Series 2, Vol. 3, pp. 98–113, 220.

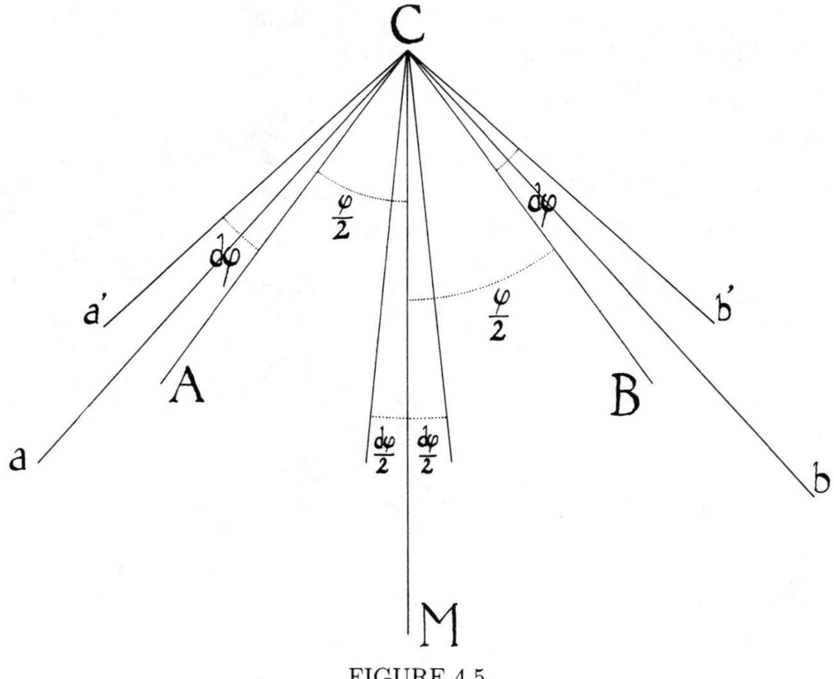

FIGURE 4.5.

the resultant z, but rather by imagining a more complex system of forces which can express some of the formal properties of the composition. In particular, the associative property requires at least three forces. In the specific case which Foncenex considers, the component forces are paired and placed symmetrically with respect to the resultant. The simplest system has only two such pairs, and the trick is to find the resultant of these four forces in two different ways. We should note that Foncenex, like some of his successors, does not find it necessary to use the commutative property. Instead, he uses a bisymmetrical relation according to which

$$(a_1 \circ a_2) \circ (a_3 \circ a_4) = (a_1 \circ a_3) \circ (a_2 \circ a_4) \tag{4.3}$$

where $\circ$ is the composition operator and a_1, a_2, a_3, a_4 are the component forces.

In the hypotheses accepted for $z = af(\varphi)$, associativity and bisymmetry with the principle of the algebraic sum for conspiring forces are necessary for finding $f = f(\varphi)$. They are also sufficient conditions if we assume that $f(\varphi)$ is continuous.

In this way, Foncenex associates a second system of forces Ca' and Cb' to the primitive system of forces CA and CB. $Ca'$and Cb' still have magnitude a and are placed symmetrically, forming an angle $\varphi + 2d\varphi$ (Figure 4.5). The resultant of CA and CB has the same direction as the resultant of $Ca'$and

Cb', so that we can sum their magnitudes to obtain the total resultant r of CA, CB, Ca', Cb' :

$$r = af(\varphi) + af(\varphi + 2d\varphi). \tag{4.4}$$

On the other hand, thanks to bisymmetry, r can be interpreted as the resultant of two forces Ca, Cb, which represent the resultants of CA, Ca' and CB, Cb'. Ca and Cb bisect the angles $\widehat{a'CA}$ and $\widehat{BCb'}$ respectively, and themselves form an angle $\varphi + d\varphi$. The magnitude of Ca and Cb is $r' = af(d\varphi)$; thus the total resultant r is given by

$$r = af(d\varphi)f(\varphi + d\varphi). \tag{4.5}$$

All that remains is to compare the right-hand sides of equations (4.4) and (4.5):

$$f(\varphi) + f(\varphi + 2d\varphi) = f(d\varphi)f(\varphi + d\varphi). \tag{4.6}$$

It seems likely that some of d'Alembert's memorable works escaped Foncenex's notice.[33] In these, with reference to another theme, d'Alembert discusses an equation related to his own. Moreover, d'Alembert's *Addition* (1750) gave a general solution of the function equation:

$$f(\varphi - \varepsilon)f(\varphi + \varepsilon) = f(\varphi)f(\varepsilon) \tag{4.7}$$

of which equation (4.6) is a specific expression. Foncenex developed his own technique for equation (4.6). Setting $df(0) = V\,d\varphi$, he observes that, for the principle of summing conspiring forces, $f(d\varphi)$ must equal $2 + V\,d\varphi$. Moreover, since $f(\varphi + d\varphi) = f + df$, and $f(\varphi + 2d\varphi) = f + 2df + d^2 f$, then, by equation (4.6)

$$2f + 2df + d^2 f = (f + df)(2 + V\,d\varphi),$$

that is,

$$d^2 f = fV\,d\varphi + V\,d\varphi\,df. \tag{4.8}$$

This last equation allows us to say that V is an infinitesimal quantity. Thus, if we set $V = H\,d\varphi$, equation (4.8) becomes $d^2 f = fH\,d\varphi^2$, whence

$$f(\varphi) = Ae^{\sqrt{H}\varphi} + Be^{-\sqrt{H}\varphi}, \tag{4.9}$$

where A and B are constants. The conditions for $\varphi = 0$:

$$f(0) = 2, \qquad \frac{df}{d\varphi} = V,$$

[33] J. d'Alembert, "Recherches sur la courbe que forme une corde tendüe mise en vibration," *Histoire de l'Académie des Sciences* of Berlin (1747), Vol. 3, pp. 214–219, 220–249; J. d'Alembert, "Addition au mémoire sur la courbe que forme une corde tendüe mise en vibration," *Histoire de l'Académie des Sciences* of Berlin (1750), Vol. 6, pp. 355–360.

give

$$f(\varphi) = 2\cos(\sqrt{-H}\varphi);$$

and from the condition $f = 0$ for $\varphi = \pi$, one obtains

$$\sqrt{-H}\,\pi = \frac{2\mu+1}{2}\,\pi \qquad (\mu = 0, \pm 1, \pm 2, \ldots),$$

that is,

$$f(\varphi) = 2\cos\frac{2\mu+1}{2}\,\varphi.$$

"Now," Foncenex concludes,[34] "it is evident that when the angle φ is less than two right angles, force z, and consequently the value of f, must always be positive; such a condition could not exist unless $\mu = 0$. Finally, there will be

$$f(\varphi) = 2\cos\frac{\varphi}{2}.\text{''} \tag{4.10}$$

We will omit the passages which lead Foncenex to a complete demonstration of the parallelogram rule in order to mention the *Scholie I*, which concludes the second section. Here the author elaborates a new solution for equation (4.7) or, more precisely, for its equivalent

$$f(\varphi) + f(\varphi + 2\varepsilon) = f(\varepsilon)f(\varphi + \varepsilon) \tag{4.11}$$

without having to differentiate it. First, consider the case in which φ is a multiple of ε, and set $f(\varepsilon) = K$, $f(\varphi) = f^{(n)}$, $f(\varphi+\varepsilon) = f^{(n+1)}$, $f(\varphi+2\epsilon) = f^{(n+2)}$. Then equation (4.11) assumes the form of a recurrent series whose ratio is $K - 1$:

$$f^{(n+2)} - Kf^{(n+1)} + f^{(n)} = 0. \tag{4.12}$$

Thus in general we obtain $f^{(n)} = Dx^n + Ey^n$, where D and E are constants, n is an exponent, and x, y are the roots of the equation $u^2 - Ku + 1 = 0$. The result is

$$f^{(n)} = D\left(\frac{K}{2} + \sqrt{K^2/4 - 1}\right)^n + E\left(\frac{K}{2} - \sqrt{K^2/4 - 1}\right)^n. \tag{4.13}$$

If we set $K = 2\cos\alpha$, a simple trigonometric interpretation leads to

$$f^{(n)} = F\cos n\alpha + G\sin n\alpha, \tag{4.14}$$

where F, G are constants. Again, the conditions for $n = 0$ and for $n = \nu$ (such that $\nu\alpha = \pi/2$) reproduce equation (4.10). The generalization for the case in which φ and ε are incommensurable becomes evident "by means of a method so familiar to geometricians, and above all in the works of the ancients," Foncenex writes, "that I shall not dwell on it here."

---

[34] F. Daviet de Foncenex, *op. cit.*, p. 308.

4.8 D'Alembert's Memoir of 1769

History has been unfair to Foncenex. Present-day writers ascribe the splendid solution which we discussed above to d'Alembert and Poisson, while Foncenex's work is only cited for its section on the equilibrium of the lever. True, d'Alembert was first to recognize the basic role of the functional equation (4.7) in his study of the vibration of a stretched cord. But Foncenex's argument in favour of equation (4.6) or equation (4.11) is original and has the merit of simply and directly linking the functional equation to more formal properties of the sum—that is, to associativity and bisymmetry. He originates an idea which was to become one of the pivots of current mathematical research: the exploration of the algebraic foundations of trigonometry and geometry in general, the creation of new entities and structures, defined by the formal properties of composition laws, and the discussion of principles of algebra and analysis. In this train of thought, Foncenex's successors include such thinkers as Gauss, Cauchy, Bolyai, Lobacevskij, Abel, Stokes, Schiapparelli, Weierstrass, and others.

Why do historians give all the credit to d'Alembert? The reason is not his Berlin essay of 1750, but rather a memoir which he presented to the Academy of Paris in 1769.[35] Its title and structure are strikingly similar to Foncenex's essay. Unfortunately, perhaps, Foncenex revered d'Alembert, citing many of his works, but d'Alembert gives little attention to Foncenex. He cites the Turin *Mémoire*, but only for the section on the equilibrium of the lever, and only to point out its flaws.[36] Yet the reasoning which d'Alembert develops in his first section (*Article premier*) on the composition of forces is almost the same as Foncenex's, reproducing the same passages with such minor changes that they escape notice at first reading. One of these slight variations is, however, worth brief discussion.

Unlike Foncenex, d'Alembert immediately considers the most general case of two forces AB and AC, both acting concurrently on A and with different magnitudes and directions (Figure 4.6). By considering a straight line AD and duplicating the forces AB and AC in the forces Ab and Ac, which are symmetrical with respect to AD, he finds again a system of two pairs of equal forces. Using (without credit) Foncenex's lemma, d'Alembert can apply the bisymmetry property (equation (4.3) in the following form:

$$(AB \circ AC) \circ (Ab \circ Ac) = (AB \circ Ab) \circ (AC \circ Ac).$$

If a, b and z are the magnitudes of AC, AB and their resultant AF respectively, it follows that

$$zf(u+m) = af(m) + bf(m+\alpha), \tag{4.15}$$

[35] D'Alembert, "Mémoire sur les principes de la mécanique," *Mémoires de l'Académie Royale des Sciences, Année 1769*, (Paris, 1772), pp. 278–286.

[36] *Ibid*, pp. 285–286.

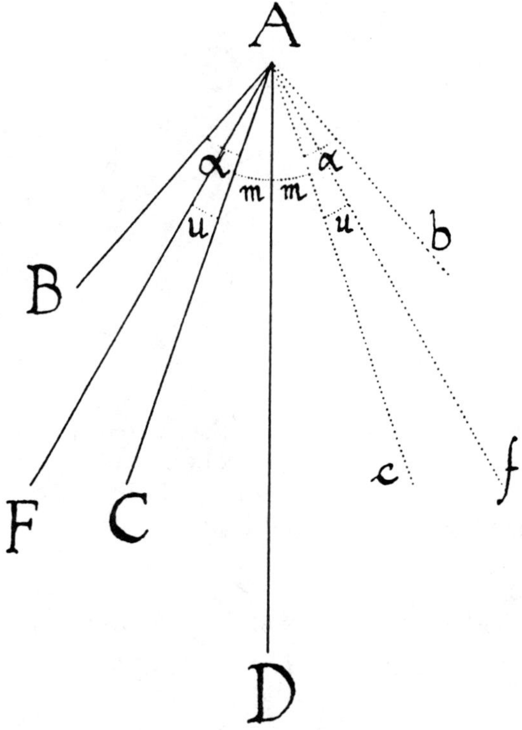

FIGURE 4.6.

where α, u, m are the angles represented in Figure 4.6. The determination of $f(\alpha)$ is similar to the above, by means of the functional equation (4.7), and leads to $f(\alpha) = 2\cos\alpha$. At this point, d'Alembert realizes that (4.15) gives an expression for the magnitude of resultant z, which contains the arbitrary parameter m:

$$z = \frac{af(m) + bf(m + \alpha)}{f(u + m)}. \tag{4.16}$$

Whatever m may be, the left-hand side of equation (4.16) has the same value. This is possible because $f(\alpha) = 2\cos\alpha$ and leads—by means of short calculations—to the proportion $\sin u : \sin(\alpha - u) = b : a$ which provides the direction of AF.

This treatment is significant in light of the epistemological considerations of Section 1.4. The magnitude z of the resultant is determined by equation (4.16) according to a general law of composition, subordinated to parameter m. The equality $f(\alpha) = 2\cos\alpha$ may therefore be understood

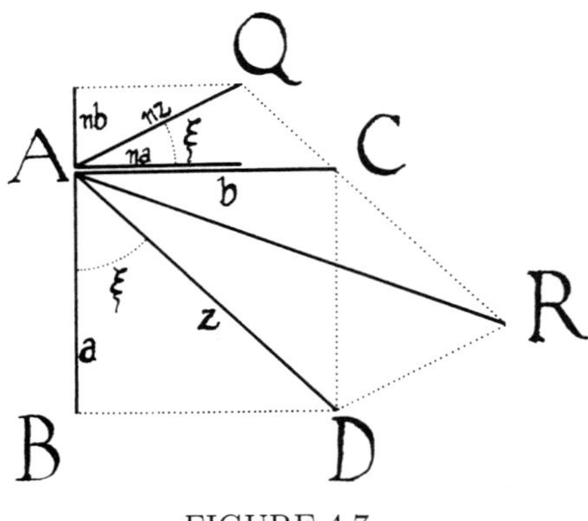

FIGURE 4.7.

as a condition which allows us to affirm the existence of the same z for every m-application of the theory produced by Foncenex's lemma. In other words, it can represent the restriction on $f(\alpha)$ which permits the prediction of Ramsey-eliminability of the function z.[37]

4.9 Further Developments: D'Alembert, Poisson, Cauchy, Dorna and Darboux

Foncenex's original research was later developed and enriched by several thinkers. D'Alembert was among the first. Just after he presented his *Mémoire* to the Academy of Paris in 1769, he published the sixth volume of his *Opuscules mathématiques*,[38] which contains "a new demonstration of the parallelogram of forces." This version is far more complex than those he had developed in 1761[39] and 1769, but it was, as always, brilliant. In this demonstration, he restricts his attention to determining the direction of the resultant of two perpendicular forces a and b; for the rest, d'Alembert refers to Daniel Bernoulli. Relying on the usual hypotheses and denoting by ξ the angle formed by the resultant z with the straight line AB (Figure 4.7), let $b/a = \zeta$, and $\tan\xi = \varphi(\zeta)$. A second system of orthogonal forces na and nb (where n is a real number) is introduced as shown in the figure. AR is the resultant of all four forces. It can be determined in two ways, either

[37] F.P. Ramsey, "Theories," in *Foundations of Mathematics* (Patterson, N.J.: Littlefield, Adams, 1960), pp. 212–236.

[38] J. d'Alembert, "Nouvelle démonstration du parallélogramme des forces," *Opuscules mathématiques* (Paris, 1773), tome 6, memoir 51, pp. 360–370.

[39] J. d'Alembert, "Démonstration du principe de la composition des forces," *Opuscules mathématiques* (Paris, 1761), tome 1, memoir 5, pp. 169–179. D'Alembert curtails and improves upon Bernoulli's treatment, using a demonstrative technique based on mathematical induction.

by compounding the orthogonal forces of magnitude $(a - nb)$, $(b + na)$, or by compounding the partial resultants z and nz. A functional equation in $\varphi(\zeta)$ results:

$$\frac{\varphi(\zeta) + \varphi(n)}{1 - \varphi(n)\varphi(\zeta)} = \varphi\left(\frac{\zeta + n}{1 - n\zeta}\right). \tag{4.17}$$

D'Alembert found a general solution for it—obviously by presuming $\varphi(\zeta)$ continuous and differentiable. Thus we have

$$\varphi(\zeta) = \tan(\alpha \arctan \zeta) \tag{4.18}$$

where α = constant. The expression $\varphi(\zeta)$ must increase with ζ, and it must be positive if $\zeta > 0$; moreover the principle of symmetry dictates that $\varphi(1) = 1$. All of this leads to the statement $\alpha = 1$; thus $\varphi(\zeta) = \zeta = b/a$, in conformity with the parallelogram rule.

Among those who developed the demonstrative method of Foncenex and d'Alembert, S.D. Poisson is one of the most widely cited.[40] Some even give him credit for the functional equation (4.7). His main contribution, presented in the *Traité de mécanique* (1811) does not deal with the static-mechanical theme, but rather with the direct solution of equation (4.7). He does this by techniques which are linked to mathematical induction. Recognizing that the solution $f(\varphi) = 2\cos\varphi$ is valid for some values of φ, he demonstrates successively that it also holds for 2φ, 3φ, $\ldots$, $\frac{1}{2}\varphi$, $\frac{1}{4}\varphi$, $\ldots$, and thus for every $x = m\varphi/2^n$ (m, n = integers).

In his *Exercices de mathématique*,[41] A.L. Cauchy uses a similar inductive procedure to demonstrate the formula

$$R = 2P\cos(\widehat{P,R}), \tag{4.19}$$

which gives the resultant R of two forces P. His approach to the parallelogram rule refers directly to Daniel Bernoulli and follows his arguments. He starts from the case of two perpendicular forces P and Q. Cauchy first finds the magnitude of the resultant R, using Bernoulli's method, and then finds the direction of R by induction. Because of symmetry, R must have the direction of the diagonal if $Q = P$, and we can easily demonstrate the same thesis if $Q = P\sqrt{2}$. (To do this, we look at the case of three mutually orthogonal forces P whose resultant is $P\sqrt{3}$ and passes through the diagonal of the cube [Figure 4.8].) At this point, Cauchy shows that, if we can prove that the thesis holds for $Q = P\sqrt{m}$ (m = positive integer) then we can also verify it for $Q = P\sqrt{m+1}$, and thus for every $Q = P\sqrt{m/n}$ (m, n = positive integers). Finally, a well-known standard argument allows him to conclude that the resultant passes through the diagonal of the rectangle, whatever the ratio of Q and P.

---

[40] S.D. Poisson, *Traité de mécanique*, 2nd ed. (Paris, 1833), Vol. 1, sections 26–27.

[41] Cauchy, "Sur la résultante."

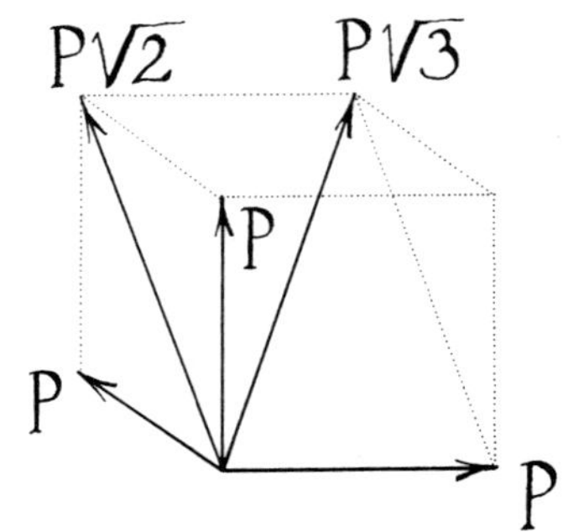

FIGURE 4.8.

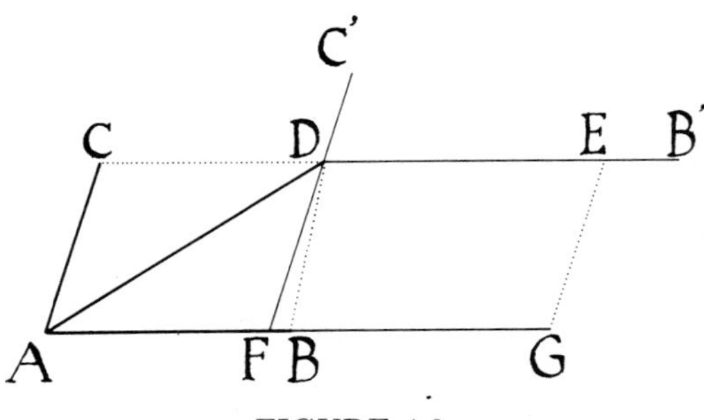

FIGURE 4.9.

Of the numerous nineteenth-century elaborations of Foncenex's demonstrations, a variant proposed by A. Dorna in his *Elementi di meccanica razionale* (1865)[42] is worth exploring.

Let AB and AC be two forces whose resultant is AD (Figure 4.9). To prove that the quadrilateral $ABCD$ is a parallelogram, Dorna applies the resultant in D, substituting a second system of components constituted by DB' and FD, parallel to and equal to AB and AC respectively. Moreover, in F he introduces a third force, FG, of arbitrary magnitude ω. Keeping $\overline{AC}$ constant, the author now assumes that the length of the side CD can be represented as a function of magnitude x of AB; thus $\overline{CD} = f(x)$. Similarly, for the quadrilateral $FGED$, $\overline{DE} = f(\omega)$. On the other hand, by compounding DB', FD and FG in a convenient way, we see that segment CE may be seen by analogy as a side of the quadrilateral relative to the composition of $AB + FG$ with AC, so that $\overline{CE} = f(x + \omega)$. In conclusion, since

$$\overline{CE} = \overline{CD} + \overline{DE} = f(x) + f(\omega)$$

we obtain the functional equation

$$f(x) + f(\omega) = f(x + \omega) \tag{4.20}$$

---

[42] A. Dorna, *Elementi di meccanica razionale*, 2d ed. (Turin, 1873), pp. 17–18.

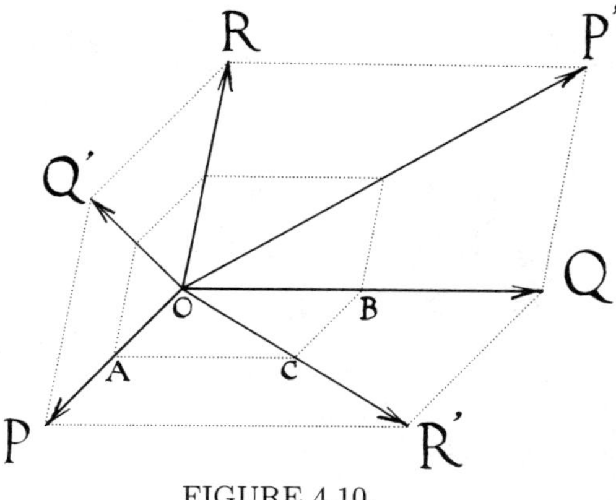

FIGURE 4.10.

for which, as we know, Cauchy provided a general solution, assuming $f(x)$ continuous. That solution is $f(x) = \mathcal{A}x$, where $\mathcal{A}$ is an arbitrary constant. The application of the principle of symmetry for $\overline{AB} = \overline{AC}$ requires that $\mathcal{A} = 1$. Therefore the quadrilateral $ABCD$ is a parallelogram.

In his notable paper of 1875,[43] Darboux underscores the importance of equation (4.20) in clarifying the hypotheses which support the law of composition of forces. We must also credit him with careful criticism of the postulates which connect all the preceding demonstrations – postulates which he proves to be necessary.

Referring to *n lines* $OA_1, OA_2, \ldots, OA_n$, rather than to forces, Darboux poses the problem in strictly geometrical terms. His aim is to determine for these lines a composition law subject to the following conditions:

I. A resultant which is unique and determined remains invariable when some of these lines are replaced by their partial resultant; it is independent of the order in which the partial compositions are made.

II. It is also independent of the orientation of the system in space.[44]

From these conditions, he shows that the resultant of two equal and opposite lines is nil, and that the resultant of two lines OA, OB is necessarily co-planar. Having established this, Darboux first examines the problem of the direction of the resultant, following a line of reasoning similar to Cauchy's. Instead of confining his attention to the plane of component lines, he adds one dimension. Consider three lines P, Q and R, which are not co-planar (Figure 4.10). Their partial resultants are R' (of P and Q); Q' (of P and R); and P' (of Q and R). It follows that the total resultant

[43] G. Darboux, "Sur la composition des forces en statique," *Bulletin des sciences mathématiques et astronomiques* (1875), Vol. 9. pp. 281–288.

[44] *Ibid.*, p. 283.

S can be derived in three ways: as a resultant of R' and R, or of Q' and Q, or of P' and P. This means that the planes PP', QQ' and RR' must all intersect on the same straight line.

Ceva's theorem concerning the trihedral angle maintains that

$$\frac{\sin R'OP}{\sin R'OQ}\frac{\sin P'OQ}{\sin P'OR}\frac{\sin Q'OR}{\sin Q'OP} = -1. \tag{4.21}$$

Now, fixing line OR perpendicular to plane POQ, and giving it length 1, we can state that the ratio $\sin P'OR/\sin P'OQ$ depends only on Q, while ratio $\sin Q'OR/\sin Q'OP$ depends only on P. If $f(Q)$, $f(P)$ give this dependence, then equation (4.21) becomes

$$\frac{\sin R'OP}{\sin QOR'} = \frac{f(Q)}{f(P)}. \tag{4.22}$$

This equation offers the solution to the problem of the direction of the resultant R' of P, Q: the direction is the diagonal OC of the parallelogram with sides OA and OB, with $OA = f(P)$ and $OB = f(Q)$. What about the magnitude of the resultant? Up to now, we have only been able to say that the length of the diagonal OC is a function of R', but the function $f(R')$ remains to be determined. To do this (Darboux says) we need a third postulate:

III. The law of the composition of lines must reduce to the algebraic sum for lines of the same direction.

In this case for P and Q, the result is:

$$f(P+Q) = f(P) + f(Q). \tag{4.23}$$

This, in fact, is Cauchy's functional equation.

In his solution of this equation, Darboux made his most conspicuous contribution, both in his work of 1875 and later on.[45] Cauchy had stated the result $f(x) = \mathcal{A}x$, assuming f to be continuous. Do we really need another postulate on the continuity of f? This sort of question would never have occurred to eighteenth-century scientists; they firmly held, on metaphysical grounds, that "nature doesn't make leaps" (*natura non facit saltus*). But Darboux's cultural milieu was quite different. The late nineteenth century saw a grand rearrangement of mathematical foundations which recognized continuity as a complex and subtle concept, while it took the formal properties of composition (associativity, commutativity, etc.) as the bases of mathematics. Subsequent developments confirmed this arrangement, creating (as in Bourbaki's work) precise hierarchies, from elementary to complex structures. In his work of 1875, Darboux shows that, to obtain $f(x) = \mathcal{A}x$,

[45] G. Darboux, "Sur le théorème fondamental de la géometrie projective," *Mathematische Annalen,* Vol. 17 (1880), pp. 51–61.

we need only to assume that $f(x)$ is integrable. If this is the case, equation (4.23) ensures that $f(x)$ is always increasing, and therefore that we can avoid the assumption of continuity. Darboux's essay of 1880, and works by H. Hankel (1870), J.L. Jensen (1878–1906), S. Pincherle and U. Amaldi (1901), R. Volpi (1903) and others confirm this. In 1905, G. Hamel found non-linear solutions, using an appropriate base of real numbers, the existence of which is connected to the axiom of choice.

4.10 Conclusion

As we know, research is now directed towards other goals. The composition of forces continued to be the subject of ever more elegant and ingenious demonstrations, but the purpose of the argument had changed. For example, a paper by P. Tchebichef again concerned the removal of any assumption about continuity.[46] A lucid essay by K. Culmann presented the parallelogram rule as an immediate consequence of a linear transformation.[47] F. Siacci's analyses belong—albeit indirectly—to nineteenth-century research into non-Euclidean geometry.[48] Both the theory of groups and vectorial calculus provide a "logical ancestor" to force (for example, the concept of Abelian group) and a proper mathematical reference—the general theory of linear transformations and substitutions.[49]

The initial question from which Bernoulli had digressed—that is, is the parallelogram rule a rational or empirical truth?—lost real interest during the course of the nineteenth century. Once the myth (still present in Kant's theory of science) of a universal and necessary truth of scientific knowledge had been destroyed, the theoretical problem was no longer that of finding whether the axioms needed for the demonstration can be established *a priori* or *a posteriori*. Rather, the problem was to list these axioms in a complete and essential way, whatever their epistemological origins.

Thus we should hardly be surprised that many of the authors who contributed to the rational demonstration of the parallelogram rule were declared empiricists about mechanics. For example, the Neapolitan scientist M. Zanotti agreed with Schnuse's empirical conception,[50] but maintained that, if the concept of force could simply be brought back to its kinematic

---

[46] P. Tchebichef, "Sur la résultante de deux forces appliquées à un seul point," *Société mathématique de Moscou* (1876), proceedings of the session of Dec. 14, 1875, *Société mathématique de France* (1878), Vol. 6, pp. 188–193.

[47] K. Culmann, *Ueber das Parallelogramm und über die Zusammensetzung der Kräfte* (Berlin, 1870).

[48] F. Siacci, *op. cit.*

[49] *See*, for example, C. Burali-Forti and R. Marcolongo, *Elementi di calcolo vettoriale ...* (Bologna, 1909); C. Burali-Forti and R. Marcolongo, *Omografie vettoriali* (Turin, 1909); C. Burali-Forti, "L'importance des transformations linéaires des vecteurs dans le calcul vectoriel général," *L'Enseignement mathématique*, 10th year (1908), pp. 414–417.

[50] Schnuse, *Die Grundlehren der Statik fester Körper* (Leipzig, 1851).

consequences, then "rational mechanics ought to be considered a mathematical novel." As a matter of fact, in his essay of 1857 on rational mechanics, he developed the interesting approach to the composition of forces given by A.F. Möbius in the Crelle's *Journal.*[51] Siacci fervently supported the demonstrability of statical principles, but he quoted Dante in support of empiricism: "...only the senses can apprehend/What is the object of the intellect."[52]

Our historical excursus on the parallelogram rule cannot claim to be complete. For a systematic collection of demonstrations, the reader should consult A.H. Westphal's essay *Ueber die Beweise für das Parallelogramm der Kräfte* and E. Georges's lecture *Die Zusammensetzung der Kräfte.*[53] Our aim was only to examine some of the epistemological problems that the different attempts at demonstration imply. At first glance, this seems to be rather idle, since no demonstration can completely avoid an empirical reference. One must at least select some specific axioms. But a more careful examination shows that the true question underlying the parallelogram rule involves the theoretical arrangement of statics and mechanics. What *is* the relationship? Is statics the specific case in which movement ceases? Is it part of the whole? Or is it the relation between theories of different ranks? The distinction between topological and phenomenological laws that so enlivens critical analyses of the formal structure of physical theories could be very useful in facing this question.[54]

In fact, we can reinterpret the repeated attempts to demonstrate the parallelogram rule according to a line of thought that leads to the opposite of their original aim—that is, not to establish statics as a meta-empirical science but, on the contrary, to set forth a list of mathematical axioms which "describe some basic facts homogeneous with our intuition" (Hilbert), and therefore to clarify the *empirical claim* that even the most abstract of theories must somehow imply. From this point of view, the clear distinction between the "mathematical model" expressed by a formal theory and its subsequent interpretation in terms of physical objects (*natural* science) finds in the history of statics one of the main routes that led to its acknowledgment. Hence follows the axiomatic approach to the foundations of mechanics that, starting from Hilbert's program, was accomplished in this century thanks to W. Noll, C. Truesdell, and their colleagues.

---

[51] M. Zanotti, *Elementi di meccanica razionale* (Naples, 1857), pp. 8–20.

[52] *See* F. Siacci, *op. cit.*, nota 2, p. 9; "*Così parlar conviensi al vostro ingegno/ Però che solo da sensato apprende/ Ciò che fa poscia d'intelletto degno*," *Divina Commedia, Paradiso* 4:40–42.

[53] A.H. Westphal, *Ueber die Beweise für das Parallelogramm der Kräfte*, inaugural disseration, Göttengen, 1867; E. Georges, *Die Zusammensetzung der Kräfte*, inaugural dissertation, Halle, 1909.

[54] *See* E. Tonti, *La struttura formale delle teorie fisiche* (Milan, 1976).

Part II

De Resistentia Solidorum

5

Galileo and His "Problem"

5.1 Introduction

In 1684 Leibniz published a fundamental essay on the resistance of solids in the *Acta eruditorum* of Leipzig. The work begins, "Mechanical science seems to have two parts: the first one concerns the power of acting or moving, the other the power of reacting or resisting, that is, the strength of bodies."[1] According to Leibniz, therefore, the resistance of bodies is not merely one chapter of the many which make up mechanics; it is half of the entire discipline. Usually, such dichotomies are hardly probatory: they belong to a metaphysical way of thinking, according to which reality is two-faced, since every concept is linked with its opposite.

Nonetheless, Leibniz's division is particularly provocative and very nearly convincing. The action/reaction dichotomy really is central to mechanics, and neatly evokes the dual nature of the concept of force, which we discussed above. Moreover, this division of mechanics suits the interpretation of static principles which we have tried to establish—one which does not subordinate statics to dynamics. Leibniz adds to the definition of statics as a science which, by dealing with reactive forces, must also concern the resistance of bodies. Although we may disagree with the pre-eminence which he gives to the resistance of solids, we shall accept Leibniz's suggestion in order to examine some of the subjects which originate from this theme.

Historically, a number of profound questions were raised about the structure of matter. These complex hypotheses involved an entire cosmology, mathematical theories based on axiomatic models, and a good deal of experimental research. The debates and discussions which concerned the very technical and practical question of the strength of materials branched out in different directions, permeating philosophical and scientific fields which were contested by irreconcilable factions, with radically different world views. Their disputes pitted Aristotle against Democritus, Galileo against Aristotle, Descartes against Galileo, Leibniz against Descartes, and Newton against Leibniz.

However disorderly the field was, we can at least be certain how the battle started. In the essay cited above, Leibniz continues as follows:

---

[1] G.W. Leibniz, "Demonstrationes novae de resistentia solidorum," *Acta eruditorum Lipsiae* (July, 1684), p. 319.

> Of these [two parts], the second [i.e., the resistance of bodies] has certainly been considered by some people. Archimedes, who alone among the ancients introduced geometry into mechanics, did not reach this point. Later, after Archimedes, almost nothing was done in geometrical mechanics until Galileo, who—endowed with an accurate understanding and a rich and thorough knowledge of geometry—first brought to light the sacred courts of science (*pomoeria scientiae protulit primus*). It was, in fact, he who began to relate the resistance of solids to the laws of geometry.

Leibniz was right. The resistance of solids is the youngest topic in the field of mechanics. When Leibniz wrote (1684), it was not yet fifty years old. It has an author who may be called its first founder, and an exact beginning—an identifiable place where it is fully formulated for the first time, Galileo's *Discorsi e dimostrazioni matematiche intorno a due nuove scienze* ("Discourses and mathematical demonstrations concerning two new sciences"), published at Leiden in 1638.

Thus we must begin by looking at this source, particularly at the first two chapters. The text is brilliant, shrewd, almost playful. It shows the excitement of a scientist who knows he has found an unprecedented truth which contradicts tradition, a truth which is almost tangible and has enormous practical benefits. Galileo's discovery helps the skilled workman ensure that his construction is sound, but it has an unspoken theme as well: it is an implicit but ferocious criticism of the deadening influence of the medieval Aristotelians. It attacks those who turn their backs on nature in order to fiddle with the holy texts of the past—who will not leave the bastions of authority. Sometimes this criticism manifests itself in sarcasm, but more often it is expressed by concealed allusions and by digressions which seem inexplicable until one looks for the metaphors they contain. Under the geometrical and mathematical mask, there is a physical concept which disquieted Galileo's contemporaries; it is a radical departure from Aristotle's cosmology, and Aristotle's cosmology was the framework within which current Catholic theology thought and taught the truth of faith.

For this reason, Galileo's work caused an uproar even inside the Galileian school. The subject of solids of equal resistance seems an improbable cause for a raging dispute, one which lasted for almost a century after the master's death. In fact, it was the "subtext" which divided the feuding parties; the discussion of resistance became a pretext for a larger battle between the heirs of the great Pisan. At the end, however, the battle spilled out into the purely scientific aspects of Galileo's theory, and the question of resistance had a fundamental role in the birth of the modern mechanics of solids.

5.2 Galileo: A Short Account

It would be superfluous to recount the life of Galileo, the scientific revolution he provoked, and his sufferings at the hands of the Catholic church. There are a great many biographies, critical and historical essays, and detailed editions of his trials by the Holy Office, as well as dramatic accounts and historical fiction. In fact, the discussion of the dispute between Galileo and his opponents is still open and lively.

Of course, the question no longer concerns choosing between the Ptolemaic and Copernican systems, as was still the case as late as the 18th century. Rather it is an epistemological discussion. Throughout the last century, interest in Galileo has reawakened. Writers of all stripes— supporters of nearly opposite notions in the history of science—have claimed Galileo as their own, superimposing their ideologies on his work. In consequence, today we can contemplate as many Galileos as there are schools of thought. There is a positivist Galileo, a Marxist Galileo, a critical-rationalist Galileo, a neo-positivist Galileo, an anarchist Galileo. Everyone claims the man.

For our purposes, none of these Galileos is very useful. We have to look directly at his text, especially at the first two chapters ("Days") of the *Discorsi.* The topics discussed in these chapters are irrelevant to the great discussion of the Copernican system which is generally at the center of historical and critical analyses. But we must make some mention of the events which immediately preceded the composition of the *Discorsi*, events which illuminate it and give it its peculiar dissembling, allusive and enigmatic qualities.

Galileo was born in Pisa in 1564, the son of Vincenzo Galilei and his wife Giulia Ammannati. In 1581 his father sent him to the University of Pisa to study medicine, but Galileo had little interest in the subject and left the university without a degree in 1585. His great love was mathematics; in 1583, he had met Tartaglia's pupil Ostilio Ricci in Florence, and Ricci introduced him to the subject and fostered his enthusiasm for Archimedes. Legend has it that Galileo discovered the law of isochronism of the pendulum late in 1582 during a sermon in the Cathedral of Pisa, by watching the swing of a chandelier. From 1589 to 1592, he taught mathematics at the University of Pisa. During this period, he developed his studies on falling bodies, set out in *De Motu.* Already his work contrasted sharply with Aristotelian doctrine. The opposition of colleagues who advocated the old system forced him to move to Padua where, in 1592, he succeeded to the university's chair of mathematics. He held this position until 1610, and these 18 years were among the calmest and most fruitful in his life. At Padua, he invented a military compass and the telescope, and made his first astronomical discoveries.

In 1609, he learned of an "eyepiece" constructed by a Dutch optician, and this led to his most famous invention, the telescope, which he first pointed skywards in the night of 7 January 1610. His astronomical observations led to extraordinary discoveries: the mountains of the moon, the phases of Venus, the moons of Jupiter, Saturn's ring, the composition of the Milky Way, and sun spots.

A growing curiosity on the part of upper-class laymen brought Galileo to court and introduced him to the aristocracy. "I have been favored by many of these very illustrious gentlemen, cardinals, prelates and various princes," he writes in one of his letters, "who wished to see my observations, and they all have been satisfied." The Jesuits organized a solemn session in his honor at their college in Rome.

In March 1610, he published the *Sidereus nuncius*, which added fuel to the growing debate between the traditionalists and the Copernicans. But the whole matter could have been kept in the purely scientific sphere had it not been for two Florentine laymen, Francesco Sizzi and Ludovico delle Colonne, who forced it into theology.

It all happened by chance. At a banquet given by the Grand Duke of Tuscany in 1613, the scriptural argument adopted by the Ptolemaists, consisting principally of a passage from Joshua ("Sun, stand thou still upon Gibeon" [Josh. 10:12]), came under discussion. Galileo learned of this, and tried to explain his own ideas in a series of letters to Benedetto Castelli, Mgr. Piero Dini and the Grand Duchess Cristina di Lorena. In these missives, he firmly asserted the necessary and legitimate separation of research and theology. But the dispute grew noisier and more heated. Luther had brandished an objection on Biblical grounds against Copernicus. In Rome, the question was judged to be of some delicacy. Moreover, as so often happens, theology was intermixed with politics. In 1616, the Church decided to condemn Copernicus' works. In the same year, Galileo was tried before the Holy Office. At the Pope's behest, he was told not to hold, teach or defend the Copernican heresy. He received a stern warning to this effect on February 26th, and a milder reprimand from Cardinal Bellarmino on May 26th.

The reason for this inconsistency is worth looking at Cardinal Bellarmino, an excellent theologian and a highly cultured man, had settled the problem with a compromise. Writing to Fr. Foscarini (who was also involved in the Copernican question) in April 1616, Bellarmino stated

> It seems to me that your lordship and Lord Galileo are prudently limiting yourselves to speaking by supposition, and not in an absolute manner as I have always believed that Copernicus has spoken: because by saying that the earth is supposed to move and the sun to stand still, all the appearances are saved; it is better than presenting eccentric curves and epicycles; it is well said and implies no danger, and this is enough for the mathematician: on the other hand, however, desiring to assert that the sun is really the center of the world and revolves on its own axis ... is a very dangerous thing.

Some over-zealous Catholic historians, preoccupied with defending the Church, have proposed that Bellarmino was right in suggesting that Galileo speak only hypothetically. After all, the Copernican theory was not proved until Herschel's and Bessel's work in the 1820s and Foucault's experiment in 1851—precisely the time when the Church dropped Galileo's works from the *Index librorum prohibitorum* (1822). In fact, Bellarmino's suggestion had quite another meaning.

The aim pursued by Bellarmino (and later by Urban VIII) was to avoid any conflict between science and faith, by recognizing that the nature of induction gives the empirical sciences a purely hypothetical and instrumental value. Thus, in the case of a clash between a scientific theory and Scripture, one must accept God's wish to act improbably (but not impossibly), as described in the Bible.[2] This solution is certainly ingenious, and authors such as P. Duhem,[3] who inclined towards the conventional, saw in it a foreshadowing of the nineteenth- and twentieth-century view of science which triumphed in the works of Mach, Poincaré and others. But this creates a deceiving God, one who amuses himself by driving man to hypotheses which are reasonable, persuasive, confirmed by experiment—and wrong. This God exercises his omnipotence in the narrow gap between the extremely improbable and the impossible. Bellarmino's compromise is theologically untenable. Curiously, Galileo's position—the very one which the Holy Office condemned—strongly anticipates the stand taken by the Church since Leo XIII's encyclical *Providentissimus*. The papal encyclical cites the same texts from St. Augustine as Galileo quoted in his "Lettera alla Granduchessa di Toscana."[4] Galileo had stated that, if an irrefutable scientific proposition seems to oppose the scriptural assertion, we should "investigate the true meaning of the Holy Letters" by correcting their interpretation, remembering that "the Scriptures, in order to adapt themselves to the understanding of everybody, say many things whose outward aspect and literal expression are different from the absolute truth."[5]

---

[2] G. Morpurgo-Tagliabue, "I processi di Galileo e l'epistemologia," *Rivista di storia della filosofia*, Vols. 2, 3 (1947), Vol. 1 (1948); reprinted Rome, 1981, p. 56.

[3] P. Duhem, *La théorie physique, son objet et sa structure* (Paris, 1906) (hereafter cited as Duhem, *La théorie physique*); P. Duhem, Σῴζειν τὰ φαινόμενα. *Essai sur la notion de théorie physique de Platon à Galilée* (Paris, 1908).

[4] A good synthesis of the Galileian theses in contemporary theological culture may be found in A. Dubarle, "Les principes exégétiques et théologiques de Galilée concernants les sciences de la nature," *Revue des sciences philosophiques et théologiques* (1966), pp. 67–87; M. D'Addio, "Considerazioni sui processi di Galileo," *Rivista di storia della chiesa in Italia* (1983), pp. 2–56; F. Russo, "Galileo e la cultura teologica del suo tempo," in P. Poupard, ed., *Galileo Galilei, 350 anni di storia(1633–1983). Studi e ricerche* (Rome, 1984), pp. 150–178.

[5] G. Galilei, "Lettera a Cristina di Lorena, Granduchessa di Toscana," in *Edizione nazionale delle opere di G. Galilei*, 2d ed. (Florence, 1929–1939) (hereafter cited as Galileo, *Opere*), Vol. 5, pp. 309–348.

We return to biography: what happened next is well known. Relying on the benevolence of Cardinal Bellarmino and of Maffeo Cardinal Barberini (later Pope Urban VIII), Galileo boldly set forth his theory in the *Discorsi delle comete* ("Discourse on comets," 1619) and in *Il saggiatore* ("The assayer," 1623), a polemic directed against the Jesuit Orazio Grassi. Violent hostilities broke out after the daring publication of the *Dialogo sopra i due massimi sistemi del mondo* ("Dialogue concerning the two chief world systems") in 1632. Some historians claim that complex political infighting—in which the sinister Cardinal Gaspare Borgia was involved—helped to embitter the conflict. In any event, Galileo's faith in Urban VIII's indulgence and the friendship of Grand Duke Ferdinand II, to whom the *Dialogo* was dedicated, turned out to be misplaced.

Galileo was ordered to appear before the Holy Office and was tried for a second time. Was his enemies' triumph the result of a diabolical intrigue plotted by the Father Commissary Vincenzo Maculano and his colleagues, as L. Geymonat suspects? Or was it the cruel exploitation of a moment's weakness, as G. Morpurgo-Tagliabue is inclined to believe?[6] Whatever happened, Galileo was caught in a trap of his own making. He had to defend himself against the charge of violating the precept of 1616, which may not have been formally valid.[7] They questioned him so mercilessly that he stumbled over their quibbles. He decided on April 12 to protest his innocence by affirming that, when he wrote the *Dialogo*, he intended "neither to defend, nor to teach the said Copernican opinion, but, on the contrary, to confute it."[8] Immediately, three of the inquisitors (A. Oreggi, M. Inchofer and Z. Pasqualigo) saw that they had him. As long as Galileo maintained that Bellarmino had not forbidden him to discuss heliocentrism, provided that he would not defend it as certain and compatible with Scripture, it would be difficult to find arguments in the *Dialogo* which could justify a condemnation. But Galileo's last declaration, that he had really wanted to "confute" the Copernican theory, was clearly a lie; he could therefore be accused of malice and deceit.

After months of suffering, on June 21, 1633, broken by his inquisitors, he gave in. The moment is recorded on the back of a folio of the record of his inquisition, signed by Galileo. To his inquisitors' question, he answered "I do not hold nor did I hold this opinion of Copernicus after having been warned to abandon it; after all, here I am in your hands, do what you like."[9] They responded: "He should tell the truth or else be put to torture." Galileo

---

[6] L. Geymonat, *Galileo Galilei*, 5th ed. (Turin, 1969), p. 184; G. Morpurgo-Tagliabue, *op. cit.*, pp. 136ff.

[7] E. Costanzi, *La chiesa e la dottrina copernicana* (Siena, 1898).

[8] G. Favaro, *Galileo e l'Inquisizione. Documenti del processo galileiano* (Florence, 1907), p. 80; see also Galileo, *Opere*, Vol. 19, pp. 27–421.

[9] Vatican. Archivio Vaticano, "Ex Archivio S. Officii—Contra Galilaeum Galilei mathematicum," Vol. 1181, fol 453v

answered, "I am here to show obedience, and I have not kept this opinion after the determination taken as I have said." They made him sign, and let him go.[10]

On June 22, the day of his conviction, Galileo knelt to make his abjuration:

> As I have been judged by this Holy Office ... as vehemently suspect of heresy, that is that I have thought and believed that the sun is the center of the world and immobile and the earth is not the center and that it moves ... with my sincere heart and unfeigned faith, I abjure, damn and detest the above errors and heresies.[11]

This humiliation did not save him from the sentence of life imprisonment. In addition, the Holy Office issued a "general prohibition of all that has been published and is going to be published," and everlasting silence about the past and future, with no exceptions, without any chance of derogation. This was cruelty, but it was slightly tempered. He was allowed to live first at the Villa Medici alla Trinità dei Monti in Rome, then at Siena, at the court of his friend Archbishop Ascanio Piccolomini, and finally in his villa at Arcetri. He spent his last years there, completing the work which serves as a foundation for modern mechanics, the *Discorsi.*

Publishing the work created problems. The prohibition on publication was too recent to be lightly disregarded. Giovanni Pieroni, a military engineer and possibly a student of Galileo's, tried in vain to publish the treatise at Venice. By sheer luck, the famous Dutch printer Lodewijk Elzevier made a trip to Italy, and by August 1636 Galileo had sent him the first two chapters through Fulgenzio Micanzio. As the printing progressed, Galileo pushed himself to finish the work. In March 1637, Elzevier sent proofs with engraved figures to Micanzio. Galileo was delighted. In a letter to Benedetto Guerrini he wrote, "the first printed folio is on the way; that comes as a token that they are working for me in Leiden at the Elzevirii, the most famous printers in Europe."

The *Discorsi* was published in mid-1638; by 1639, the first French translation of it appeared in Paris under the impressive title *Les nouvelles pensées de Galilei, mathématicien et ingenieur du Duc de Florence. Où par des inventions merveilleuses, et des demonstrations inconnuës iusqu'à présent; il est traitté de la proportion des mouvements, tant naturels que violens, et de tout ce qu'il y a de plus subtil dans les méchaniques et dans la phisique* ("The new thoughts of Galileo, mathematician and engineer of the duke of Florence. In which, by marvellous inventions, and by demonstrations unknown until now, are discussed the proportion of movements, both natural

---

[10] D. Berti, *Il processo originale di Galileo Galilei*, 2d ed. (Rome, 1878), p. 216.
[11] Galileo, *Opere*, Vol. 19, pp. 402–407.

and violent, as well as all that is most subtle in mechanics and physics.") This is, perhaps, a fair description.

Distressed by blindness but surrounded by his affectionate students, Galileo died at Arcetri on January 8, 1642.

5.3 The Subtext: Galileo's Atomism

This brief narrative of Galileo's life may seem to have little to do with the topic under discussion, the resistance of solids. It seems hard to believe that the *Discorsi* could give rise to controversy. The book makes no reference to the Copernican question, nor is there any hint of Galileo's conflict with the Church or any mention of theology or scriptural interpretation. The analysis seems limited, harmless, incapable of giving rise to any questions of faith or doctrine. But underlying this straightforward discussion of mechanics there are troubling questions and hypotheses. To address these, we must first summarize the intellectual climate in which Galileo worked.

The concepts of material reality and of the intrinsic properties of bodies were, in Galileo's lifetime, governed by the Aristotelian theory of *hylemorphism* and by a good deal of metaphysical prejudice. Theology had created a comfortable image of the world, and had inextricably tied faith to a description of physical reality, to such an extent that some Christian dogma was based on cosmology and its laws. But the rise of Renaissance humanism and the great flowering of physical science suddenly brought into question all of the tidy assumptions upon which this cosmology was based. This explains why the growth of new sciences often came into conflict with traditional thought, in spite of the fact that the newcomers had roots in late scholasticism. The Copernican theory was by no means the only, or even the most important, such challenge. We should remember that the vexing question of the center of the universe, and whether other worlds might exist, had been debated for a long time. The Bishop of Paris, Étienne Tempier, even considered the thesis that "God cannot make many worlds" heretical, including it in his famous *Syllabus* of condemned propositions (March 7, 1277).

Copernican cosmology was a flank attack, but the debate on the nature of matter itself was a frontal assault. During this time, the atomistic hypothesis reappeared on the horizon of science after having been branded as atheistic and materialistic for centuries. To question Aristotelian assumptions about matter itself was far more serious than to fiddle with the orientation of the earth and sun; such questions were seen as particularly insidious in their ability to cast doubt on received theology and to trouble men's minds. The hypothesis could, of course, be stripped of its associations with Democritus, Epicurus and Lucretius, but it was still deeply threatening. Donato Rossetti tells us that the students of the University of

Pisa were obliged "to flee the atomists as ethnic pagans and publicans."[12] The concept had to be considered as a direct attack on faith and truth. The Counter-Reformation only intensified the Church's hostility to atomism when the Council of Trent enshrined the doctrine of transubstantiation of the Host. By distinguishing *substantia* from *species* or "sensory appearance," theologians could claim that the consecrated bread and wine, while apparently unchanged, had been transformed into the actual body and blood of Christ. According to atomism, on the other hand, atoms were seen as both causes and carriers of sensations, and as constituting principles of reality, at the same time substance and species. This doctrine was irrevocably opposed to the doctrine of transubstantiation.

In his youth, Galileo had been attracted by the new perspectives which atomism opened in physics. In his *Discorso sulle cose che stanno su l'acqua* ("Discourse on floating things") (1612) he tried to verify, with reference to hydrostatics, the reliability of some Democritean hypotheses on water and also on heat. But above all, in his *Saggiatore* (1623) he developed a general theory which described the elements of nature and perceptible phenomena (except for sound, which he saw as being wavelike) by means of the movement of particles of matter. One of P. Redondi's essays,[13] published to commemorate the 350th anniversary of Galileo's conviction, argues convincingly that the real cause for the Church's attack on Galileo was his atomism. Redondi maintains that the scientist's heliocentrism was a pretext, adopted to keep the true reason for the conviction from the public. He bases this claim on the discovery of an anonymous document, found in the Vatican archives, on the *Saggiatore* and its Democritean origins.[14]

Whether or not Redondi is right, it is certain that atomism occupies an important place in Galileo's thought, surfacing over and over again (if not always consistently) in his writings. In the *Discorsi*, the atomic hypothesis is directly connected with the themes of the resistance of solids, but the connection is never formally stated and remains rather enigmatic. Redondi, addressing this, speaks of a "substitution of theory" to characterize the change from the explicit positions of the *Saggiatore* to the reformulation of the *Discorsi*. "The conceptual substitution," he says,

> takes place under the eyes of the readers, but it is almost imperceptible, like a clever game of cards .... Through Salviati, Galileo speaks of points, spaces and lines. One listens, fascinated, to that audacious infinitesimal solution to a difficult geometrical paradox, then realizes that Salviati has begun to

[12] D. Rossetti, *Composizione e passioni de' vetri ....* (Livorno, 1671), introduction, p. 4.

[13] P. Redondi, *Galileo eretico* (Turin, 1983).

[14] Vatican. Archivio Vaticano, MS, G3, Archivium Sacrae Congregationis de Doctrina Fidei, AD FE series, ff 292r, 292v, 293r.

> speak of particles instead of points, of hollows instead of spaces, of bodies instead of lines, and that the terms of each couple are synonymous.[15]

In other words, Redondi suggests, Galileo made a purely semantic substitution in order to replace the metaphysical and materialistic speculations in the *Saggiatore* with new and more powerful metaphysical speculations of a mathematical character.

5.4 The Primacy of Geometry over Logic in the Discorsi

How can we account for the difference between the terms and references used in the *Saggiatore* and those in the *Discorsi*? It would seem reasonable to suppose that Galileo, having burnt his fingers, was being cautious, taking his revenge on his inquisitors without leaving himself open to renewed attack. But the question is more complex, and is connected to the concept of science which Galileo gradually evolved, a concept which reaches its full expression in the *Discorsi*.

Some scientists, as good Christians, tried to reconcile atomism and received theology, but Galileo's position was very different. Unlike many of his contemporaries and successors (Rossetti, for example) he never pretended to justify his hypotheses and results "with the Holy Scriptures, with Councils and with Holy Fathers,"[16] nor did he aim to substitute a new all-embracing philosophy for the Peripatetic one. On the contrary, he intended to limit himself to strict scientific examination and to rid it of inconclusive metaphysical baggage. In a fragment of unknown date,[17] we find a wonderfully sarcastic simile. The Peripatetic is like a mule-driver who sees that his cart is badly balanced. Instead of shifting the load, he adds a heavy rock to the light side; when this overbalances the load in the other direction, he adds a rock to the first side. And so on and so forth, until the poor mule's knees buckle under him. Galileo's aim is to strip the load of all unnecessary weight—that is, of all the metaphysical and theological arguments which add nothing to the mathematical description of reality.

This attitude is confirmed in Galileo's letters to the Bolognese professor Fortuno Liceti, his last public statement on subjects connected with atomism. In a letter dated September 15, 1640, we find the following passage: "Truth is surely achieved if experience is placed before any kind of abstract theory, as we are certain that in such theory there will be, at the least, a covert fallacy, because a meaningful experience [*sensata esperienza*] cannot

[15] P. Redondi, *op. cit.*, p. 22.
[16] D. Rossetti, *loc. cit.*
[17] Galileo, *Opere*, Vol. 8, p. 640.

be contrary to the truth."[18] This anticipates an epistemological concept like that presented by Karl Popper in this century. According to Galileo, every theory based on general principles of natural philosophy is bound to be "at least covertly" wrong. With the progress of experimental knowledge, the time may come when theory and data collide. And if this happens, we must discard the theory, not the data; a single conflict with experience is enough to prove that a theory is fallacious.[19] What Popper adds is the observation that it is this very risk—that of being proved wrong—that makes a *theory* empirically meaningful; if a theory cannot be disproved, it is useless.[20] For Galileo, it is *experience* that must be meaningful. The brute fact is not enough; it shows us *that* something happens, but not *what* or *how*.[21] Science aims to make experience meaningful, to interpret it, to use it as a tool for understanding nature.

The *Saggiatore* shows that Galileo had already recognized how a pure datum should be enriched to make it significant. The great book of nature, he says, "is written in a mathematical language and its letters are triangles, circles and other geometrical figures. Without these, it is impossible for us human beings to understand a word; without them, we wander futilely in a dark labyrinth."[22] What lends significance to experience is, therefore, its transformation by mathematics. The mathematical language of nature decides the truth of every theory, for it is the only demonstrative basis of those principles which philosophy poses *a priori*.[23]

The *Discorsi* take this argument to its limits. Mathematics (or, in Galileo's terms, geometry) is not only the pre-eminent instrument for understanding; it is the *only* such instrument. It perfectly satisfies the real needs of research, unlike "logic," whose vague arguments lead nowhere. In the Second Day of the *Discorsi*, for example, after a brilliant geometrical demonstration set out in contrast to a superficial "logical" pseudo-deduction, Galileo makes his Aristotelian Simplicio admit, "Indeed, I begin to understand that while logic is an excellent guide in discourse, it does not, as regards stimulation to discovery, compare with ... geometry."[24]

By differentiating logic and geometry, Galileo means to suggest a profound epistemological alternative. The datum of experience cannot be given significance by the ancient logic of Peripatetic philosophy, which would swamp the fact in metaphysics. Instead, we need mathematical hypotheses

---

[18] *Ibid.*, Vol. 18, p. 249.

[19] "Dialogo sopra i due massimi sistemi del mondo," in *ibid.*, Vol. 7, p. 148.

[20] K. Popper, *Logik der Forschung* (Vienna, 1934).

[21] Galileo, *Opere*, Vol. 18, p. 208.

[22] *Ibid.*, Vol. 6, p. 232.

[23] *Ibid.*, Vol. 7, pp. 75–76.

[24] "Discorsi e dimostrazioni matematiche, intorno a due nuove scienze. Attenenti alla mecanica e i movimenti locali ...., in *ibid.*, Vol. 8, pp. 49–318 (hereafter cited as Galileo, *Discorsi*), p. 175.

which can be confirmed by experience. In a famous passage from the Third Day, Galileo perfects this proposal: every hypothesis, however reasonable and convincing, must be taken "as a postulate, the absolute truth of which will be established when we find that other inferences from it, correspond with and agree point by point with experience."[25] The postulate cannot be justified simply because it agrees with a pre-existing larger theory. It must be mathematically applicable, so that its inferences can be tested.

This is the foundation of modern scientific method. Science is divorced from theology and philosophy; they cannot interfere in the definition of its principles, for science has its own criteria for validity. It is not in the least surprising that the seventeenth-century Jesuits pounced on this point in the *Discorsi* and denounced it roundly.

Aside from this, the style of the *Discorsi* is extremely prudent. There is never a hint of positions which might clash with the Church's teachings, never a handhold for the inquisitors. The tone of the work differs from that of the *Dialogo sopra i massimi sistemi.* The three characters are the same, but they are presented differently. In the *Dialogo*, Simplicio was a comic representative of a ruined speculative system, an easy target for critics; in the *Discorsi*, he poses serious problems and objections to Filippo Salviati, a fine experimenter and mathematician, and to Francesco Sagredo, an enlightened and cosmopolitan Venetian patrician and enthusiast for the new learning.[26] When the argument turns to the thorny question of atomism, Galileo is the first to point out the risks of yielding to Democritus' materialism, and evades the issue as a matter of courtesy (perhaps with a touch of irony): "Further references to these affairs," he makes Simplicio say to Salviati, who conveys Galileo's own thoughts, "I omit, not only as matter of good form, but also because I know how unpleasant they are to the good-tempered and well-ordered mind of one so religious and pious, so orthodox and God-fearing as you."[27] He leaves nothing for anyone to protest.

Not surprisingly, the *Discorsi* was fully accepted by the Church. It obeyed Bellarmino's warning to keep the hypothetical formulation of principles within the strict limits of mathematical treatment. In fact, the work won praise from seventeenth-century Jesuit scientists. Frs. A. Terill, H. Fabri, I.G. Pardies, P. Casati and F. Eschinardo all recognized the beauty and power of Galileo's theory of the resistance of solids. But not everyone was so uncritical. Bellarmino's compromise had left Galileo a way out; the cardinal's injunction to speak hypothetically was intended to diminish the value

[25] *Ibid.*, pp. 207–208.

[26] A. Banfi, *Vita di Galileo Galilei* (Milan, 1979), pp. 184ff.

[27] Galileo, *Discorsi*, p. 72.

of an empirical theory, but in Galileo's hands, the hypothetical and deductive method was transformed into a powerful instrument for neutralizing metaphysics and theology.

For this reason, Fr. Terill decided to rewrite the first two Days of the *Discorsi*, using logic alone and stubbornly ignoring geometry. And for this reason, Fr. Eschinardo wrote, at the end of his fine commentary on Galileo's resistance of solids, "all these things cannot be rejected by scholars. The only objection is that the author distinguishes logic from geometry ... as if they were two discordant sciences."[28] The two Jesuits went straight to the heart of the matter. Galileo's claim for the autonomy of science compromised the traditional conception of a hierarchical unity of knowledge with theology at its head. Therefore, the Church's control of culture could be dangerously weakened. Once again, however, Galileo was right, even on this specific theological and philosophical level; he anticipated a concept of science that today is firmly maintained by the Church itself, and which is an unquestionable condition for a confident dialogue between science and theology.

One final remark: the fact that geometry reigns supreme in the *Discorsi* may make the work sound pedantic. Quite the opposite: the dialogue teems with digressions. Instead of a trip along a dull highway, we are in for adventures along a good many side roads, each of which is seen as worth lingering over. Sagredo makes the pure fun of this clear in a passage from the First Day:

> We are not committed to any closed and concise method, but meet only for our own pleasure. If we digress now, it is in order not to lose information; who knows, if we let this occasion pass, that we shall meet with it again some other time? In fact, how do we know that we shall not discover curious things that are more interesting than the answers we originally sought?

The digressions and suggestions that abound in the *Discorsi* have given rise to criticism (sometimes quite harsh) from Galileo's contemporaries. Frs. Mersenne and Fabri tried to "improve" the treatise in their French and Latin translations; they changed the brilliant dialogue into a tiresome series of theorems and demonstrations. More than the others, Descartes declared himself annoyed by Galileo's exuberance. In a letter to Mersenne dated October 11, 1638, he praised the *Discorsi* but added:

> It seems to me that he [Galileo] is quite defective in his continuous digressions and that he never dwells on a thorough explanation of his subjects; which shows that by no means has

[28] F. Eschinardo, *De impetu* (Rome, 1684), section 120.

> he examined them with order and that, without having considered the first causes of nature, he has only sought the reasons for some particular effects, and thus he has built without foundations.[29]

But Descartes was a mischief-maker, especially with regard to Galileo. Furthermore, he may have realized that, in the *Discorsi*'s digressions, Galileo offered a vision of the world, presented by parables, that completely opposed Descartes' own.

Science needs the discipline and order that the author of the *Discours de la méthode* preferred—there is no doubt of that. But Galileo's exuberance, curiosity and freedom are even more useful: "The work is play for mortal stakes," as Frost put it.

5.5 The First Day of the Discorsi

> Frequent experience of your famous arsenal, my Venetian friends, seems to me to open a large field to speculative minds for philosophizing, and particularly in that area which is called mechanics, inasmuch as every sort of instrument and machine is continually put in operation there. And among its great number of artisans there must be some who, through observations handed down by their predecessors as well as those which they attentively and continually make for themselves, are truly expert and whose reasoning is of the finest.[30]

With these words, the first Day of the *Discorsi* begins. They express the attention which Galileo pays to technical problems and their solutions, handed down from the past or found in day-to-day work.

This is true, Sagredo tells Salviati. It is often useful to ask questions of those whose work has given them experience and skill, especially foremen (*proti*), because of their superiority to ordinary laborers. This, despite the fact that

> what we were told a little while ago by that venerable workman is something commonly said and believed, despite which I hold it to be completely idle, as are many other things that come from the lips of persons of little learning, put forth, I believe, just to show they can say something concerning that which they don't understand.

---

[29] R. Descartes, *Epistolae* (Amsterdam, 1668), Vol. 2, p. 276.

[30] All quotations in this and the next three sections, unless otherwise attributed, are from Galileo, *Discorsi*, pp. 49–150.

The old man, we learn, insisted that bigger structures were weaker than small ones "because many devices succeed on a small scale that cannot exist in great size." This thesis irritated Sagredo's geometrical mind.

> Now, all reasonings about mechanics have their foundations in geometry, in which I do not see that largeness and smallness make large circles, triangles, cylinders, cones, or any other figures [or] solids subject to properties different from those of small ones; hence if the large scaffolding is built with every member proportional to its counterpart in the smaller one, and if the smaller is sound and stable under the use for which it is designed, I fail to see why the larger should not also be proof against adverse and destructive shocks that it may encounter.

In Salviati's opinion,

> The common notion is indeed an idle one, so much so that with equal truth its contrary may be asserted; one may say that many machines can be made to work more perfectly on a large scale than on a small one. For example, take a clock that is both to show the hours and to strike; one of a certain size will run more accurately than any smaller one....
>
> [But] recourse to imperfections of matter, capable of contaminating the purest mathematical demonstrations, still does not suffice to excuse the misbehavior of machines in the concrete as compared with their abstract ideal counterparts. Nevertheless I do say just that, and I affirm that abstracting all imperfections of matter, and assuming it to be quite perfect and inalterable and free from all accidental change, still the mere fact that it is material makes the larger framework, fabricated from the same material and in the same proportions as the smaller, correspond in every way to it except in strength and resistance against violent shocks [*invasioni*]; and the larger the structure is, the weaker in proportion it will be. And since I am assuming matter to be inalterable—that is, always the same—it is evident that for this [condition] as for any other eternal and necessary property, purely mathematical demonstrations can be produced that are no less rigorous than any others.
>
> ...For it can be demonstrated geometrically that the larger ones are always proportionately less resistant than the smaller. And finally, not only artificial machines and structures, but natural ones as well, have limits necessarily placed on them beyond which neither art nor nature can go while maintaining always the same proportions and the same material.

As other authors have noted, this passage clearly anticipates model theory, which states that the ratio of the sizes must be varied if the model and the full-scale artifact are to be equally strong. But Salviati does more; he implies a new interest in how the strength of a structure relates to its size, not only to its geometrical form.

The development of structural design and construction techniques falls into two great periods. The first such period lacked a precise concept of stress; therefore the safety of a structure was thought to depend on its form and the composition of its parts. The example of a masonry arch or vault is a good one; the architect's task was to find a profile for the intrados and to define a shape for the single voussoirs and buttresses that would prevent any movement, giving the whole structure the behavior of a single rigid block. In the second period, the problem of structural design changes: the form is given, and the architect's job is to ensure that fracture cannot occur, determining the strength of materials and hence the dimensions that would keep stress within acceptable limits.

This passage from the *Discorsi* marks the transition from the first period to the second. From this point, we can look backward and forward at once. If form alone were responsible for strength, we should conclude (says Galileo) that we can scale up or down without risking collapse or failure, just as in geometry the properties of a circle are independent of its size. But this does not happen in reality.

> Who does not see that a horse falling from a height of three or four braccia [1 braccio = 58.4cm] will break his bones, while a dog falling from the same height, or a cat from a height of eight or ten, or even more, will suffer no harm? Thus a cricket might fall without damage from a tower or an ant from the moon.

The same paradox applies to masonry: "Thus, for example, a small obelisk or column or other solid figure can certainly be laid down or set up without danger of breaking, while very large ones will fall apart at the slightest provocation, and that purely on account of their own weight."

Thus: if we increase the dimensions, we diminish the resistance. Geometric proportion is not enough. The ideal, purely mathematical model which the ancients used to change the proportions of masonry structures, a model considering only surface and volume, will not do. We must also consider the material and its resistance. Yet—and this is the essential point—we cannot dismiss these new aspects as random imperfections; they must be studied by way of "geometrical demonstrations."

The subtle Salviati knows how to be an iconoclast; he takes pleasure in shocking his interlocutors (especially Simplicio, the faithful Aristotelian) with extreme examples.

> Here I must tell you of a case really worth hearing about, as are all events beyond expectation.... A very large column of marble was laid down, and its two ends were rested on sections of a beam. After some time had elapsed, it occurred to

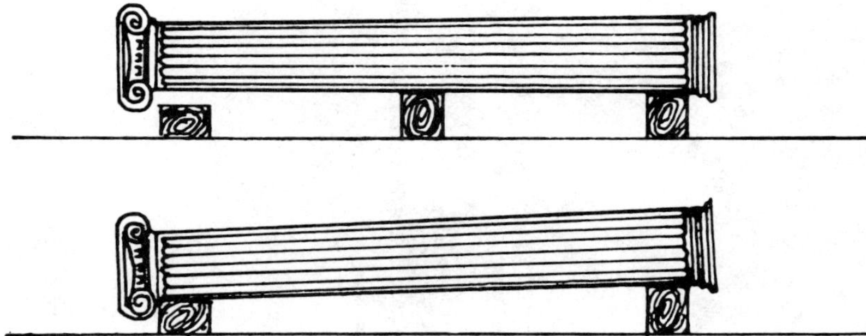

FIGURE 5.1.

> a mechanic that in order to insure against its breaking of its own weight in the middle, it would be wise to place a third similar support there as well. This suggestion seemed opportune to most people, but the result showed quite the contrary. Not many months passed before the column was found cracked and broken, directly over the new support at the center.

Simplicio explodes incredulously: "A truly remarkable event, and most unexpected, if indeed this was due to the addition of the new support in the middle." Salviati responds,

> It surely did result from that, and to recognize the cause of the effect removes the marvel of it. For the two pieces of the column being placed flat on the ground, it was seen that the beam-section on which one end had been supported had rotted and settled over a long period of time, while the support at the middle remained solid and strong. This had caused one half of the column to remain suspended in the air; and, abandoned by the support at the other end, its excessive weight made it do what it would not have done had it been supported only on the two original [beams], for if one of them had settled, the column would simply have gone along with it.

Galileo's observation is exact and more important than it might seem at first sight. In any simply stiff structure—that is, in any structure with constraints which strictly prevent rigid movements or displacements—incidental subsidence or imperfections will not cause stress to the structure itself, which will merely settle into the new configuration (Figure 5.1).

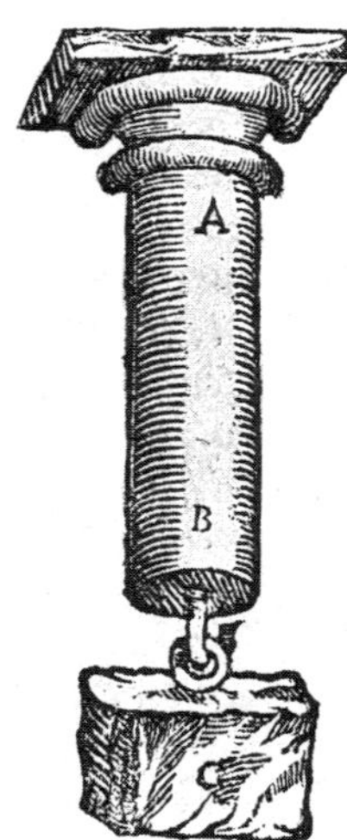

FIGURE 5.2. Galileo's figure for defining the "absolute" resistance of a solid. From the *Discorsi* of 1638.

Sagredo wants to know more about the new science, and presses his friend Salviati:

> Then smooth out for us these rough spots, Salviati, and clear up these obscurities, if you have any way of doing so, for indeed I am beginning to think that this subject of resistance is a field full of beautiful and useful considerations. And if you are willing that it be made the subject of our discussions today, that will be most welcome to me, and I believe to Simplicio.

Thus Salviati prepares to illustrate what he has learned "from our Academician [Galileo], who made many speculations about this subject, all geometrically demonstrated, according to his custom, in such a way that not without reason this could be called a new science." Immediately, however, the dialogue stumbles over the question *why*? Why is a solid more or less resistant to fracture? What is the cause of resistance?

> Let us draw the cylinder or prism AB, of wood or other solid and coherent material, fastened above at A, and hanging plumb; at the other end, B, let the weight C be attached [Figure 5.2]. It is manifest that whatever may be the tenacity and mutual coherence of the parts of this solid, provided only that that is not infinite[ly strong], it can be overcome by the force of the pulling weight C, of which the heaviness [*gravità*] can be increased as much as we please, and that this solid will finally break, just like a rope. And just as we understand that the resistance of a rope is derived from the multitude of hempen fibers

> that compose it, so in wood there are seen fibers and filaments stretched out lengthwise which render it even more resistant to breakage than hemp of the same length would be. In a stone or metal cylinder, the coherence of parts seems still greater, and depends on some other cement than that of filaments or fibers. Yet even these [cylinders] are broken by a sufficient pull.

5.6 Attempts to Explain the Cause of Resistance

With these words, Salviati throws out a bit of speculative bait. The fish he hooks is, of course, Simplicio, the representative of Aristotelian enquiry into causes. Simplicio wants to know more: more about filaments (in wood, the fibers are as long as the piece; in rope, they are short and twisted) and about non-filamentous materials like stone and metal. Salviati veers away from atomism; these questions, he says, are not necessary to our purposes and would require rather a long digression. Note the importance of this remark: we can reach an exact and useful theory of the strength of materials without understanding the physical reasons for cohesion and fracture. A sound mathematical hypothesis describing some of the phenomena of resistance—what we should now call a constitutive equation—can replace the results of complex and problematic inquiries into the physics of the solid state, at least for Galileo's purposes. It would be only a slight exaggeration to call Salviati's statement the beginning of modern rational mechanics of materials.

But Salviati is not to be let off so lightly; Sagredo takes Simplicio's part in demanding to know "what that cement may be that so tenaciously holds together the parts of solids.... Moreover, this knowledge is necessary for an understanding of the coherence between the parts of those very filaments of which some solids are composed." In fact, such a question could not be answered since all involved lacked the physical information needed to deal with it. But in this case, the fact that the goal was unreachable drove Galileo to put forth the most daring and interesting conjectures, guesses that were to be useful in the development of science in spite of being insufficient to solve the specific problem. Salviati's response is in two parts. The first regards the case of a cord of 100 braccia, composed of short fibers spun together, "each not more than two or three braccia in length." The strength of the individual fibers is not enough to explain the resistance of the cord as a whole. Instead, we have to take into account the pressure which each filament exerts on the others.

> But the very act of twisting [in making] rope binds the threads mutually in such a way that later, when the rope is pulled with great force, its filaments will break rather than separate from one another. This is manifestly known by seeing

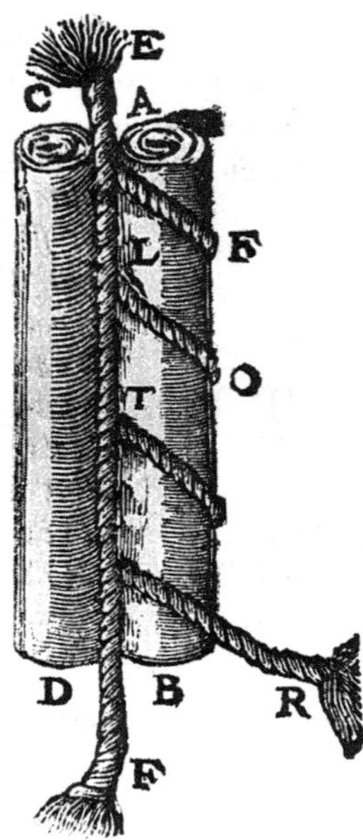

FIGURE 5.3. Galileo's example for clarifying the different resistance opposed by filaments, according to their interweaving. From the *Discorsi* of 1638.

> that the filaments at the broken ends are very short, and not one braccio or more in length, as would be seen if the parting of the rope were made not by a breaking of its filaments, but by their mere separation one from another, and their slipping.

Elaborating this example leads Salviati close to a truth of primary importance: "That tightly held thread, which does not obey the person who pulls on it with some force and tries to draw it out from between the fingers, resists because it is retained by a double compression, for the upper finger presses against it no less than the lower, the one pressing against the other"— an anticipation of the principle of action and reaction.

But this hint is dropped; Salviati's aim is directed elsewhere. He wants to be sure of the behavior of different filamentous materials, considering not only their number but their interweaving as well. For example, let us look at filament EF (Figure 5.3). It behaves differently when it is pressed between two small contiguous cylinders AB and CD than it does when it slides freely or when it is coiled helically around one of the cylinders. In the latter case, "the greater the force that pulls the filament in order to unwind it from the cylinder, the more tightly will the filament be pressed against the cylinder." What matters is not only the strength of the material, but how it is arranged.

> Now who does not see that such is the resistance of those filaments which, together with thousands of like windings, make

> up the thick rope? Indeed, such binding by twisting cement things so tenaciously that from a few rushes, and not very long ones, woven with but few turns, very strong cord is made that I believe is called pack twine [*susta*].

The first part of Salviati's explanation is apparently innocuous; who could object to a description of rope? But it hides a much more important concept, one which he might not have been able to express without incurring very serious censure. What, really, are these filaments which, when spun together, form a body able to resist extension? What was Galileo thinking about when he described the rushes in pack rope, each of which is held only by its adjacent elements? How can we not see a marked likeness between the interwoven rushes or fibers and the "first bodies" celebrated by the atomist Lucretius?

There is more: Lucretius himself tackled the problem of explaining the variety of bodies and their mechanical behavior, including their comparative strength. He proposed a solution in the second book of *De Rerum Natura*:

> ...some rest
> Within immense space the first bodies [the atoms]
> Never have; but, more and more excited
> By inner restless varying force
> A part of them bumps and bounces
> Pushed to and fro throughout vast space;
> Yet another part in a tiny place
> Forms groups by means of blows
> And all those more densely linked
> Together coiled and fixed
> By their interwoven forms,
> And thus restricted in their moving,
> Form bitter oak and strong oak
> And hard iron of heavy stroke
> And rocks of adamant.

The analogy is undeniable. Through these pages of the *Discorsi*, the ghost of atomism strolls silently. But the rope which evokes the image also makes it unrecognizable, at least for the moment. However short the filaments may be (and Galileo's filaments are anything but short; his three-cubit filament would be at least four and a half feet long), they cannot possibly be identified with the short-lived first elements of material. On the contrary, one characteristic of the fibers is that they break under strong traction, and this contradicts the notion of indivisibility which defines the atom. At best, the rope is a macroscopic model for far smaller interactions which may occur among the different particles of a composite material.

In fact, Sagredo seems not to catch the hidden meaning of Salviati's explanation and moves on to the wonderful force of winches, and to a

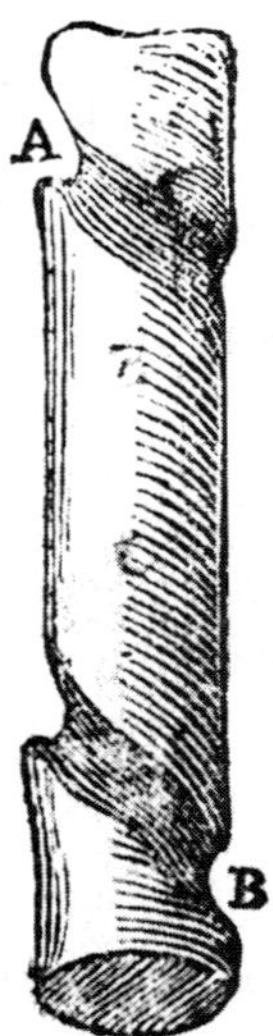

FIGURE 5.4. A "simple but clever device" to slide down a cord from a window: it shows the role played by friction between the interwoven elements of a solid. From the *Discorsi* of 1638.

"simple but clever device" invented by one of his young relatives to slide down a cord from a window without cutting his hands (Figure 5.4). It is Salviati's turn to speak again, and he decides to reveal the second part of his answer to Simplicio on the resistance of bodies.

> Since you want to hear my thoughts about resistance to breakage on the part of other bodies, whose texture is not of filaments, ... but whose parts cohere by reason of other causes... [I say that] these... may be reduced to two kinds, one of which is the celebrated repugnance that nature has against allowing a void to exist. The other, when this of the void is deemed insufficient, requires the introduction of some sticky, viscous, or gluey substance that shall tenaciously connect the particles of which the body is composed.

5.7 For and Against the Power of the Vacuum

This appeal to nature's *horror vacui* is a sop thrown to Aristotle, but it has, according to Salviati, some experimental evidence to support it. If you stack two "exquisitely smoothed, cleaned, and polished" slabs and pick them up by the upper slab, the upper one will carry the lower one with it and keep it lifted indefinitely, even when the latter is big and heavy. This would prove in fact the aversion of nature for an empty space, even during the brief moment required for the outside air to rush in and fill up the region between the two slabs.

But the best of the argument goes to Sagredo. Perhaps Galileo wanted an intelligent amateur to lead the attack on ancient prejudices, or perhaps he was merely being discreet, by keeping his student and spokesman Salviati in the background. His noble Venetian friend, as a non-expert, could take a bold approach. Sagredo attacks with increasingly deadly strokes. First, he remarks, we can accept that the *horror vacui* explains coherence only if we *deny* the reason behind the *horror vacui* as an explanation of the resistance of solids. Aristotle believed that in a vacuum all motion would be instantaneous, and therefore that a vacuum was impossible. But consider our two slabs. If we assume that the abhorrence of a vacuum holds them together, then we must also assume that the vacuum created by their eventual disjunction cannot be instantaneously filled by the incoming air. "Hence we must say that by force (or contrary to nature) a vacuum is sometimes to be admitted—though in my opinion nothing is contrary to nature save the impossible, and that never happens."

Note the epistemological accuracy of this remark. It opens the way to a new concept of nature, one no longer intended as an unchangeable view of a pre-constituted order which judges what is possible or impossible, but rather one which allows us to consider any possibility. While Diodoro Crono, with his "invincible argument," circumscribed the possible within the sphere of the real, Galileo includes the real in a broader, undefinable sphere of the possible—nature herself. Only a deductive science, based on provable hypotheses, can explore this sphere. We have stopped imposing our order upon nature; we are willing to let nature manifest ever new aspects of the possible.

Back to the vacuum: "But here another difficulty arises," Sagredo goes on,

> and this is that although experience assures me of the truth of the conclusion, my mind is still not entirely satisfied about the cause to which the effect is to be attributed. For the effect of separating the two surfaces occurs prior to the [existence of this] void, which consequently follows the separation. Now, it seems to me that the cause should precede the effect, in time at least, if not in physical existence [*natura*]; also, that for a positive effect, there should be a positive cause. Hence I cannot see how the cause of adherence of the two slabs and their repugnance to being separated—effects that are actual—can be a void that does not exist [first], but which must follow. And there can be no action by things that do not exist, according to the definite statement of the Philosopher [Aristotle].

This is a low blow. It hits the mark on a metaphysical level—on the level where the Aristotelians, who denied the existence of a vacuum, were better informed. It is natural for Simplicio to step in and defend the traditional thesis, using a new metaphysical principle: "nature does not undertake to do that which refuses [*repugna*] to be done." Galileo uses this bit of repartee to let the reader know that blind obedience to the metaphysicians is useless. Their theories—as Bartoli was to state a few years later—tack against the winds of reality, like "winds upon which a ship is hauling; they drive into the intended direction even if the winds blow in the opposite direction."[31] We should abandon the metaphysicians' way in favor of concrete scientific argument.

Sagredo shows this. So be it, he says; let Simplicio have his victory. But there is another objection: how can we explain that bodies behave differently, and that some are more resistant than others? A vacuum is a vacuum. "If for one effect there is only one cause... (or if many are assigned, they are reducible to one), then why won't this one of the void, which surely does exist, suffice also for all resistances?" Galileo had already touched on the difference in behavior of different solids in his argument about filaments, and had come to an atomistic model. Now he sets the multiplicity of phenomena against the unity of a metaphysical explanation. In fact, Galileo was to come to a unitary explanation himself—atomism—but this was to be based on mathematics, not metaphysics.

Salviati had been silent, but now he joins in, rather unwillingly:

> I do not wish at present to enter into a contest as to whether the void is in itself enough, without any other retainer, to hold united the separable parts of coherent [*consistenti*] bodies. But I will say that the void which fights and is vanquished between two plates is not in itself enough reason for the firm bonding [*collegamento*] of the parts of a solid marble or metal cylinder.

We need still more causes. Salviati even declares that he knows how to distinguish the part of resistance which can be attributed to the *horror vacui*; by measuring it, we can show the need for additional causes. Sagredo is looking absently at him, absorbed by a new idea. Instead of taking up Salviati's argument, he sets off on an obscure digression:

> I was wondering whether, since it takes more than a million in Spanish gold every year to pay the army, something besides small coins must be provided for the soldiers' pay. But go on, Salvati; assume that I grant your argument, and show us how to separate the operation of the void from all other [actions]; then, measuring this, make us see that it is inadequate for the effect we are discussing.

---

[31] D. Bartoli, *La tensione e la pressione, disputanti qual di loro sostenga l'argentovivo ne' cannelli* (Bologna, 1677), p. 10.

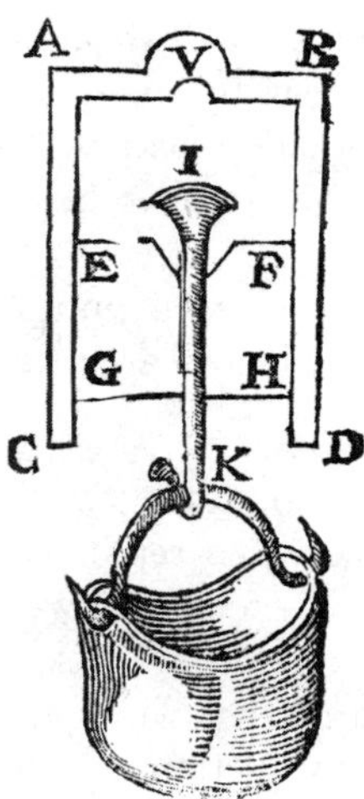

FIGURE 5.5. Experiment proposed by Galileo to define a physical measure for nature's abhorrence of a vacuum. From the *Discorsi* of 1638.

5.8 First Intimations of an Atomistic Theory of Resistance

What on earth have Spanish coins to do with the theory of resistance? What has Sagredo got up his sleeve? Galileo knows how to keep us on the hook— how to infuse mystery into what seems like a straightforward discussion. Salviati answers cryptically, "Your daemon is guiding you," and goes on to describe an experiment which could establish the insufficiency of a vacuum. Partly fill a hollow cylinder AD (Figure 5.5) with water, insert a piston EH, and evacuate the air through valve I. Now close the valve, suspend the piston, and attach a container to the hook K. Fill the container with weights until the piston is pulled free. The total weight added is a measure of the vacuum created. Naturally, Simplicio cannot wait to raise objections about the perfect fit of the piston, the porosity of the cylinder, and the evaporation of water. (To these could be added other objections made by the Venetian military engineer Antoine de Ville, in a letter to Galileo dated March 3, 1635.) In fact, given the technical limitations of the period, this remarkably ingenious system could not be constructed; for this reason, it seems to confirm Koyré's thesis[32] that Galileo's experiments are only imaginary. Geymonat and Caruso disagree; they affirm that in fact Galileo's interest in the theory of the vacuum was almost certainly

---

[32] A. Koyré, *Études Galiléennes* (Paris, 1939), Vol. 1, pp. 71ff

prompted by practical problems of hydraulic engineering.[33] Perhaps the invention here illustrated refers to some attempt by Galileo to build a similar mechanism in his workshop next to his studio at Padua.

The discussion between Salviati and Sagredo takes yet another turn. Sagredo links the preceding experiment to one which he had clearly made some years back, on a piston suction pump used to draw water from a cistern. When the water goes below a certain level, the pump no longer works.

> I thought that the apparatus was worn out; but when I found a master [mechanic] to repair it, he told me that there was nothing at all wrong except the water [level] which, having gone down too far, did not allow itself to be lifted to such a height. He added that neither with pumps nor with any other device that lifts water by suction is it possible to make this rise a hairbreadth more than eighteen braccia; whether pumps are [of] large [bore] or small, that is the measure of this absolutely limited height.
>
> Well, up to now I have been so dull-witted that although I understood that a rope, a wooden staff, or an iron rod can be lengthened until its own weight breaks it when attached from above, it never occurred to me that the same thing will happen, and much more easily, with a rope or rod of water.

Galileo had treated this topic in a letter of August 6, 1630 to Giovan Battista Baliani, a Genoese physicist, and his follower:

> If hemp or steel cords break under excessive weight, what doubt could we have that a cord of water will likewise break? In fact, it will break more easily, insofar as the parts of water, becoming separated from one another, do not have to overcome resistances other than that of the vacuum that is created at the moment of division. This is because in the case of iron and other solids, there is a very strong and tenacious attachment of the parts that is absent in water.[34]

Baliani responded in a remarkable letter of October 24, 1630, approaching the problem much more reasonably by referring to atmospheric pressure. For Galileo, the eighteen braccia which represented the limit of the height of a rod of water finally disproved the old Aristotelian notion of the *horror vacui* as an incommensurable force of nature. And if the *horror vacui* was disproved—that is, if the incommensurable turned out to be eighteen braccia—there had to be other causes for the resistance of solids.

---

[33] *See* the critical edition of the *Discorsi* edited by L. Geymonat and A. Caruso (Turin, 1958), p. 616.

[34] Galileo, *Opere*, Vol. 14, pp. 127ff.

Salviati, responding to Sagredo's remarks about the pump, uses another example:

> Take, for example, a copper wire of any thickness and length, fix one of its ends on high, and to the other end add greater and greater weight until it finally breaks. Let the maximum weight that it can sustain be, for example, fifty pounds. It is obvious that fifty pounds of copper, over and above the weight of this wire, say one-eighth of an ounce, drawn into a wire of the same thickness, would be the maximum length of wire that could maintain itself.

The result is that

> [a] copper rod that is able to sustain itself up to a length of 4,801 braccia encounters a resistance dependent on the void that, in comparison with its other resistances, is as much as the weight of a rod of water eighteen braccia long and as thick as the copper; and if we find, for example, that copper is nine times as heavy as water, then the resistance to breakage of any copper rod, so far as the void is concerned, will be as the weight of two braccia of the same rod.

For a deeper understanding of the dialogue between Sagredo and Salviati on this subject, we must allow ourselves a Galilean digression. The relation between a rod of water and a vacuum had been debated in the thirteenth and fourteenth centuries by such authors as Albertus Magnus, Roger Bacon and Egidio Colonna. Galileo's interpretation is similar to Egidio's; the latter favored the existence of a positive force, the *tractatus a vacuo*, which keeps bodies in contact. Egidio, in his time, had been a fierce opponent of Aristotelianism and a supporter of the atomistic concept of material reality. Surely there must be a link between a water rod, a vacuum and atomism. And the link includes other problems discussed by the later Scholastics, such as those pertaining to the mathematical infinite and the existence of other worlds. The latter applies to the question of the existence of an empty space; if other worlds exist, there must be a void between them. If this is the case, then the Aristotelian notion of the universal contiguity of all bodies is untrue (see Part 1, Section 1.6). Late authors of the thirteenth and fourteenth centuries, such as Richard of Middleton and Walter Burley, went so far as to assert that God's omnipotence obliges us to recognize both the multiplicity of worlds and the existence of a vacuum. The Parisian bishop Étienne Tempier, who condemned the monoterrestrial view as heresy, also condemned a proposition concerning the immobility of the world, which was based on the impossibility of a vacuum.

The principle of divine omnipotence, professed with particular emphasis by the Franciscan school (Duns Scotus and William of Ockham are representatives), created a group of theories and a scientific orientation that together gave shape to a new vision of the world. From it, Richard of Middleton and Ockham deduced the potential infinity of the world and the possibility of infinitely unifying or dividing quantities. The consequent discussion of the paradoxes of infinity sometimes reached high levels of logical subtlety, laying the foundation for the future mathematical research. A.C. Crombie traces the concepts in *conclusion* 17C of *Centiloquium theologicum* (attributed to Ockham, but of unknown origin) down to the nineteenth and twentieth centuries, to authors like Cantor, Dedekind and Russell. One such concept is the paradox of the part and the whole for infinite sets:

> Nothing opposes the fact that a part be equal to the whole or that it be not smaller, because this is observed not only intensively but also extensively, as in the whole universe there is not a larger number of parts than in a bean, insofar as in a bean there are an infinite number of parts.[35]

The discussion between Salviati, Sagredo and Simplicio has started out on analogous speculative ways. After deciding that not all cohesion can be attributed to the *horror vacui*, Sagredo brings up the central question: what, then, holds things together? The hypothesis of "sticky stuff" is untenable; there is no proof of such a glue in reality, and even if we accept the notion, we still have to decide what keeps the parts of glue together. As Salviati tells Sagredo, "A little while ago, I told you that your daemon was guiding you; now I find myself in the same straits." Salviati knows that the *horror vacui* is insufficient and the existence of a glue is incredible. He must try to do the impossible, to discover that the vacuum *is* a sufficient cause. Simplicio is skeptical:

> If you have already demonstrated that in the separation of two large pieces of a solid, the resistance of the large void is very small in comparison with that which holds together the minimum particles, then why do you not wish to admit it as certain that the latter [resistance] has a cause quite different from that of the former?

Unknowingly, by repeating an adjective ("large pieces," "large void") Simplicio has opened up the way to Salviati's solution. Salviati now takes up Sagredo's thought about the pay of the army.

> To this, Sagredo replies that every individual soldier was paid with pennies and farthings collected by general levies, although a million in gold was not enough to pay the whole army. Who knows that there are not other tiny voids operating on the most minute particles, so that the same coinage as that with

[35] A.C. Crombie, *Augustine to Galileo* (London, 1952). This quotation is taken from the Milan edition (1970), p. 245.

> which the parts are joined is used throughout? I shall tell you what has sometimes passed through my mind on this; I do this not as the true solution, but rather as a kind of fantasy... that I subject to your higher reflections.

There follows rather a fantastical image: as a metal cools from the liquid to the solid state, the fire coming out of it leaves very small empty pores. "Although each individual vacuum is exceedingly minute, and therefore easily overcome, yet their number is so extraordinarily great that their combined resistance is, so to speak, multiplied almost without limit."

This is Galileo's solution to the problem of the resistance of solids. The concepts presented in this passage of the *Discorsi* are not altogether new; they appear in the 1612 *Discorso intorno alle cose che stanno su l'acqua o che in quella si muovono* and in a subsequent discussion with Vincenzo di Grazia (*Considerazioni sopra il discorso di Galilei intorno alle cose ...*).[36] Di Grazia accused Galileo of deducing atomism; "which I cannot believe," de Grazia wrote, "as this repudiates his mathematics, which do not admit that a line is composed of points."[37] The reply to di Grazia was in the name of Benedetto Castelli, but with Galileo's direct contribution.[38] The elements of fire which form the vacuums are "*corpi quanti*," finite in number and size, and therefore they "have nothing to do with raising questions about whether lines or other *continua* be composed of indivisible elements. On the other hand, where have you ever found that the composition of lines by points repudiates mathematics?"

5.9 Democritus or Plato?

An echo of the old dispute with di Grazia enters the *Discorsi*, which goes into new and near-frantic digressions on mathematical problems about the continuous and the discrete. As Sagredo puts it,

> *Sagredo*: There is no doubt that as long as a resistance is not infinite, it can be overcome by the sheer multitiude of minimal forces. Thus a number of ants might bring to land a ship loaded with grain, for our eyes daily show us that an ant can readily transport a grain, and it is clear that in the ship there are not infinitely many grains, but some limited number. We can take a number several times as great, and put that number of ants

[36] Vincenzo di Grazia, *Considerazioni sopra il discorso di Galilei intorno alle cose ....* (Florence: 1613).

[37] Galileo, *Opere*, Vol. 4, p. 417.

[38] B. Castelli, *Risposta alle opposizioni del Sig. Ludovico delle Colonne e del Sig. Vincenzo di Grazia, contro al trattato del Sig. Galileo Galilei intorno alle cose che stanno su l'acqua, ecc.* (Florence, 1615); Galileo, *Opere*, Vol. 4, pp. 693–788.

> to work; and they will bring to land not only the grain, but the ship along with it. It is true that the number would have to be large, but in my opinion so is that of the voids that hold together the minimum particles of a metal.
>
> *Salviati*: But if an infinitude were required, you would perhaps hold this to be impossible?
>
> *Sagredo*: No, not if the metal were infinite in bulk, [but] otherwise....
>
> *Salviati*: Otherwise, what? Well, since paradoxes are at hand, let us see how it might be demonstrated that in a finite continuous extension it is not impossible for infinitely many voids to be found.

With these exchanges, the discussion enters the field of pure mathematics, where that "substitution of theories" of which Redondi writes is skilfully applied. The example which Salviati uses to demonstrate his thesis, Aristotle's wheel, concerns the rolling of two concentric circles on two tangential horizontal straight lines. Galileo explains the ensuing paradox in this way:

> And just so, I shall say, in the circles (which are polygons of infinitely many sides), the line passed over by the infinitely many sides of the large circle, arranged continuously [in a straight line], is equal in length to the line passed over by the infinitely many sides of the smaller, but in the latter case with the interposition of as many voids between them. And just as the "sides" [of circles] are not quantified, but are infinitely many, so the interposed voids are not quantified, but are infinitely many....

The transition from this model to physical reality is both obvious and surprising:

> but if we take the highest and ultimate resolution [of surfaces and bodies] into the prime components, unquantifiable and infinitely many, then we can conceive such components as being expanded into immense space without the interposition of any quantified void spaces, but only of infinitely many unquantifiable voids. In this way there would be no contradiction in expanding, for instance, a little globe of gold into a very great space without introducing quantifiable void spaces—provided, however, that gold is assumed to be composed of infinitely many indivisibles.

Simplicio intervenes, alarmed by what seems very like Democritus' atomism. But Galileo's ideas (not very clearly defined, as we have seen) are in fact closer to Plato's *Timeus* than to Democritus' materialism. Linguistically and conceptually, Galileo vacillates between *minimi quanti* (finite number of indivisible entities) and *infiniti indivisibili* (infinite indivisible entities). This indecision gives rise to problems of interpretation which Galileo's student Vincenzo Viviani had already observed: "If [the atoms] are infinite, it is necessary that they are not *quanti*, but a little further on, [Galileo] makes them *quanti*."[39] This ambiguity is favored by the mathematical image. The term "indivisible" (which Galileo may have adopted from the geometrical work of the English mathematician and theologian Thomas Bradwardine[40]) contains two distinct but overlapping concepts: that of the physical atom, the smallest unit of a substance, which cannot be split without radically altering the properties of the substance itself; and that of the geometrically indivisible, which belongs to pure mathematics. Both concepts seriously questioned the Peripatetic doctrine; the former because of its obvious reference to Leucippus' and Democritus' atomism, and the latter because it raised a different concept of geometry. The emergence of this conflict with Aristotelianism was precisely what unified the two distinct concepts. And this unity was confirmed by tradition. We should remember that, from the era of Roger Bacon, Duns Scotus and Bradwardine on, one of the chief arguments against Democritean atomism consisted of "demonstrating" the impossibility of believing that a line is composed of points, proposing as proof the apparent incommensurability of the diagonal and the side of a square.

Simplicio confirms this alliance of physics and mathematics:

> Also, this composing the line of points, the divisible of indivisibles, the quantified of unquantifiables—these reefs seem to me to be hard to pass. And not absent from my difficulties is the necessity of assuming the void, so conclusively refuted by Aristotle.

These are the objections which Antonio Rosso had made to Galileo in his *Esercitazioni filosofiche* ("Philosophical exercises"), published in Venice in 1633. Galileo now devotes a good part of the discussion to them, alternating between the physical and mathematical models to shore up their defenses.

We should leave the three disputants to their digressions, which Galileo's extraordinary imagination multiplies to excess. The discussion moves swiftly through the paradoxes of the infinite, Archimedes' mirrors, the velocity of light, the theorems of geometry, and the condensation and rarefaction of bodies, to falling bodies, which allow Galileo to exhibit his fundamental

---

[39] Florence (Firenze), Biblioteca Nazionale, Mss. Gal. pt. 5, Vol. 9, p. 17r.

[40] *See* Geymonat and Caruso, *op. cit.*, p. 625.

results. From this last to the movement of a pendulum is only a short step, and the pendulum leads naturally to the consideration of the non-intuitive but undoubted law of the isochronism of small oscillations. Galileo then moves from the free oscillations of the pendulum to forced oscillations, caused by rhythmically puffing at a sphere, and to the ability of a bell-ringer to handle a very large bell by means of a rope.

The discussion moves quickly, as the curiosity of the three interlocutors grows livelier and more far-reaching. By the end of the first day, the discussion has reached the "wonderful phenomenon" of musical sound. Consider a lute or cembalo: on what does the pitch of a vibrating string depend? Galileo had given the problem some thought. He experimented with an astonishing variety of common things, or he could simply intuit the hidden harmony in them. So he guesses that the pitch of a sound depends on its frequency, which in turn depends on the length, thickness, density and tension of the string, according to a constant ratio. He was right, although the discovery of this ratio had to wait for the next century.

5.10 The Second Day

Salviati arrives at the meeting a little late, as usual in academic meetings. Sagredo and Simplicio are looking forward to reviewing the discussion of "that resistance which all bodies have to fracture, and [which] depends on that cement which holds the parts glued together so that they would yield only under considerable pull," and of the cause of that coherence which is "mainly in the vacuum."

Salviati comes straight to the point.

> Taking up the original thread, then, whatever may be the resistance of solid bodies to parting under a violent pull, its presence in them is beyond any doubt. This resistance is very great against a force that pulls them in a straight line, but is observed to be much less when the force is across them. Thus we see that a steel or glass rod, for example, supports a weight of a thousand pounds lengthwise, but when fixed horizontally in a wall, it is broken by attaching only fifty [pounds] to it. We must speak of this second resistance, seeking the proportions in which it is found in prisms and cylinders of the same material, whether similar or dissimilar in shape, length, and thickness. In such speculations I take as a known principle one which is demonstrated in mechanics about the properties of the rod which we call the lever: that in using a lever, the force is

> to the resistance in the inverse ratio of the distances from the fulcrum to the force and to the resistance.[41]

The scientific horizon of the discussion has come clearly into view. We are no longer concerned with *why* something happens, only with *how* it happens. Experiment shows that a beam loaded at right angles to its length (for example, a cantilever) can only carry a fraction of the weight it can sustain if the load is set on its end (for example, a pier). Galileo believes that he can find a full explanation by applying the equilibrium law of the lever. Therefore his treatment would reach the exactness of geometry if he could give the law of the lever the aspect of a geometrical theorem. For this reason, he takes up Archimedes' demonstration and introduces some improvements to it, according to the line of thought of Valerio and Benedetti (see Part 1, Sections 2.1, 2.2). In addition, he sees the necessity of (sometimes) considering the weight of the arms of the lever— something which classical theory generally ignored. He thereby introduces a distinction between two ways of considering weight:

> When we consider an instrument in the abstract, i.e., apart from the weight of its own material, we shall speak of taking it in an absolute sense [*prendere assolutamente*]; but if we fill one of these simple and absolute figures with matter and thus give it weight, we shall refer to [it] ... as a "moment" or "compound force" [*momento o forza composta*].

Unfortunately, Galileo's faith in his ability to explain the resistance of solids by means of the equilibrium law alone is somewhat misplaced. A different hypothesis comes into his discussion of a loaded beam and the properties of rupture. As we shall see, the discovery of the inadequacy of the static model required a long and difficult journey and a new, richer vocabulary. For the moment, let us continue with Galileo's text, appreciating its simplicity and beauty as well as its limitations.

After identifying the lever as the model for the entire forthcoming analysis, Galileo formulates and (in his own way) resolves the problem which, from a historical viewpoint, is the single most discussed and important in the science of structures. To this day, it is usually called Galileo's problem. It involves the resistance to fracture of a cantilever beam, loaded at its free end.

> Now, getting back to our first purpose, it will not be difficult to understand the reason whence it comes about that: A solid prism or cylinder of glass, steel, wood, or other material capable of fracture, which suspended lengthwise will sustain a very

---

[41] Unless otherwise attributed, all quotations in this and the next four sections are from Galileo, *Discorsi*, pp. 151–189.

FIGURE 5.6. Galileo's figure for his fundamental problem. From the *Discorsi* of 1638.

heavy weight attached to it, will sometimes be broken across (as said earlier) by a very much smaller weight, according as its length exceeds its thickness.

Let us imagine the solid prism $ABCD$ fixed into a wall at the part AB; and at the other end is understood to be the force of the weight E (assuming always that the wall is vertical and the prism or cylinder is fixed into the wall at right angles). It is evident that if it must break, it will break at the place B, where the niche in the wall serves as support, BC being the arm of the lever on which the force is applied. The thickness BA of the solid is the other arm of this lever, wherein resides the resistance, which consists of the attachment that must exist between the part of the solid outside the wall and the part that is inside. Now, by what has been said above, the moment of the force applied at C has, to the moment of the resistance which exists in the thickness of the prism (that is, in the attachment of the base BA with its contiguous part), the same ratio that the length CB has to one-half of BA. Hence the absolute resistance to fracture in the prism BD, (being that which it makes against being pulled [apart] lengthwise, for then the motion of the movers is equal to that of the moved) has, to resistance against breakage by means of the lever BC, the same ratio as that of the length BC to one-half of AB, in the prism; or, in the cylinder, to the radius of its base. And let this be our first proposition.

5.11 Opening Remarks

Little scientific literature has deserved more comment, provoked more theoretical discussion, generated more hypotheses, or raised more questions than this one. Let us reconsider it in a little more detail. The structure of reasoning is simple and clear, but its development is sketchy, giving only a glimpse of the physical meaning of its premisses. Choosing an elbow lever ABC with fulcrum at B has interesting implications for the mechanics of solids and the science of structures.

First, the geometrical constraint—the intersection of beam and wall—is represented in statical terms as a *force* expressing resistance, analogous to the "power" operating on the end of a lever. The concept of equilibrium unifies facts that are distinct and almost alien to each other: the geometrical form of a constraint; the resistance produced, perhaps by the *horror vacui*, perhaps by "glue"; the force exerted by something or somebody on a lever arm; and finally the weight on the opposite arm. Since the law of the lever must hold, this law becomes the means for expressing the interrelationships between these different facts. It allows us to establish or discover their equivalence; we can reduce each one in terms of another. This unifying reduction lies at the base of the whole mechanics of solids, but it has become so natural that its relevance is often neglected, even though it raises problems at the methodological level, for example with regard to the principles of scientific operationalism (*vide* P.W. Bridgman [42]).

Second, this reduction is not defined abstractly or in general terms. Rather, it refers to the limit case—impending collapse. As such, it can be verified. We can in fact compare the weight which cracks a cantilever to the ultimate weight which can be supported end-on by a similar beam. In this way, the concept of resistance is endowed with a definite measure—that is, the maximum weight that a beam, axially loaded, can carry. We have therefore set premisses for a future limit analysis of structures and for criteria of resistance.

Third, Galileo assumes that his beam is rigid and fragile—that all points of the section AB of the beam where it meets the wall crack at once. He has, in fact, chosen to describe one particular class of bodies, but this is not his aim. Rather, he sees this ideal behavior as a perfect mathematical model which can "make the experiment meaningful," yielding a truth which is valid as long as we disregard the "imperfections of matter" and proceed only by geometrical demonstration. The case is ideal, but not too ideal. It provides an *absolute reference* that allows us to interpret the real differences as being the effect of possible secondary causes, just as the principle of inertia permits us to ascertain the presence of friction. This orientation was to survive in works by future authors from Vincenzo Viviani to Robert

---

[42] P.W. Bridgman, *The Logic of Modern Physics* (New York, 1927).

Hooke. These writers applied this approach to the search for what we would now call constitutive laws. Such laws do more than record empirical results; they permit the creation of mathematically consistent theories, in the light of which engineers can interpret their experimental data and the variety of material behavior.

We shall try to follow some of these developments, bearing in mind what Galileo's first proposition implies. The first criticism was made by the Genoan Giovan Battista Baliani, in a letter to Galileo dated July 1, 1639. He asks respectfully for a major clarification: "I would like your Grace to have explained at greater length in the first proposition how the moment of the force at C to the moment of resistance is like CB to the half of BA."[43] This question would raise considerable discussion, mostly outside Italy, during the late seventeenth and early eighteenth centuries. Why half? Why not a third or some other fraction? Galileo answers in a letter to Baliani (August 1, 1639). This proportion, he explains, comes from the supposition that "resistances of equal moments are circumfused" around the center of the fixed base of a prism or cylinder,

> in which operation the same accident occurs that intervenes in the straight line AB whose support is in C, where there are placed on the minor distance CB as many equal weights as you like, [suspended] at equal distances. They will show the same resistance to the force placed in A, as if all the aforementioned weights, reduced into only one, were hanging from the half of BC.[44]

For Galileo, if the different points of the clamped section are equally resistant, we can immediately apply the lemma on which Archimedes' demonstration of the law of the lever is based (see Part 1, section 2.1).

Baliani replied on August 19. He fully accepts the experimental plausibility of Galileo's thesis, but insists that it seems not to have been arrived at deductively, and requires a new postulate. Of course Baliani was right: Galileo's results could not be reduced to a simple corollary of the law of the lever. But Galileo, in a letter of September 1, insists that his first proposition could indeed be reduced to the principle of the lever, according to which "the force of resistance is inversely proportional to the distance from the point of support."[45]

---

[43] Galileo, *Opere*, Vol. 18, p. 70.
[44] *Ibid.*, p. 77.
[45] *Ibid.*, p. 94.

We can express Galileo's problem in rather more lucid terms. When a beam is subjected to simple traction or compression, Galileo admits implicitly, the stress cannot exceed a certain limit, σ_{lim}, or the beam will break. Therefore we must also have a limit value N_{lim} for the normal force:

$$N_{\text{lim}} = \sigma_{\text{lim}} S \tag{5.1}$$

where S is the area of the cross-section. We will define the absolute resistance P_c as the axial load which causes collapse; in terms of the normal force:

$$N_{\text{lim}} = P_c. \tag{5.2}$$

The justification of equation (5.1) can be found in the demonstrative arguments of proposition 4, where two cylinders A and B are compared. They have the same length, but different diameters (DC and EF, respectively). If $EF > DC$, then B will be stronger than A.

> For first, consider the absolute and simple resistance that resides in the bases (that is, in the [areas of] circles EF and DC), when these are to be broken by exerting a force that stretches them lengthwise. There is no doubt that the resistance of cylinder B is [in that case] greater than that of cylinder A by as much as circle EF is greater than [circle] CD, because so many the more are the fibers, filaments, or holding elements that keep the parts of such solids together.

This is the only passage in the Second Day which mentions the physical causes of resistance discussed in the First Day. It is by no means coincidental that Galileo links the notion of fibers to equation (5.1), which is critically important to the themes of the Second Day. The geometrical demonstrations of proposition 1 can proceed independently of any physical hypothesis, since they depend only on the principle of the lever (if we accept Galileo's interpretation). But we need to link these demonstrations to the reasons which explain resistance, and this is precisely what equation (5.1) does. We shall see how some of Galileo's successors—especially Viviani and Marchetti—used peculiar methods to eliminate this relation, and to make the arguments in the Second Day consistent and self-conclusive according to a strict axiomatic system.

Consider a cantilever beam (Figure 5.7), rectangular in section and with length l, thickness B and height H. One end is inset into a wall; the other is loaded with weight Q_c until the cantilever breaks. Galileo believes that the behavior of the system at the moment of collapse can be reduced to that of an angular lever. The fulcrum is the bottom of the intersection of beam and wall; one arm lies along the axis of the beam to the loaded end, and the other is the part inset into the wall. Galileo implicitly assumes the uniform distribution of the limit stress σ_{lim}, equivalent to the axial limit

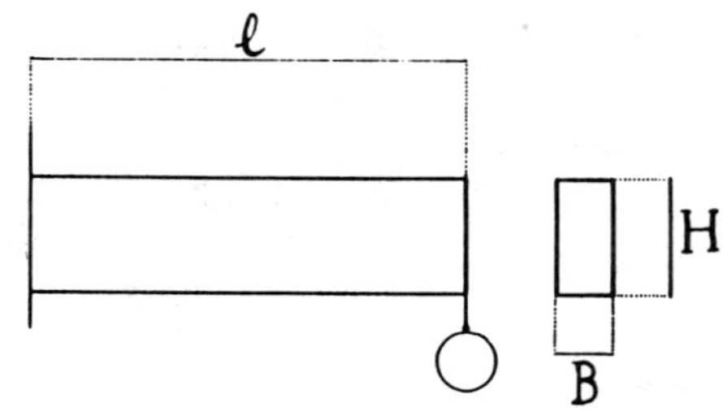

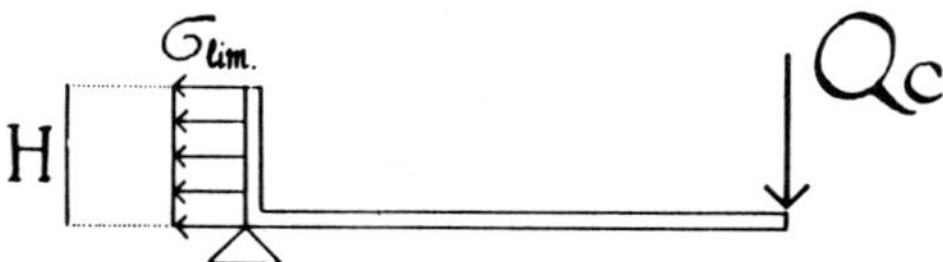

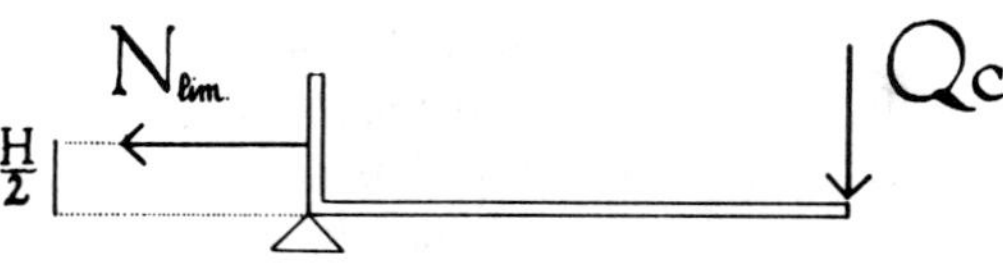

FIGURE 5.7.

force N_{lim} applied at the middle of the height. The equilibrium law of the lever gives us

$$N_{\text{lim}} : Q_c = l : \frac{H}{2} \tag{5.3}$$

and this is the first proposition precisely as Galileo states it. Although equation (5.3) is not perfect, many of its consequences, as Galileo himself established, are true and important. Meanwhile, we observe that

$$N_{\text{lim}} = \sigma_{\text{lim}} BH$$

so that, according to Galileo, the resisting moment of the beam cannot exceed the limit value

$$M_{\text{lim}} = \frac{1}{2}\sigma_{\text{lim}} BH^2. \tag{5.4}$$

Correspondingly, the limit value Q_c of the applied load cannot exceed the value

$$Q_c = \frac{M_{\text{lim}}}{l}. \tag{5.5}$$

This formula is exact, in the sense that it defines the load which will cause the beam to collapse as long as no significant deformation occurs before

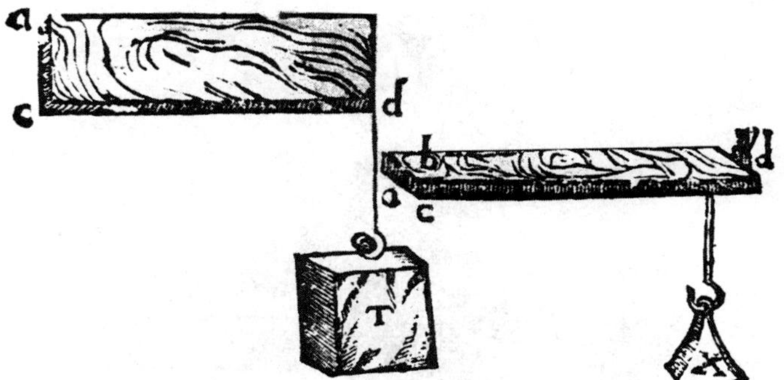

FIGURE 5.8. Different resistance to fracture opposed by the same prism when stood on edge and when laid flat. From the *Discorsi* of 1638.

the moment of fracture. But the truth of equation (5.5) depends on a correct evaluation of M_{lim}. In fact, equation (5.4) depends on the (erroneous) assumption that the stress σ_{lim} is uniformly distributed at the clamped section, and the equation therefore cannot be accepted. As we now know, the correct expression (which can be deduced from the hypothesis of perfect plasticity) is

$$M_{\text{lim}} = \frac{1}{4}\sigma_{\text{lim}} BH^2. \tag{5.6}$$

By comparing equations (5.4) and (5.6), we realize that, qualitatively at least, Galileo's solution accounts for the essential elements which determine the limit resistance of a beam. From equation (5.5), it follows that the breaking load is inversely proportional to the length of the beam, while from equation (5.4), we can show that the resistance increases with the square of the height. The remaining problems which Galileo resolves in the Second Day are essentially corollaries to these fundamental results.

5.12 Corollaries

Salviati presents the second proposition as follows:

> Now we can immediately understand how, and in what ratio, a rod, or rather a prism of greater breadth than thickness, more greatly resists breaking when loaded across its breadth than across its thickness. For an understanding of this, imagine a ruler *ad* whose breadth is *ac* and whose thickness, much less, is *cb*. It is asked why, when we wish to break it on edge as in the first figure, it will resist the great weight *T*; but placed flat, as in the second figure [Figure 5.8], it will not [even] resist *X*, which is less than *T*.

FIGURE 5.9. Galileo's drawing for demonstrating that the resistance of a heavy beam decreases when its length increases. From the *Discorsi* of 1638.

The reason is obvious, as the limit resisting moment in the first case is

$$M_{\text{lim}}^{(1)} = \frac{1}{2}\sigma_{\text{lim}}(cb)(ac)^2$$

and in the second case,

$$M_{\text{lim}}^{(2)} = \frac{1}{2}\sigma_{\text{lim}}(ac)(cb)^2.$$

Therefore

$$M_{\text{lim}}^{(1)} : M_{\text{lim}}^{(2)} = (ac) : (cb).$$

Salviati concludes: "the same ruler or prism, broader than it is thick, more greatly resists being broken when on edge than when flat, according to the ratio of its breadth to its thickness."

Pursuing his examination, Galileo wants to examine separately the effects of length and thickness (i.e., cross section) on the resistance to fracture of a beam. In the first case, he takes into account the fact that increasing the length of the beam will, of course, also increase its weight. This yields the third proposition: "The ratio in which the moment of heaviness of a horizontal prism or cylinder increases, in relation to its own resistance to being broken by elongation, I find to be in squared proportion to the lengthening." (Figure 5.9.) In fact, the moment of weight in A for the beam (1) (Figure 5.10) with length $AB = l_1$ and unit weight q is

$$M_A^{(1)} = q\frac{l_1^2}{2}$$

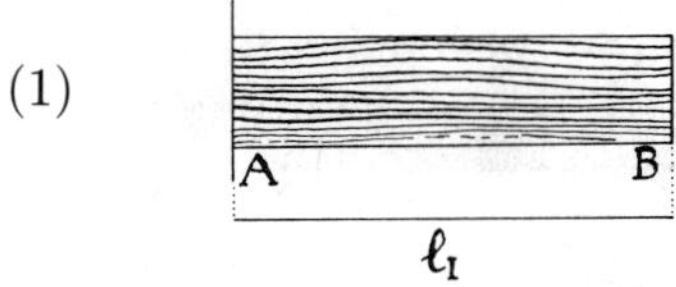

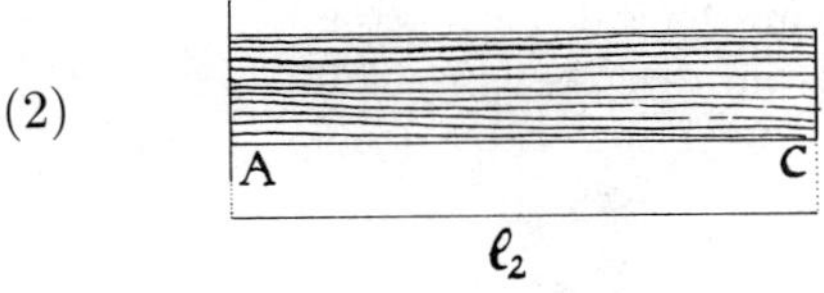

FIGURE 5.10.

while that for beam (2) with the same unit weight but with length $AC = l_2$ is

$$M_A^{(2)} = q\frac{l_2^2}{2}$$

so that

$$M_A^{(1)} : M_A^{(2)} = l_1^2 : l_2^2.$$

> We shall now show, in the second place, the ratio according to which resistance to being broken increases in prisms and cylinders of the same length, when they are increased in thickness. Here I say that: In prisms and cylinders of equal length but unequal thickness, resistance to fracture increases as the cubed ratios of the thickness or the diameters [respectively] of their bases.

As we increase the thickness, we keep a constant ratio α between diameter B and height H:

$$B = \alpha H.$$

From equation (5.4), we find that

$$M_{\text{lim}} = \frac{1}{2}\sigma_{\text{lim}}\alpha H^3. \tag{5.7}$$

That is, the resistance is proportional to the cube of the height, as proposition 4 confirms. In the case of a cylinder with a circular base of diameter H (Figure 5.11), we can derive an analogous expression:

$$M_{\text{lim}} = \frac{1}{2}\sigma_{\text{lim}}\beta H^3 \tag{5.8}$$

where β is a convenient numerical factor.

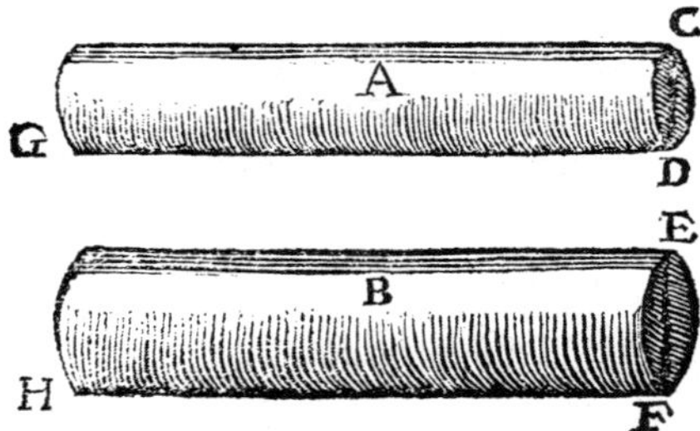

FIGURE 5.11. Galileo's figure for calculating the ratio of the resistance of two cylinders of equal length. From the *Discorsi* of 1638.

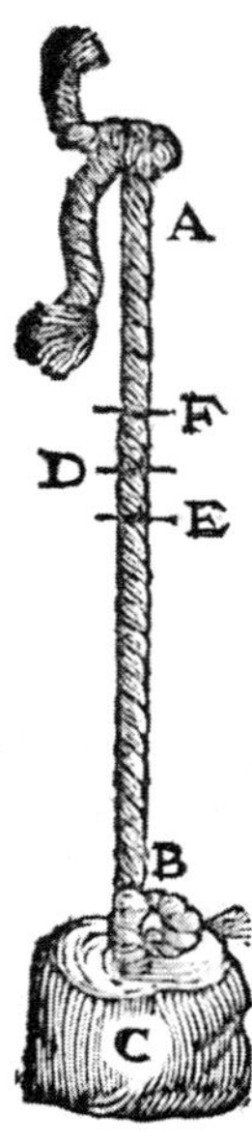

FIGURE 5.12. Galileo's figure for demonstrating that the resistance of a cord does not depend on its length. From the *Discorsi* of 1638.

At this point, Simplicio asks Salviati to remove a doubt which troubles him. He has understood very well that the breaking load Q_c of a cantilever depends on its length. But he was convinced by tradition that in the case of an axial load, the breaking load should also depend on the distance between the wall and the point at which the load is applied. Salviati points out that if a cord breaks, for example at point D (Figure 5.12), this is due only to the applied weight C and to the weight of the cord segment BE. The cord will still break if it were fixed at point F instead of A or if the weight C were suspended at E instead of B. Salviati charitably notes that Simplicio has shared this error with many others, "even very intelligent people."

The emphasis with which Galileo presents his thesis shows that he was aware of its latent significance: the resistance to traction, which in the proof of proposition 4 he recognizes as depending on the cross section of a taut cylinder, is here defined *only* in terms of the cross section. In this, as Geymonat and Caruso remark,[46] Galileo follows Leonardo ("every suspended body stretches the whole cord and equally in each of its parts"[47]) and, even more, Guidobaldo del Monte, whom Galileo fails to mention. In a paper discovered and published by Libri,[48] Guidobaldo had observed, rather more lucidly than Galileo, that

> a cord which sustains a weight, sustains as much when it is short as it does when it is long. It is quite true that in the long one, either because of its own weight or because of its length, there can be many weak parts, and perhaps it breaks more easily ... but if the cord had been sustained even a bit higher than the place that breaks because of its stretching, without any doubt it would have broken just the same, because it would have been deformed in the same manner.

(This passage hints at a criterion of fracture based on stretching, a topic that will be more fully discussed in Chapter 8, below.)

The next proposition presented by Salviati deals with the comparison of cylinders of different lengths and diameters. What is the relationship between length and diameter and the breaking load Q_c? The answer is easily obtained: M_{lim} obeys equation (5.8) and is related to Q_c by equation (5.5). In short,

$$Q_c \propto \frac{H^3}{l}. \tag{5.9}$$

That is, in prisms and cylinders which differ in both length and thickness, the resistance to fracture is directly proportional to the cube of the diameter of the base and inversely proportional to the length (proposition 5).

Galileo's propositions and demonstrations seem rather convoluted and over-elaborate for such elementary results. But he was hampered by his geometrical reasoning (based essentially on the classical theory of proportions), lacking the algebraic formalism on which we can now rely. The establishment of this mathematical idiom was to be the most important achievement of post-Galileian science. As Galileo had succeeded in divorcing science from the tiresome abstractions and prejudices of Aristotelian tradition, making it accessible in ordinary language, so the post-Galileian improvement of mathematical formalism allowed science to unfetter itself

---

[46] Geymonat and Caruso, *op. cit.*, pp. 730–731.

[47] Paris, Bibliothèque Nationale, Ms E, 32v.

[48] G. Libri, *Histoire des sciences mathématiques en Italie, depuis la renaissance des lettres jusqu'à la fin du XVII siècle* (Paris, 1838–1841), Vol. 4, p. 398.

from ordinary language and to transform the most complex chain of syllogisms into the simple consequences of calculation.

5.13 The Problem of Solids of Ultimate Dimensions

The propositions demonstrated thus far offer all the premisses needed to answer the problems formulated in the First Day which had stirred the curiosity of the interlocutors: do dimensional limits of structures exist? If resistance decreases when we increase the size of a body, preserving its proportions, is there a size which cannot be exceeded? And how can we determine this maximum?

Now we come to proposition 6:

> The compound moments of [two geometrically] similar cylinders or prisms, resulting from their own weights and [from their own] lengths serving as levers, have to one another the ratio that is the three-halves power of the ratio of the reistances of their bases.

The demonstration consists of expressing the ratio between the limit moment and the moment due to the proper weight in two similar cylinders, taking into account the fact that the absolute resistance is proportional to the square of the diameter. (But in fact the demonstration is anything but clear; it led later authors (Viviani included) to refute Galileo's thesis and to state that the ratio between the moments due to their own weight in two similar cylinders equals the square of the ratio between their absolute resistances. This misunderstanding of proposition 6 was discussed at the end of the nineteenth century by R. Caverni.[49]) We might note Simplicio's comment:

> This proposition strikes me as not only new but surprising, and at first glance very remote from the judgement I had conjecturally formed. For since the shapes are similar in all other respects, I should have thought it certain that their moments against their own resistances would also be in the same ratio.

Evidently Galileo was very proud of this proposition, which showed geometrically how the resistance of solids differed substantially from the geometrical properties of the solids themselves. The properties affecting resistance do not depend only on the ratios between the dimensions of a body, but

[49] R. Caverni, *Storia del metodo sperimentale in Italia* (Florence, 1895), Vol. 4, pp. 475ff.

also on the dimensions themselves. This allows us to predict a limit to the size of a body—which is precisely the goal of the following propositions.

Salviati says:

> After long thought about this, I found what I am about to put before you, in proper order, concerning this point. And first I shall demonstrate that: Among [geometrically] similar prisms or cylinders having weight [*gravi*] there is a single and unique case of the critical [*ultimato*] state between breaking and remaining whole when [the solid is] pulled down [*gravato*] by its own weight, such that if greater, unable to resist its own weight, it will break; and if smaller, it resists with some force whatever is done to break it.

The demonstration of this assertion proceeds, as usual, by way of geometrical rules and the repeated use of proportions. Despite Sagredo's claim that it is "short and clear," it is tortuous, but we can reduce it in this way: consider a set of prismatic cantilevers with rectangular bases and equal specific weight γ per volume unit. The cantilevers are similar in the sense that

$$B = \alpha H,$$
$$H = \alpha' l,$$

where α, α', are constants. The total weight Q of any cantilever is

$$Q = \gamma BHl = \gamma\alpha\alpha'^2 l^3$$

and can be applied ideally in the center of gravity of the prism at the distance $l/2$ from the fixed base (where the cantilever meets the wall). Moreover, the limit moment as given by equation (5.4) is

$$M_{\text{lim}} = \frac{\sigma_{\text{lim}}}{2}\alpha\alpha'^3 l^3.$$

For equilibrium at the onset of fracture, it must be

$$M_{\text{lim}} = Q\frac{l}{2}$$

and therefore

$$\frac{\sigma_{\text{lim}}}{2}\alpha\alpha'^3 l^3 = \frac{1}{2}\gamma\alpha\alpha'^2 l^4.$$

From this relation comes the following value for the maximum permissible length:

$$l = \frac{\sigma_{\text{lim}}}{\gamma}\alpha'. \tag{5.10}$$

Sagredo is struck by a result which is, in his opinion, so interesting and so improbable.

> It will therefore be necessary, in order to achieve that neutral state between holding and breaking, to alter greatly the ratio between length and thickness of the greater prism by thickening or shortening it. The investigation of that state, I think, might require equal ingenuity.

Salviati responds by stating the central problem:

> Even more, and more labor too; I know, for I spent no small time finding it. But now I wish to share it with you. Given a cylinder or prism of the maximum length that is not broken by its own weight, and given also a greater length, to find the thickness of some cylinder or prism which, at this given length, is the unique and maximum that resists its own weight.

The solution to this question requires a "long demonstration," and one "very difficult to keep in mind" at a single hearing. Actually, all we have to do is to solve equation (5.10) for the height H, taking into account that, by definition, $\alpha' = H/l$. If l_1 is the "greater length" associated with the new H_1, we have

$$H_1 = \frac{\gamma}{\sigma_{\text{lim}}} l_1^2.$$

It may be interesting to remember that Galileo developed this proposition (proposition 8) during his house arrest at Siena. During this imprisonment, Galileo managed frequent interchanges with his friends and pupils; he presented his expected results to them, and proposed questions on the resistance of solids. Proposition 8 in particular was studied by Galileo's student Mario Guiducci, Guiducci's cousins Andrea and Niccolò Arrighetti, and by Frs. Fabiano Michelini, Niccolò Aggiunti and Dino Peri. The most important contribution was that of A. Arrighetti, who, in a letter of September 25, 1633, succeeded in demonstrating that "among the infinite solids similar to a given one, only one is intermediate between fragility and consistence." The demonstration made use of the postulate that the resistances of cylindrical and prismatic solids are proportional to the squares of their thicknesses, whereas the moments of resistance are proportional to the cubes. Two days later, Galileo replied to his disciple, praising him lavishly ("the progress of your Grace is majestic and rises above a common geometrician") and presenting him with his own demonstration.[50]

Returning to the *Discorsi*, we come to proposition 9, which is connected to the preceding ones:

> Given the cylinder AC, of any moment whatever against its own resistance, and given any length DE, to find the thickness

[50] *See* for example Geymonat and Caruso, *op. cit.*, p. 731.

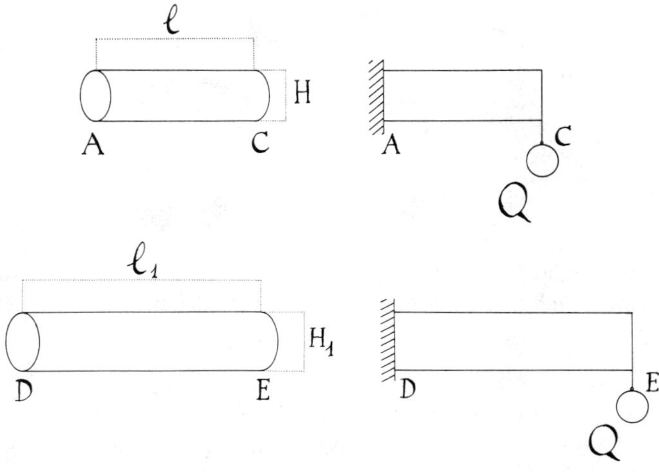

FIGURE 5.13.

> of the cylinder of length DE, such that its moment against its resistance shall have the same ratio as that of the moment of cylinder AC against its [resistance] [Figure 5.13].

Let us set l equal to the length AC of the first cylinder, and l_1 equal to the length DE of the second. Let H and H_1 be their respective diameters. If we ignore the weight of each cylinder and consider only load Q at the free end, the reactive moments at the mortise are Ql and Ql_1 respectively. The *limit moment* is given by equation (5.8) and is, therefore, proportional to βH^3 for cylinder AC and to βH_1^3 for cylinder DE. Thus we have l, l_1, proportional to the cubes of H and H_1, from which we derive the solution that the diameters are proportional to the cube roots of the lengths, that is:

$$H_1 = H\sqrt[3]{l_1/l}.$$

Galileo's comments on these results are really delightful:

> You now see how, from the things demonstrated thus far, there clearly follows the impossibility (not only for art, but for nature herself) of increasing machines to immense size. Thus it is impossible to build enormous ships, palaces, or temples, for which oars, masts, beamwork, iron chains, and in sum all parts shall hold together; nor could nature make trees of immeasurable size, because their branches would eventually fail of their own weight; and likewise it would be impossible to fashion skeletons for men, horses, or other animals which could exist and carry out their functions proportionably when such animals were increased to immense height—unless the bones were made of much harder and more resistant material than

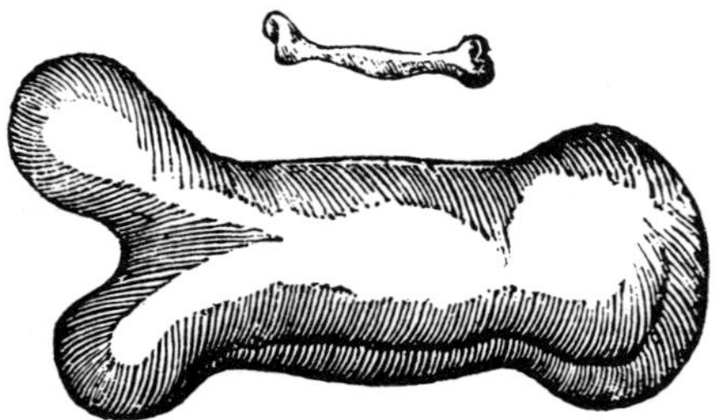

FIGURE 5.14. Galileo's sketch for demonstrating the monstrosity of an enlarged bone which performs the same function as the small one. From the *Discorsi* of 1638.

> the usual, or were deformed by disproportionate thickening, so that the shape and appearance of the animal would become monstrously gross. Perhaps this was noticed by our very alert poet [L. Ariosto] when, in describing a huge giant, he said:
>
> His height is quite beyond comparison,
> So immeasurably gross is he all over.
>
> To give one short example of what I mean, I once drew the shape of a bone [Figure 5.14], lengthened only three times, and then thickened in such proportion that it could function in its large animal relatively as the smaller bone serves the smaller animal; here are the pictures. You see how the shape becomes in the enlarged bone disproportionate.

Galileo exaggerates a little in his examples and conclusions, but his lesson is important: he knew how to shatter the unfounded assumption that similar structures have like resistance, regardless of their dimensions. Anyone who believes that architects have entirely assimilated Galileo's teaching is too optimistic. Some years ago, when blind faith in technology was still possible, various utopians presented schemes for urban planning that were as astonishing and radical as they were fruitless. Wright's mile-high skyscraper, Kenzo Tange's plan for Tokyo, Arata Isozaki's air city, Yona Friedman's three-dimensional grid, Paul Maymont's suspended town, Walter Jonas' "intra-town"—these and other "megastructural" daydreams could exist only in Gulliver's world, where a cup may be as small as a midge or as large as a castle (Figure 5.15).

Galileo would have remarked that in such a world gravity and the weight of bodies had been forgotten. In fact, proposition 10 says that "Given a prism or cylinder and its [own] weight, and given the maximum weight it sustains [at one end], we can find the maximum length beyond which the prism itself, if prolonged, would break of its own weight." Galileo is interested not in the problems of architecture, but rather in the mechanical description of animals, their body structure and how they move. In these

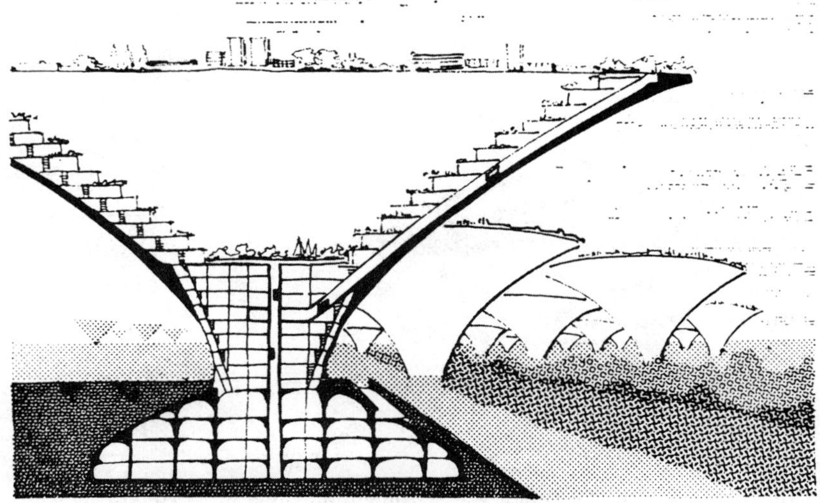

FIGURE 5.15. W. Jonas' "intra-town."

short passages, Galileo prefigures a new science which was to be highly successful in the second half of the seventeenth and the early eighteenth centuries. Giovanni Alfonso Borelli, an active and intelligent member of the Accademia del Cimento, proposed the bases of the new doctrine in his posthumous *De motu animalium*, published in Rome in 1680–1681.

Three questions follow on beams which are subject to different constraint conditions.

> Thus far there have been considered the moments and resistances of solid prisms and cylinders of which one extremity is assumed to be fixed, and only at the other end is the force of a pressing weight applied.... Now I wish some discussion of the same prisms and cylinders, but when they are sustained at both ends, or are supported on a single point taken between their extremities.
>
> I say first that the cylinder pressed [*gravato*] by its own weight [alone] and brought to that maximum length beyond which it can no longer sustain itself, either on a single support exactly at its middle, or supported by two at its extremities, can be twice as long as when fixed in a wall or sustained at one end only [Figure 5.16].

The demonstration is simple: the moment at the built-in end A for a cantilever with length l is

$$M_A = ql^2/2$$

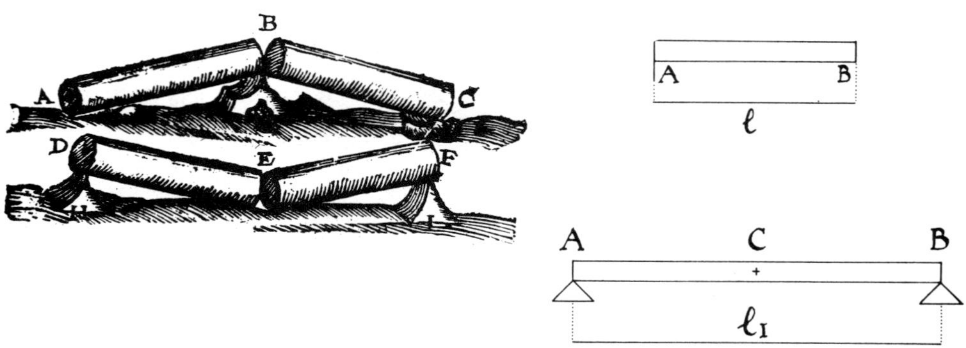

FIGURE 5.16. Galileo's figure for his treatment of beams supported in the middle or at both ends. From the *Discorsi* of 1638.

where q is the weight per unit length. The maximum moment at the middle C of a beam with span l_1 supported at both ends is

$$M_C = ql_1^2/8.$$

Therefore, if we set

$$M_A = M_C = M_{\text{lim}}$$

it follows that

$$l_1 = 2l.$$

This system is similar (as Sagredo notices) to a question posed in the Aristotelian *Mechanical Problems*. Moreover, it also relates to the law of the lever, which can be used to determine the reactions of a doubly supported beam. Propositions 12 and 13 consider similar subjects, but we shall omit them for the sake of brevity.

5.14 The Problem of Solids of Equal Resistance

At last we come to a new and important problem. Galileo was the first to state it accurately, and to work out a correct solution for a particular case. He outlines the theory which we now call structural optimization. Of course, Galileo only refers to the cantilever, but he recognizes the non-uniformity in the behavior of material under stress. If a beam is prismatic (Figure 5.17) then when it is loaded, the cross-sections next to the mortise (such as a–a) are stressed more severely than the sections at the free end (such as b–b). At the moment the beam breaks, when the mortise section is at its maximum resistance, the middle sections could actually bear more

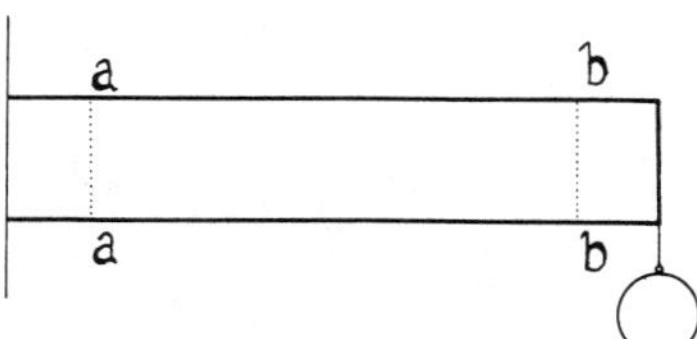

FIGURE 5.17.

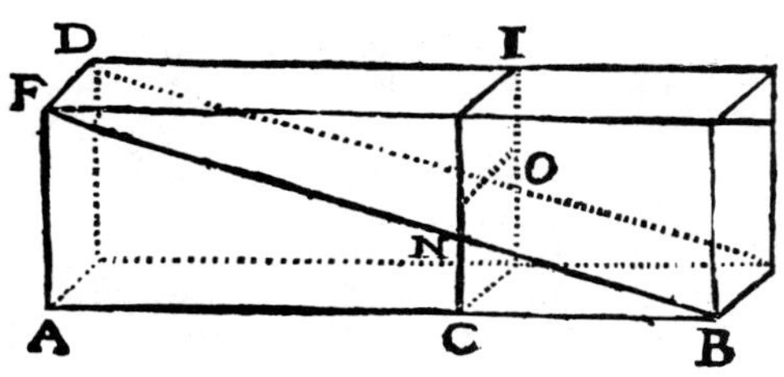

FIGURE 5.18. Galileo's figure for demonstrating that in the triangular prism the strength at C is less than at A, while the opposite occurs in the rectangular prism. From the *Discorsi* of 1638.

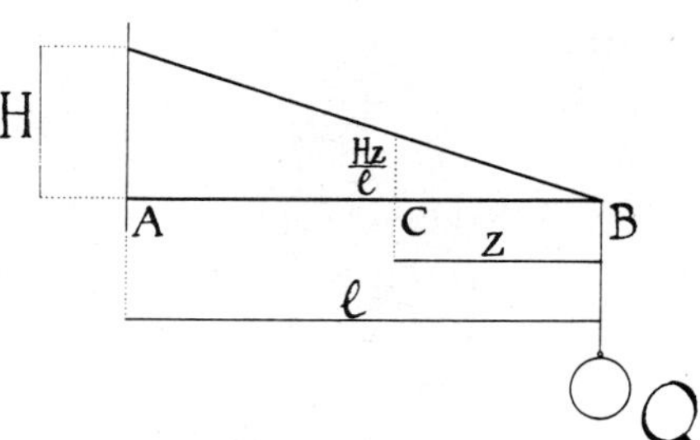

FIGURE 5.19.

weight. But the beam breaks anyway. Therefore a prismatic beam is not the best possible configuration of the material.

The situation changes if we shape the cantilever in a more appropriate way. As Galileo puts it,

> Here DB is a prism [Figure 5.18] in which the resistance to fracture by a force pressing on end B is, as previously demonstrated, less at the end AD than is the resistance at CI, by as much as length CB is less than BA. Next, consider the same prism sawed through diagonally along the line FB, so that the opposite faces form two triangles, one of which, FAB, is facing us. This solid has a nature contrary to that of the prism, since it less resists being broken over the point C than over A, by a force applied at B, in proportion as CB is less than BA.

The thesis can be demonstrated by comparing the limit moments at C and A. The height of the beam at C is given by $H(z/l)$ (see Figure 5.19). Therefore

$$M_{\text{lim}}^{(C)} = \frac{1}{2}\sigma_{\text{lim}} BH^2 \left(\frac{z^2}{l^2}\right), \qquad M_{\text{lim}}^{(A)} = \frac{1}{2}\sigma_{\text{lim}} BH^2.$$

The moments of load Q at C and at A are $M_C = Qz$, $M_A = Ql$ respectively. Therefore

$$M_{\text{lim}}^{(C)} : M_{\text{lim}}^{(A)} = z^2 : l^2, \qquad M_C : M_A = z : l,$$

or

$$\frac{M_{\text{lim}}^{(C)}}{M_{\text{lim}}^{(A)}} = \frac{M_C}{M_A}\frac{z}{l}.$$

Returning to the text:

> Thus we have taken away from the beam or prism DB a part, in fact one-half, by cutting it diagonally, leaving the wedge or triangular prism FBA; and these two solids are of contrary condition, the former being more resistant the more it is shortened [in the direction of B], and the latter losing robustness as it is shortened. Now, this being the case, it seems quite reasonable and even necessary that a cut can be made after which, the superfluous part being removed, there remains a solid of such shape that it is equally resistant in all its parts.
>
> *Simplicio*: Indeed, it is necessary that where we pass from the greater to the less, we also meet with the equal.
>
> *Sagredo*: But the point now is to find how to guide the saw as to make this cut.

Finally Salviati sets forth his solution:

> This understood, let the parabolic line FNB, whose apex is B, be drawn on the face FB of prism DB, and let the prism be sawed along this line, leaving the solid that lies between the base AD, the rectangular plane AG, the straight line BG, and the surface $BGDF$, which has the curvature of the parablic line FNB. I say that this solid [taken absolutely] is equally resistant throughout [Figure 5.20].

To make this intelligible in mathematical terms, all we need to do is to state that in the section with abscissa z and height $H(z)$, the equilibrium equation at the limit of fracture yields

$$M_{\text{lim}} = Qz$$

and therefore

$$\frac{1}{2}\sigma_{\text{lim}} BH^2(z) = Qz. \tag{5.11}$$

This determines a parabola whose height $H(z)$ at z is

$$H(z) = \sqrt{\frac{2Q}{\sigma_{\text{lim}} B}z}.$$

The reader may notice that Galileo's original drawing (see Figure 5.20)

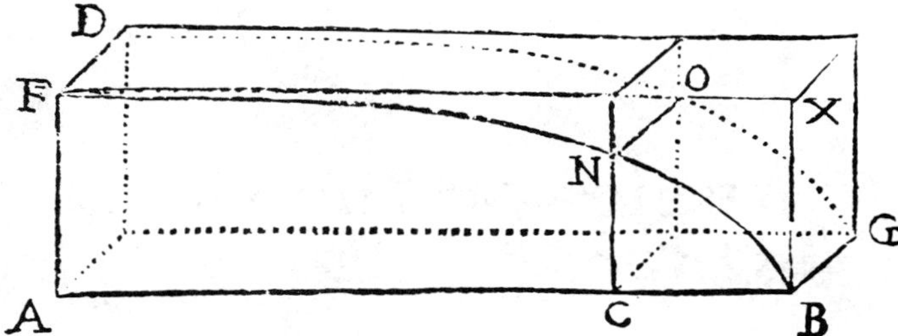

FIGURE 5.20. Galileo's solid (cantilever) of equal resistance. From the *Discorsi* of 1638.

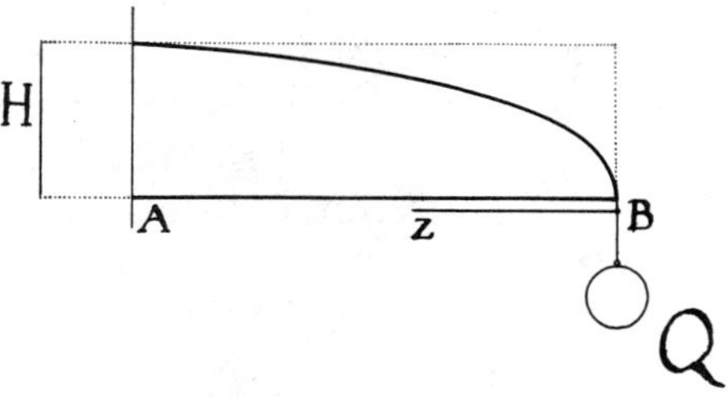

FIGURE 5.21.

is slightly wrong; the parabola given by equation (5.11) has its vertex on the loaded end section and its tangent is there vertical (see Figure 5.21).

Salviati concludes: "From this it is seen that beamwork can be built, without any reduction in strength, while diminishing weight by over thirty-three per cent. In large ships, especially to support the decks, this may be quite useful, since lightness is extremely important in such structures."

As early as the seventeenth century, the problem of solids of equal resistance led to a memorable scientific controversy in which P. Würtz, F. Blondel, A. Marchetti, V. Viviani and (later) G. Grandi, Varignon and Parent took part. But we shall deal with this in greater detail later. As Todhunter observes, "Closely as the problem of solids of equal resistance is associated with the growth of the mathematical theory of elasticity, it is nevertheless the problem of the flexure of a horizontal beam which may be said to have produced the entire theory."[51]

After discussing the fifteenth proposition in geometrical terms, to make sure that the decrease of weight really is a third, the interlocutors concentrate on other possible techniques for making bodies lighter without sacrificing strength. The progress from a solid beam to a hollow one (or, in Galileo's terms, from a cylinder to a "cane") is spontaneous. The Second Day concludes with three last theses concerning that subject, bringing the total to eighteen propositions. We shall pass over these three, since they add little in the way of new concepts.

---

[51] Todhunter and Pearson, *op. cit.*, Vol. 1, p. 4.

6

First Studies on the Causes of Resistance

6.1 Experimental Confutations: The *Horror Vacui*

Who was first to develop the themes of the First and Second Days of the *Discorsi*? According to Fr. Guido Grandi, a monk of Camaldoli, it was Vincenzo Viviani, Galileo's favorite pupil. As we shall see, the matter is not quite so simple, and Grandi cannot be considered impartial. In fact, Viviani had published nothing before the discussion on the resistance of solids exploded, with considerable eclat, in Italy and France. All the same, Grandi's reasons for this claim are significant, and they may provide a good introduction to the debate that arose as early as the 1640s about the physical (or even metaphysical) causes of resistance.

Commenting on Viviani's *Treatise of Resistances* in its first (posthumous) edition of 1718, Grandi takes his cue from a passage in which Viviani attributes resistance, at least in part, to the "abhorrence of a void." Grandi observes:

> From many parts of the work, it is expressly clear that the author was busy at it as early as 1644, but if somebody's malignity or envy should suggest that the manuscript had been purposely and falsely back-dated, there is an evident confirmation of the truth in [Viviani's] proposition, where he mentions the *force of the vacuum*, following the style of the ancients, the very style used by his own teacher Galileo in the first dialogue. This proves that his treatise was written before the year 1644, when Torricelli, by means of the famous experiment conceived by him and first executed by Viviani himself, found the real cause of what was attributed to the force of the vacuum, and proved that this was due only to air pressure. It is not credible after such a famous and successful discovery, which he knew very well before any others, that the author should continue to speak of "force of the vacuum" with the vulgar term ....[1]

Grandi thus assigns to the year 1644 a crucial transition from the way of thinking of the ancients to a new scientific thought, in which prejudices

---

[1] Cf. Grandi's commentary on Viviani's treatise, published in G. Galilei, *Opere*, 2d ed. (Florence, 1718), Vol. 3, p. 201.

about the vacuum were eliminated. The "force of the vacuum" was finally unmasked as what it really is: "a mere nothing," as Grandi called it. Grandi puts Torricelli's experiment at the center of the transition, and its meaning is treated in relation to the criticism of the *horror vacui* voiced in the First Day of the *Discorsi*. Grandi's interpretation is confirmed by Tommaso Bonaventura's biography of Torricelli, published as the introduction to the posthumous 1715 edition of the *Lezioni accademiche*. T. Bonaventura states that Torricelli

> considered what Galileo wrote in his first Dialogue about the resistance of solid bodies to their being broken, that water in pipes, said to act by attraction, does not reach further than about 18 armslength in height, and that when it goes beyond, it soon breaks, leaving the space about it empty. As a result of this, he had the thought that, by taking a body much heavier than water [i.e., mercury] ... and enclosing it in a glass cylinder, a vacuum could be created in a much smaller space .... He communicated this conjecture to Vincenzo Viviani, who was the first to make such an important experiment .... When Torricelli was satisfied about the truth of the experiment ... he gave account of it to some of his friends in Rome, including Michelangelo Ricci to whom he wrote the following in a letter on June 11, 1644: I already mentioned that some kind of philosophical experiment was being made with reference to vacuum, not just to produce a vacuum, but to make an instrument to show mutations of the air, sometimes heavier, sometimes thicker, now lighter and now thinner. Many have said that it would not be possible, others, that it would, but only going against nature, and by force. I do not know yet whether anybody has said that it would be possible without such force, and without nature's resistance. I reasoned thus: if I should find an obvious and manifest cause whence such resistance derives, and which is found in the attempt to create a vacuum, I think it would be useless to try to attribute to vacuum that result, which obviously has another cause. In contrast, having made some very easy calculations, I find that the cause that I maintain—that is, that of the weight of the air—should in itself produce stronger opposition, which it does not do in attempting to create a vacuum .... We live submerged at the bottom of the ocean of elementary air which, as a result of irrefutable experiments, we are aware has weight ....[2]

Although they are not directly related to the resistance of solids, Bonaventura's narrative and Torricelli's letter are interesting for the considerable

---

[2] E. Torricelli, *Lezioni accademiche* (Florence, 1715), preface, pp. xxv–xxviii.

methodological value of the explanation offered for the mercury experiment. From a historical point of view, Bonaventura is wrong; Giovan Battista Baliani had already suggested such an explanation. In his letter to Galileo of October 24, 1630, Baliani clearly sets forth the hypothesis that atmospheric pressure was responsible for the behavior of a water column (see section 5.8, above). Perhaps his experience of diving into the sea at Genoa suggested his explanation:

> I imagine myself at the bottom of the sea ... and if it were not for the need to breathe, I think I would stay there, even though I would feel more compressed and squeezed from all sides than I do now .... But if I were ... I do not say in the vacuum, but in the air, and if from my head upwards there was water, then I should feel the weight, which I should not be able to bear .... It occurs to me that the same happens to us in the air, since we are at the bottom of its immensity, and do not feel its weight which compresses us on every side. This is because our bodies have been created by God in such a way as to be able to resist such pressure very well without incurring any damage .... If, however, we were in a vacuum, then we should feel the weight of the air bearing down on our heads, which [pressure] would, I believe, be extremely heavy.[3]

Baliani has hit the nail on the head, both with regard to the pressure exerted by fluids in all directions, and with respect to measuring this pressure. Moreover, his central image of the sea bottom is exactly the same as that used by Torricelli in his letter to Ricci. Historically, even Torricelli's experiment may not be the first of its kind; a similar experiment had previously been carried out by Gaspare Berti, a pupil of Benedetto Castelli, using water instead of mercury.[4]

But Torricelli's intention is to explain: the weight of the water or mercury in a column is not so much the measure of a hypothetical "force of the vacuum," as Galileo thought, as it is the elimination of the vacuum as a possible force in nature. If we put water or mercury in a tube hermetically sealed at one end and immersed in the fluid at the other, the fluid does indeed rise, but not because of the vacuum at the sealed end. The equilibrium between atmospheric pressure and the weight of the column of fluid is responsible. Therefore, nature does *not* abhor a vacuum, and Galileo's arguments relating the resistance of bodies to the force of vacuum must be revised, because (as Grandi put it) such a force is "a mere nothing."

Torricelli was the first to reach this conclusion, which is the endpoint of a path of research developed in Galileian circles. But he was by no means

[3] Galileo, *Opere*, Vol. 14, pp. 157–160.

[4] *See* for example C. de Waard, *L'experience barométrique* (Thouars, 1936).

the only one to address the problem; for example, Isaac Beeckman had foreseen as far back as 1618 that air must have some weight, and that this weight could force water to rise in the tube of a vacuum pump, exerting an expansion force in all directions. In a letter to Fr. Marin Mersenne, dated October 1, 1629, Beeckman maintained that there was nothing silly in the existence of a vacuum; rather the foolishness was in trying to use the vacuum (or nature's abhorrence of it) to explain natural phenomena.[5] In the tiny village of Bugue, in Périgord, the physician Jean Rey had independently reached the conclusion that air penetrates into caverns and wells by virtue of its own weight, and exerts pressure in all directions. His ideas, published in 1630, were the subject of a correspondence with Fr. Mersenne in 1631. Mersenne's reply practically confirmed Rey's thesis: the *horror vacui* cannot be considered a cause of phenomena. The true cause "derives from the equilibrium that Nature takes on," always returning to equilibrium "by the shortest and easiest route"[6] In these same years, Descartes was nurturing similar concepts, though from an opposite starting point, absolutely denying the existence of a vacuum and maintaining the existence of a "subtle matter." Mersenne, who was in constant correspondence with Descartes, dealt with this point in his treatise on musical acoustics, *Harmonie universelle* (Paris, 1636). He refutes Descartes' reasoning *a priori*, but reserves judgment until after an acoustical experiment. If a vacuum really exists, transmission of sound through it should be impossible.

Pierre Duhem was to observe, in considering these converging accounts, that "around 1630, the same thoughts on the heaviness of air and on atmospheric pressure were under discussion in vastly different countries, by physicists having no medium of communication with one another .... The idea awaits the hour of its advent"[7]

In fact, there was such a medium of communication, a human one, and endowed with rare brightness: Mersenne. A recent essay by Robert Lenoble[8] examines this remarkable man. His scientific curiosity, tolerance and foresight were truly exceptional in his day, especially in a cleric. Marin Mersenne was born at Oizé in 1588 and attended the newly founded Jesuit college of La Flèche, where Descartes was to study some years later. Mersenne had the aptitude for the highest ecclesiastical preferment, but at twenty-two he chose instead to enter the austere Order of the Minims, founded by Francis da Paola in 1436. Mersenne taught philosophy, first at Nevers, later at Paris, cultivated a broad range of interests, including biblical studies, and disputed with supporters of hermetic and cabalistic

[5] M. Mersenne, *Correspondance du P. Marin Mersenne, réligieux Minime* ..., ed. C. de Waard (Paris, 1936), Vol. 2, pp. 282–283.

[6] Letter to Jean Rey, Sept. 1, 1631

[7] P. Duhem, "Le Père Marin Mersenne et la pesanteur de l'air," *Revue générale des sciences* (1906), p. 774.

[8] R. Lenoble, *Mersenne ou la naissance du mécanisme* (Paris, 1971).

theories. He firmly defended the new science, in the conviction that "the cause of science and the cause of God are one and the same." The most remarkable feature of Mersenne's intense activity was, however, the establishment of a vast network of friends and correspondents among the philosophers and scientists of his time.

Mersenne was totally unbiased by traditional philosophy, but a skeptic of *a priori* systems of all types. He was a true friend to Pierre Gassendi, the priest who had dared defend Epicurus and profess atomism.[9] He was, however, on equally good terms with Descartes, whom he considered one of the best minds of the day,[10] in spite of the fact that Descartes' and Gassendi's views of the physical universe were poles apart. Mersenne was open to all opinions, selecting out those which he judged best and trying to reconcile them to Christian dogma. At the same time, however, he was both reserved and pragmatic. This explains his surprisingly amicable relationship with the rough and rude Gilles Personne de Roberval. The two men shared the empiricist's temperament, a frame of mind hostile to the Cartesian tendency to recreate a system of "demonstrative physics" founded on philosophical principles.

It was, therefore, natural that Mersenne should particularly favor Galileo. In 1629, he offered to help publish one of Galileo's treatises in France, in case he should meet with difficulties in Italy. In 1634, barely a year after Galileo's conviction at Rome, Mersenne published *Les méchaniques de Galilée* in Paris, writing in the introduction:

> I should be happy if I could be instrumental to Galileo's giving us all his speculations on movement and on everything pertaining to mechanics, because anything coming from him is excellent: to this end I beg all those who are in correspondence with Florence to exhort him by letter to publish all his observations, for I hope that by now he has the requisite time and opportunity to do so at his villa in the country.

Is this last statement merely naive, or is it ironical? Evidently Mersenne refers to the recent condemnation to silence and the perpetual imprisonment issued by the Holy Office. When the *Discorsi* appeared in Leiden in 1638, Mersenne, who already knew the manuscript delivered to the Duc de Noailles, made a free translation of Galileo's work; it was anonymously published in 1639 under the title *Nouvelles pensées de Galilée* (cf. page 135).

Mersenne's circle of friends and correspondents was enormous. Besides the scientists already mentioned, we should mention Constantijn and Christiaan Huygens, Pierre Fermat, Nicolas Fabri de Peiresc, Thomas Hobbes,

---

[9] P. Gassendi, *Epicuri Philosophia. Animadversiones in decimum librum Diogenis Laertii* (Lyons, 1649).

[10] M. Mersenne, *Harmonie universelle* (Paris, 1636), first preface.

Martin Ruarus, Jean Rey, and Étienne and Blaise Pascal. During the last years of his life (he died in 1648) most of Mersenne's scientific work dealt with the final clarification of the barometric experiment, of which there were two opposing interpretations, Torricelli's and Roberval's. The latter ascribed the rise of mercury or water to an "attracting force."

What put an end to these differences was Pascal's famous experiment of September 19, 1648. He compared the heights of a mercury column on a plain and at the top of Mont Puy de Dôme. In reality, Pascal's experiment was the fruit of Mersenne's collaboration —in different ways—with Descartes, Roberval, and the Pascals, father and son, and of Mersenne's own contribution.

6.2 Mersenne and the Problem of Resistance

Mersenne made frequent contributions to the problem of the resistance of solids. His experimental results sometimes confirmed the expectations of common sense, but sometimes contradicted not only them but also the rational predictions of Galileian science. In one of his last writings, Mersenne declared his doubts about the possibility of doing justice to the results he had obtained (in this case, about the resistance of cylinders): "Let him who wishes to experiment do so before claiming that this [result] is contrary to common sense and reason; I will listen more willingly to reason if it fits the phenomena, especially those that, as it seems to me, reason has always rejected."[11]

Mersenne doubted whether the natural phenomena pertaining to the resistance of solids could indeed be *explained* by geometrical reasoning—a doubt shared by others, most especially the eighteenth-century scientist Pieter Musschenbroek, who, perhaps more than anyone else, put this branch of mechanics on an experimental footing. In his *Introductio ad cohaerentiam corporum firmorum* ("Introduction to the cohesion of solid bodies") (1729), Musschenbroek cites Mersenne's results with reverence. In one case, regarding the resistance of threads to traction, he criticizes F. Lana-Terzi for having failed to cite Mersenne in his celebrated book *Magisterium naturae et artis*, and for having simply invented ("*ex cerebro suo extruisse*") the data he proposes. On the contrary, says Musschenbroek, "long ago, Mersenne

[11] M. Mersenne, "Reflectiones physico-mathematicae," in *Novarum observationum physico-mathematicarum F. M.M. Minimi*, Book 3 (Paris, 1647) (hereafter cited as Mersenne, *Reflectiones*), p. 150.

experimentally determined the coherence of metals in Book Three of *Harmonices*, Prop. 7.... It is proper to repeat Mersenne's conclusions because they confirm my experiments."[12]

Mersenne had demonstrated that the breaking strength of wires and cords of equal diameters varied with the material used to make them.

> A gut taken from the bowels of a sheep is weaker than a metallic cord of the same thickness. Experiments show that a gut which is made of only one filament whose thickness is 1/6 of line [1 line = 0.228 cm; 1 pound = 454 g] breaks at 7 pounds, while a gold cord of equal thickness breaks at 23 pounds, as the silver and the bronze [wire] at 18 1/2 pounds and an iron one at 19. And these results are deemed to be astonishing by many people.[13]

Why were these results so surprising? Perhaps Mersenne wishes here to allude to the scientific debate burning in France at the time (especially among his friends) about the explanation of the causes of resistance. He believed that objections could be made to all existing attempts to explain the phenomenon.

But Mersenne was not especially decisive; he preferred to leave questions open, presenting alternative hypotheses at the same time. In particular, he did not hide his mistrust of the Galileian claim for a geometrical demonstration of the strength of solids. He accepted the proportion between absolute and relative resistance which Galileo had established in the case of the cantilever (Figure 6.1), but his acceptance was reserved, almost reluctant: "This proportion is accepted with difficulty by some of our geometricians; on the other hand, so far I have not been able to find anybody who has convinced me it is wrong."[14] Mersenne examined all the current explanations for resistance—the *horror vacui*, fibers and filaments, the atomism of Democritus and Lucretius (i.e., branches and hooks interwoven)—and objected to all of them. The vacuum "almost does not deserve any consideration in the present problem"[15] but fibers and filaments are no help either:

> Among all the difficulties, the most serious is that almost all bodies are distinguished in a peculiar way because of the different arrangements of the different fibers: where some happen

[12] P. Musschenbroek, "Introductio ad cohaerentiam corporum firmorum," in *Physicae experimentales, et geometricae, de magnete, tuborum capillarium vitreorumque speculorum attractione, magnitudine Terrae, cohaerentia corporum firmorum dissertationes: ut et ephemeri des meteorologicae ultrajectinae* (Leiden, 1729), p. 506.

[13] M. Mersenne, *Cogitata physico-mathematica, in quibus tam naturae quam artis effectus admirandi certissimis demonstrationibus explicantur* (Paris, 1644), Book 1, Art. 4, Prop. I, p. 271.

[14] Mersenne, *Reflectiones*, p. 150.

[15] *Ibid.*, p. 149.

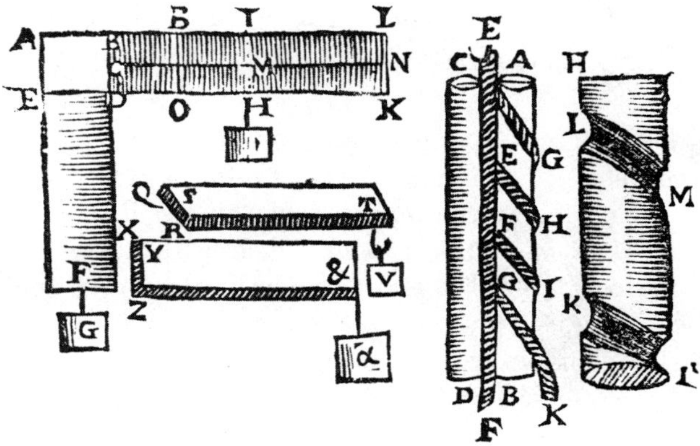

FIGURE 6.1. Mersenne's illustration of the principal subjects concerning the resistance of solids. From the "Reflectiones..." (Paris, 1647).

> to have many laminae [*schidia*], as for instance an oak, others present few or none, e.g., iron, marble and glass. It may be added that iron, copper and other metals, even single bodies, subject to force or weight, curve and bend to the form of an arch before breaking. This produces a new difficulty which escaped even Galileo's notice. If there is a physico-mathematician who is able to find a solution, he will deserve more credit than the inventor of quadrature.[16]

An accurate prediction: Euler was to solve the problem of lateral buckling in thin rods subject to axial load. As for atomism, Mersenne does not exclude it, but he is not yet satisfied: "Those who deem they can attribute robustness of cylinders to hooked atoms must explain why the hooks break or disconnect, being deformed [*evolvi*] under the action of a weight; only in this manner will they be able to throw light on this difficulty."[17]

These objections are no mere quibbles. Mersenne's point is important: he and his contemporaries had no idea what the root cause of resistance might be. The physical reasons for resistance were still wrapped in the deepest mystery. But this did not exclude the possibility of gathering useful information about the measure of resistance—a purely phenomenological description, but hardly a trivial one. Mersenne is explicit on that point, and promises to return to the subject if the Lord gives him strength.[18]

---

[16] *Ibid.*, p. 150.

[17] *Ibid.*

[18] *Ibid.*

6.3 Descartes' Concept: Stasis as the Best Adhesive

The year 1644 was critical for the problem of the resistance of solids. In it, Torricelli eliminated the vacuum from the forces of nature; Mersenne tested the resistance of materials experimentally; and Descartes published his remarkable *Principia philosophiae*, in which he dedicates a good many pages to the research of physical and metaphysical causes of the resistance of solids. Descartes agreed with Mersenne that the *horror vacui* was not a reasonable explanation for the strength of materials. In a letter to Mersenne of October 11, 1638, Descartes wrote:

> [Galileo] gives two causes for the cohesion that holds the parts of a body together: one is the abhorrence of a vacuum; the other is a certain "glue" or a sticky substance that keeps them together, and this second one he still explains by means of a vacuum. I think that they are both completely wrong. What he attributes to a vacuum cannot be attributed to anything but the heaviness of the air; and it is certain that if it were the abhorrence of a vacuum that prevents two bodies from being separated, no force would be able to divide them.[19]

Descartes' objections seem both curiously modern and antiquated at the same time—modern when he mentions air pressure, antiquated when he claims that the *horror vacui* should exert infinite resistance. This ambiguity is an accurate reflection of Descartes' position with regard to the new science. He was an innovator in mathematics and philosophy, but a conservative in both the methodology and content of physics. For instance, when he deals with the *horror vacui*, he is more Aristotelian than Aristotle; not only does he deny the existence of a vacuum in nature, but even its conceptual intelligibility. The Peripatetic arguments against the vacuum had concerned the field of physics; one of the main objections was that, since the velocity of moving bodies is inverse to the density of the medium in which they move, their velocity in a vacuum should be infinite. Descartes' argument, on the other hand, concerned the concept of a material body itself: "The nature of the *matter* or of a body generally considered, does not consist in anything that is *hard* or *heavy* or *colored* or that touches our senses in any other manner, but only in a *substance extended in length, width and depth.*" Other properties, like hardness or weight—so Descartes claims—are not only non-essential; they may be absent without depriving a body of "what makes it a body." Descartes defends this identification of "the nature of a body" with "extension in space" by arguments on condensation and rarefaction. Bodies are like sponges: "when we see a sponge

[19] R. Descartes, *Epistolae* (Amsterdam, 1668), Vol. 2, p. 277.

swollen with water . . . we do not understand that each part of it has, for this reason, a major extension, but only that the pores or interstices between its parts have become larger than when it was dry."[20] In this scheme of things, a vacuum is a contradiction in terms, and Descartes says as much:

> As far as a *vacuum* is concerned, in the meaning that philosophers give the word, that is, as *space in which there is absolutely no substance*, it is manifest that its existence is impossible. In fact, a *pure space* cannot be found in the universe, because the extension of space does not differ from the extension of a body and because of the fact that a body is substance depends only on its property of extension . . . and because it is absurd to believe that the extension is nil, we must conclude that space, by the fact that it has an extension even if we suppose it empty, is necessarily supplied with substance.[21]

This argument, at first sight, looks remarkably unreliable. But this is largely because we assign different meanings to words like "substance," "extension" and "body," not because of the structure of the argument. The "subtle substance" which fills space naturally suggests the image of a fluid, as it did to Descartes and his followers. But this subtle substance is, in itself, only the difference between the extension and the pure nil. And it is not unreasonable, in the sphere of philosophical realism, to state that such a difference belongs to the realm of material reality.

Descartes dedicates several sections of his work (sections 54–64) to his attempt to explain the resistance of solids on the level of philosophical principles. First he defines a solid by contrasting it to a fluid: "a *fluid body* is one that is divided into many particles which move separately . . . ; a *solid* is one whose parts are in mutual contact in the resting state, without any tendency to separate." This puts us at the heart of the Cartesian argument:

> I do not think that you can imagine a glue which is more suitable to attach the parts of hard bodies together than their own *stasis*. What else could it be? Not *substance*, because given that the parts themselves are *substance*, why should they be united by virtue of another *substance* rather than by themselves? And neither will it be a different *manner* of *stasis*, because there is no *manner* more contrary to motion that could separate these parts than the *stasis* which is in them. And, on the other hand, we know no other kinds of things besides *substances* and *manners*.

---

[20] R. Descartes, *Principia philosophiae* (Amsterdam, 1644), Part 2, sections 4–7.

[21] *Ibid.*, section 16.

FIGURE 6.2. Descartes' figure for his metaphysical explanation of the resistance of solids. From the *Principia Philosophia* (Amsterdam, 1644).

This is a splendid fossil, a monument to a way of "making science" that had been used for centuries. This is what it means to use "logic" instead of "geometry," in Galileo's terms.

Descartes is determined to apply his notion of stasis to the resistance of solids. Given a body B, immersed in a fluid whose particles are all in motion (Figure 6.2), he imagines complex circuits for the particles, both when B is moving and when it is at rest. This recalls the Aristotelian concept of the universe as a *plenum*, where every action is transmitted by contact, and where motion exists "in the transport of a body which is moved by the closeness of other bodies that touch it" (section 62). For this reason, the rules for impact, which Descartes had already established, take on an essential role, rather like that of the axioms of a deductive system. Unfortunately, Descartes' rules sometimes contradict experience and, for this reason, his explanations are contorted. For example, in section 63 he examines the following question:

> If it is true that the parts of a solid are not held together by any "glue" but rather by *stasis* in which each part is in a state of rest relative to the others ... if it is, moreover, true that a slowly moved body always has enough force to move another smaller one from a state of rest, according to Rule Five [of impact], we could justly ask ourselves how it is possible that, for instance, we do not succeed in breaking a nail with only the force of our hands, or any other little piece of iron still smaller than the hands themselves.

The reason, according to Descartes, is that the surface of contact between hand and nail is small, and "it can be separated more easily from the rest of the hand than a part of the nail from the rest of the nail."

Perhaps Descartes realized that his physical explanations were inadequate. But his speculative pride verged on arrogance; his principles *must* be sufficient to explain all the phenomena of nature, because they are absolutely rational. And if experience should contradict them? Then experience is wrong. "The demonstrations of all this are so certain that, even if experience should show us the contrary, we would nevertheless be obliged to give credit to our reason more than to our senses."

FIGURE 6.3. A Batavian drop. From: Thomas Aquinas a Nativitate, *Institutiones philosophicae*, Vol. III (Venice, 1762).

6.4 The Atomist Rossetti and His Explanation of Resistance

We can abandon Descartes to his *a priori* certainties. The problem of the resistance of solids itself resisted the speculative efforts of one of the greatest intelligences in the history of philosophy. The physical reason for resistance remained enigmatic. The *horror vacui* had been thoroughly discredited, and Galileo's elaborate hypothesis of minute interstitial vacuums had to be thoroughly reviewed.

To complicate matters even further, around the middle of the seventeenth century a strange object—almost a toy—came to the attention of scientists. This came from Batavia and consisted of small drops "of glass or crystal of any paste or color, white, red, turquoise, or yellow" (Figure 6.3). They were made by dropping molten glass into cold or tepid water, taking care that the drops came out with a bulb and a thin tail, sometimes very long. On being removed from the water, some of these shattered at once; "others blow up shortly afterwards, others endure for some hours or some days, and others last for months." Francesco Redi, from whom these words are taken, tells of receiving a hoard "of those wrought at Hamburg" from the Grand Duke of Tuscany. One drop promptly exploded. But the most interesting experiment consisted of breaking one of the drops near its end; then the whole drop shattered into pieces, some very small, some larger. The size of the pieces, Redi found, depended on the temperature and composition of the liquid in which the glass had been submerged. He tested wine, oil, vinegar, salted and sugared water, muddy water mixed with sand, and water saturated with nitrate, alum, vitriol, melted wax, and so forth.[22]

What could cause such curious behavior? Of course, the seventeenth century knew too little to arrive at an adequate explanation of such a complex

[22] F. Redi, "Osservazioni ... intorno a quelle gocciole o fili di vetro, che rotte in qualsisia parte tutte quante si stritolano," in Rossetti, *op. cit.*

phenomenon. In this case, just as in others considered by Galileo, experimentation acquired an unusual role: rather than confirming hypotheses or establishing theories, it became the source of questions that animated speculation. Some of the explanations spilled over into fantasy. The hypotheses offered were completely erroneous, but they were fruitful. A scientist thought he had answered a question; his answer was wrong for *that* question but correct for one that had yet to be posed. His hypothesis was incorrect, but it led to unrecognized truths which would be confirmed by future experiments.

This is what happened in the case of the Batavian drops. The scientist in question was Donato Rossetti, a canon of Leghorn and professor of the University of Pisa. He was one of those "learned men" whom the Grand Duke Ferdinand II of Tuscany consulted "when experiments were to be made or those already made were to be discussed." It was at one such discussion that the Grand Duke presented some drops "of a certain greenish glass" from Hamburg. The astonishment of the audience was extreme; "there was nobody who, at first, would risk, even when ordered by his Grace, to suggest any reason for so strange and wonderful an effect, each one preferring to reflect on the matter and give his opinion later."

Rossetti immediately set out to use the principles which he thought to be "the foundations of all natural things"—that is, the principles of atomism. He had been working enthusiastically for years, giving free rein to his imprudence. This is evinced by a long quarrel with the Modenese astronomer and physician Geminiano Montanari about the causes of the phenomenon of capillary action.[23] Rossetti immediately convinced himself that consideration of the atomic structure of bodies would lead to a perfect clarification. But this was a dangerous course to follow: "There are some," he wrote,

---

[23] *See* G. Montanari, *Pensieri fisico-matematici sopra alcune esperienze fatte in Bologna, etc. Intorno diversi effetti de' liquidi in cannucce di vetro e altri vasi* (Bologna, 1667); G. Montanari, *Prostasi fisico-matematica circa gli equilibrj, e dispareri per essi insorti tra il Dott. Geminiano Montanari e il Dott. Rossetti etc.* (Bologna, 1669); D. Rossetti, *Antignome fisico-matematiche con il nuovo orbe e sistema terrestre ...* (Livorno, 1667); D. Rossetti, *Dimostrazione fisico-matematica delle sette proposizioni che promesse D. R.* (Florence, 1668); D. Rossetti, *Lettera del Dott. D.R. al Sig. Carlo Fracassati*; D. Rossetti, *Riposta del Dott. D.R. alle opposizioni del Signor Geminiano Montanari ...* (Florence, 1668); D. Rossetti, *Insegnamenti fisico-matematici ... ad Ottavio Finetti, scolare del Dott. Geminiano Montanari ... sopra la prostasi che quegli stampò per questi* (Livorno, 1669). On the subject of Batavian drops G. Montanari gave some contribution in two letters to the Grand Duke Ferdinand II (April 22, 1670) and to Girolamo Savorgnano (December 31, 1670) (*See* G. Montanari, *Speculazioni Fisiche* (Bologna, 1671). Montagnani resumed the previous atomistic explanations suggested by R. Boyle, T. Hobbes and other authors. On the same subject, *see* also A. Marchetti, *Lettera nella quale si ricerca donde avvenga che alcune perette di vetro ... tutti si stritolino, al Granduca di Toscana Ferdinando II* (Florence, 1677).

> not knowing what atoms are, and not having the capacity to recognize them even if they were as visible as pilasters and towers, and not knowing what the atomists say nowadays, and in particular those who work at Pisa, where people think first of all of satisfying piety and religion, who are diffusing the idea that, without their chimerical principles and their master's [Aristotle's] metaphysics, it is not possible to defend some mysteries of faith. Therefore the doctrine of atoms, insofar as it is totally contrary to those principles and to that metaphysic, must be abhorred and shunned by the faithful.

These are the same problems as Galileo had encountered. But Rossetti made up his mind to go ahead and to try to be a Catholic atomist. He declared at the outset that if he saw any danger to faith, he would renounce atoms "even if they could be distinguished one by one with the naked eye and with the hand." He had, he thought, completed a consistent "physico-mathematical work," fully defensible from the objections of the ignorant, and that he could insert his explanation of the Batavian drops into a broader, more convincing system. This work, *Composizione e passioni de' vetri* ("The composition and properties of glasses,") was published in Leghorn by Giovan Vincenzo Bonfigli in 1671.

Rossetti's central idea was that atoms have "poles" which are the centers of fields of attractive or repulsive forces. He calls himself a "polist," not an atomist. Poles can explain anything, he believes.

> If you do not take the poles of atoms and of molecules for granted, at least admit their existence as possible, and you will accordingly find that all properties of bodies are demonstratively explained through them, including also those which have not yet been explained by any philosopher. Let us mention but a few: the major and minor resistance of solids; the properties of a spring and of a bow; the ease and readiness with which water is made from air and the air from water; and finally how and why all the compounds and all the mixtures vary. Am I then to be blamed if I claim to be called a *polist*, and if I declare that with the poles I will develop a physico-mathematical *corpus*, unless you have an argument or experiment which demonstrates them to be impossible or false?

As we can see, the instance of the Batavian drops is only secondary; what Rossetti wants is to start from the "natural principle of things" (chapter 1) in order to illustrate successively "the poles of atoms and molecules" (chapter 2), the properties of attraction and repulsion of atoms (chapter 3), those of molecules in themselves (chapter 4) and their aggregates (chapter 5). Finally, chapter 6 gets to the heart of the matter: what happens "when the molecules and their series suffer violence"? This is precisely the question of the resistance of bodies, dealt with in a new explanatory key. The following chapters (there are twenty-four in all) concern the enigma of the Batavian drops themselves, trying to interpret Redi's experimental results. These interpretations are highly fanciful, and their "confirmation" by Redi's data is the weakest part of the treatment.

Previously, in his *Insegnamenti fisico-matematici* ("Physico-mathematical teachings") (1669), written to counter Montanari's tenets, Rossetti had divided atoms into two species, the dark (*tenebrosi*) and the bright (*lucidi*). He did this in order to explain the phenomenon of light, which he gave an atomistic interpretation. In *Composizione e passioni de' vetri* he illustrates this distinction more clearly: the "bright atoms" are continuously "thrown off by force from the solar globe so that they give us all the light and fire we have." The "dark atoms" are, instead

> those which, in their various types and properties, are destined to form our earthly mass, and with the bright or without the bright, with their different adherences, unions, connections and interlacings, to form all that we are used to call physical matter, as far as it is becoming generated, corrupted, altered, and in some manner becoming different from what it was before.[24]

What distinguishes Rossetti's atomism from that of the ancients was the reason for the union and disunion of atoms. The ancient philosophers believed that atoms moved at random in the immensity of empty space. A *parvum clinamen*—a sudden, arbitrary deviation of motion—was responsible for their collisions, unions and aggregations. In contrast, Rossetti believed that atoms are governed by their reciprocal "natural appetence," and "that this appetence is greater and more vehement in some atoms and less and more moderate in others, according to their greater or less energy." In particular, he believed in a general law: the dark crave the bright, the bright abhor the dark, and among the dark, some crave and others abhor. Thus Rossetti substitutes, in place of the ancient causality, an "essential principle," a "natural necessity" which recreates the cosmic order. The author is explicit: his vision of the world is just as consistent, organic and orderly as the Aristotelian one. The interatomic laws of "appetence" and "abhorrence" are "precisely in the way that a Peripatetic would say it, that earth and water crave for the center of the world because they are, by their nature, directed there, and that fire and air abhor the same center because naturally they move away from it."

Rossetti had anticipated some of the definitions of these laws of attraction and repulsion in the *Antignome* (1667), but he now sets about formulating mathematically structured hypotheses. First, he establishes that appetence and abhorrence manifest themselves only within a specific interatomic distance, so that we can define a sphere of energy. Furthermore, "the nearer

[24] Rossetti, *Composizione*, p. 1.

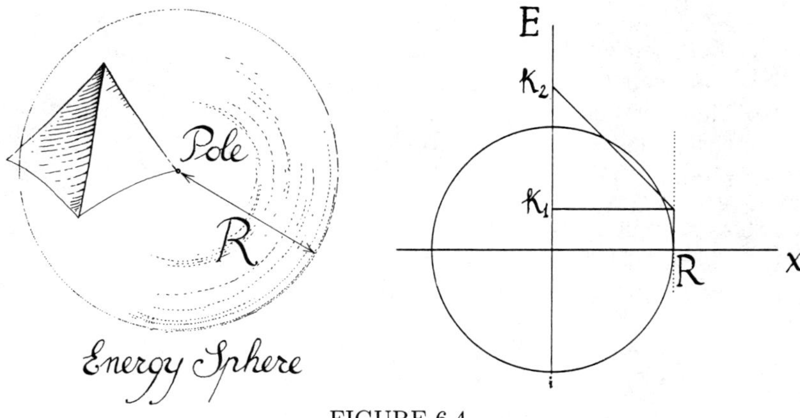

FIGURE 6.4.

the appetent atom is to appetence and the abhorring to abhorrence, that is, the nearer it is to the center of the sphere, the greater the energy."

One problem remains: to define the law of variation of the energy as a function of the distance from the center. But a direct experiment, Rossetti says, is impossible; all we can do is to choose the simplest and most convenient solution. Notice the perfect methodological consciousness of his statement:

> And now, as I think that it is not possible to confirm experimentally by what proportion such energy will increase, I think it can be said that this happens with the simplest and best-known proportion; and therefore, that the maximal energy is in the center of the sphere; the minimal at the ends of the diameter; and that the energies in the middle, which we shall call medium ones, are to be taken in their differences with the proportion of the distances." (Figure 6.4)[25]

In other words, Rossetti proposes the energy/distance law; given E = energy, R = radius of the energy sphere, x = the distance from the center, and K_1, K_2 = minimal and maximal energy, respectively, then

$$E = \begin{cases} (K_1 - K_2)\dfrac{x}{R} + K_2 & \text{if } x \leq R, \\ 0 & \text{if } x > R. \end{cases}$$

In this way, the empty space of the old atomists is "filled" with energy, or, more precisely, with action at a distance which the atoms exercise among themselves within their respective energy spheres. Having established the general law, Rossetti now examines its particular expressions for the bright and dark atoms.

---

[25] *Ibid.*, p. 2.

> I say that the dark atoms with each of their parts and with each of their extremities have appetence for the bright, so that they are going to join them at any point and always at the one to which they are nearest. And that the abhorring atoms with [all] their parts and extremities will flee from each part and extremity of those abhorred.[26]

Upon this principle Rossetti intends to base his atomistic interpretation of the phenomena of light and heat (fire). The discussion of the dark atoms is better developed, and here the concept of polarity comes into play:

> The dark among themselves have certain determined points, or poles, like the terms used for a magnet, by which means they join together. In fact, if, because of appetence, an atom should find another one with a blind impetus, and if they should join together without any rule, it is certain that we would see every day the continuous creation of new things and the cessation of others because of a lack of their individual parts, since for any given compound the same components and the same construction of them and the same order are necessary. And how can it happen, e.g. in plants, that there are always the same roots, always the same trunks with the same skin and bark ...? And how, in conclusion, in all these things, and in any other, [is there] the same distribution of the parts, and the same structure of the parts, if nature did not always order and compose them with the same hand and with the same art? ... From here it must result that, for the atom A with appetence for B, a pole C is the one that has to join a determined pole D of the atom B for which A has appetence and, therefore, the straight line CD that connects the said poles is the line atom A must describe approaching point C. And if atom B is considered to be immobile, this is just as if point D were the center of the earth, point C the center of gravity of body A, and the straight line CD the line of the barycenter and of the direction. [Figure 6.5][27]

This passage is astonishingly modern, not in tone, but in its anticipation of concepts of physical chemistry. It foreshadows the notion of valence, which assures the constant formation of the same chemical compounds. The metaphor at the end prefigures Bohr's and Rutherford's atomic model; they too saw the atom as a microcosm. Rossetti goes further in his speculative construction, imagining atoms supplied with one or more poles of attraction, and studying the possible aggregations in terms of equilibrium. This is how the molecules composed of dark atoms are created. Later,

[26] *Ibid.*, p. 4.
[27] *Ibid.*, pp. 4–5.

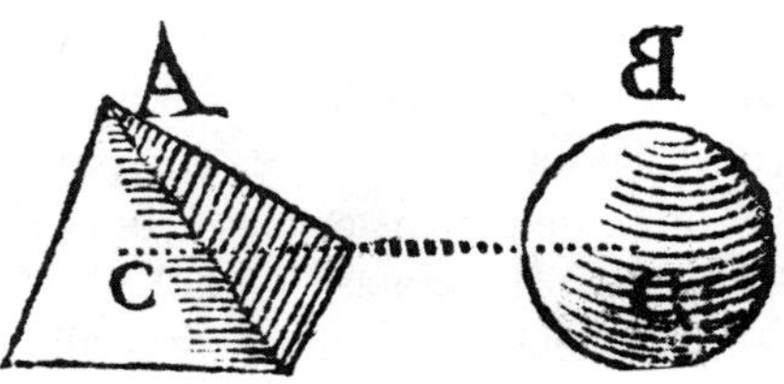

FIGURE 6.5. Rossetti's figure for explaining his concept of pole. From the *Composizione e passioni de' vetri...* (Leghorn, 1671).

these molecules come to include bright atoms, which are kept there "by force like a prisoner." Bright atoms "do not join, unless around the solar globe."[28] Molecules in turn can attract other molecules (shades of polymer chemistry!) and "they tie themselves in knots" in a more or less stable configuration, depending on the position of the poles.

> If the molecules have their poles outside, that is, in the surface common to the molecule as well as to the atom in which the pole is, one will adhere firmly to the other and will become attached to the same poles, and in the case of hindrance, at the minimum distance .... And, therefore, in the first case, when the poles of the molecules are outside, and become attached firmly together, it will be difficult to separate them from one another, and they will greatly resist separation, the more and more the appetence increases.[29]

We have reached the central point: the cause of resistance of solids is about to be clarified. It depends on the force of attraction that keeps the poles of the molecules together. It is maximal when the poles directly touch one another and minimal when they are unconnected and remain at the minimum distance. The law of variation of energy relative to distance thus allows us to explain why different materials are more or less resistant. This explanation is completed in chapter 6 ("When Molecules and their Series Suffer Violence"). The first case studied is that of

> spherical molecules that have their poles at the center, when as a consequence of an extrinsic cause, without being separated from one another and thrown out of the sphere of energy, they are moved from their site or position. When changed, they will not return to their primary order and constitution, since they will become attached at any point of contact, because the poles are always at the least possible distance.[30]

[28] *Ibid.*, p. 10.
[29] *Ibid.*, p. 13.
[30] *Ibid.*, p. 16.

Rossetti sees mercury as the best, and possibly the only, example of such a material. It has no resistance. The second case is that of spherical or near-spherical molecules whose poles are on the outer surface: "removed from the first adhesion by some extrinsic cause, they will not always return to it, but often make a new attachment; and this will happen when two poles different from the first ones meet at a minor distance."[31] According to Rossetti, this explains numerous phenomena: the ductility of gold and silver, the characteristics of viscous and plastic materials; the properties of glues and pastes.

The third case deserves particular mention. It concerns spherical and non-spherical molecules possessing a non-central pole: "the molecules, being changed in such a way that the poles do not come out of the energy sphere, will return to the first attachment, because between the points not corresponding to the center and to the surface of the sphere, there is a single line which is the shortest."[32] This is the elastic behavior of bows and springs. Rossetti gives us the law:

> If, when separating the appetent poles, it happens that some abhorrent atoms approach, and if the maximal energies of appetence are equal to the abhorrent energies, the same force is required for every distance, and from every distance the molecules do return to the former attachment with the same energy. But less force suffices for a major distance, if the maximum energy of appetence is greater than that of abhorrence, and more and more force will be necessary, if the maximum energy of appetence is less than that of abhorrence.[33]

The text is rather obscure, but Rossetti makes a noteworthy attempt to correlate the external force with the difference of the internal "efforts" or energies which result from the deformation produced by the external force.

Chapter 7 examines another promising idea: the "spring force" with which a molecule resists deformation and returns to its resting state is proportional to the deformation itself.

> With a series of molecules ... it is possible to build up a small physical cylinder, which can stretch in accordance with the motion of the molecules placed along its length and axis, but opposing a greater resistance to a greater stretching than to a smaller one; and, when lengthened, it returns by itself as if it were a spring to its natural length; and it ... can be bent or curved, but it opposes greater resistance to a greater curvature than to a smaller; and by itself, as if by the force of a spring or

---

[31] *Ibid.*, p. 17.
[32] *Ibid.*, p. 18.
[33] *Ibid.*

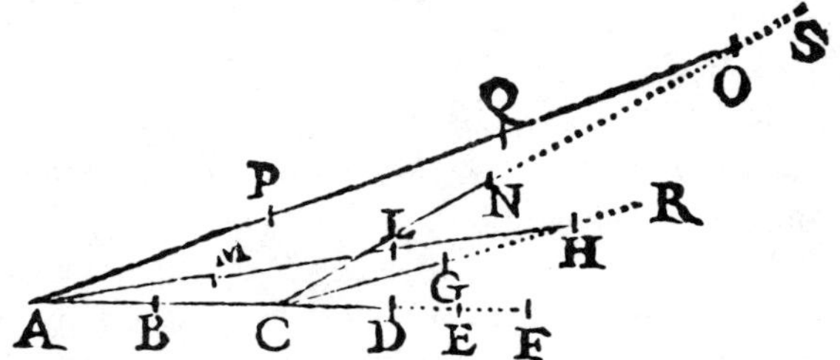

FIGURE 6.6. Rossetti's figure for his molecular explanation of elasticity. From the *Composizione e passioni de' vetri...* (Leghorn, 1671).

> bow, it returns to its former straight shape. If [there are] some little cylinders AB, BC, CD, attached to each other at B and C [Figure 6.6], they have more resistance to their maximum stretching and lengthening of eight degrees, with eight degrees of resistance, and to the stretching of seven degrees, with seven degrees of resistance; and to that of six with six, etc.; and if they have a resistance of six degrees to being detached—which resistance we could call the force of the knot—and if knot C is fixed, the little cylinder CD will be stretched towards E; CE arrives at the sixth degree, and CE, that is, CG will not become detached from the knot by its stretching; but if it is stretched more towards F, that is, towards H, it will become detached.[34]

It is a pity that Rossetti's work received so little attention. Many of his ideas suggest important concepts in chemistry, physics and mechanics: valence, polarity, molecular interactions, the proportionality between deformation and reaction, the criterion of rupture. We attribute the elastic law to Robert Hooke and the criterion of fracture to Edmé Mariotte, but Rossetti should deserve some of the credit for both. On the other hand, a scientific discovery lies not only in the intuition of a relation, but also in the ability to communicate it in the most efficient and useful manner possible. At this latter, Rossetti—more quarrelsome than reasonable, more an inventor than an interpreter of experiments—failed completely.

6.5 Atomism and Vacuum: Newton, Leibniz and Clarke

We shall try, however belatedly, to compensate for history's neglect of Rossetti by at least attributing to him what follows. The atomistic concept of reality can be freed from the old notion of random, aimless motion of

[34] *Ibid.*, p. 21.

elementary particles (*parvum clinamen*; see Section 6.4, above) and can become instead a new vision, as coherent as those of the Aristotelians and a good deal more sensible. In this vision, the world is no longer arranged by a "natural" disposition of heavy bodies (earth and water) and light bodies (air and fire), such that any deviation causes violent perturbation. Instead the world exists in harmony, with a well-defined interplay of attractive and repulsive forces which, from a distance, govern all matter and natural phenomena.

The merit of familiarizing science with the notion of attraction at a distance belongs, of course, to Sir Isaac Newton. He more than any other was to be the reference for later scientists interested in the atomistic explanation for elasticity and the strength of materials. But Newton, unlike Rossetti, fully understood the implications of the concept of attraction and the philosophical questions inherent in it. He explicitly limits his approach in the famous *Quaestio* 23 of his *Optice*:

> Here I do not seek which the efficient cause of these Attractions is. What I call *Attraction* is perhaps caused by some *impulse*, or by *something else* unknown to us. I use the word *Attraction* in general, to denote the force with which the bodies tend one towards the other, whatever the cause of this force may be. As a matter of fact, it is necessary that we learn from the phenomena of Nature *what bodies* attract each other, and *what* are the *laws* and the *properties* of such an Attraction, before it is convenient to search for the *efficient cause* of the Attraction.[35]

And, a little later:

> I consider these Principles, not as *occult Qualities* which are supposed to originate from the *specific Forms* of things, but as *universal laws* of Nature, according to which the things themselves have been formed .... To sustain that every distinct species of things contains *specific occult Qualities*, by means of which things have certain active forces, to sustain—I say—such a doctrine is not to say anything.[36]

The emphases in the preceding passages are not Newton's, but Samuel Clarke's. Clarke, one of Newton's followers, added them in his fifth reply to Leibniz, the last (because of Leibniz's death) in a long interchange of essays. Among the many subjects discussed are atoms, the vacuum, and attraction at a distance. In this dialogue, Clarke comes off as the modern scientist, while Leibniz often gives way to vague metaphysical reasonings.

---

[35] I. Newton, *Optice: sive de reflexionibus, refractionibus, inflexionibus et coloribus lucis*, (London, 1706), book 3, p. 322.
[36] *Ibid.*, p. 344.

Leibniz's criticism of atomism starts in the *Apostille* (Notes) of his "Fourth Writing":

> All those who are in favor of the vacuum let themselves be led more by imagination than by reason. When I was young, I also believed in the vacuum and atoms, but reason had freed me of it .... We should like nature to be as limited as our spirit, but this means refusing to recognize the greatness and the majesty of the Author of things. The smallest of the corpuscles is actually infinitely subdivisible, and contains a world of new creatures, of which the universe would be deprived if that corpuscle were an atom.[37]

These arguments are weak, and can be summarized as follows: there is no sufficient reason to admit the existence of imperfection in nature. A vacuum, as devoid of matter, is imperfect. So is the atom, since it is indivisible, and to be indivisible is to be imperfect. His problems with the notion of attraction at a distance are more serious; how can such an attraction differ from magic or a miracle?

Clarke concedes this last point:

> It is true [that] if a body attracts another without the intervention of any means, that is not so much a miracle as a contradiction; indeed this is equivalent to supposing that something acts where it is not. But the means by which two bodies attract each other can be *invisible* and *intangible*, and of a different nature from that of bodies considered by mechanics.[38]

Gravitation can therefore be produced by "*regular* and *natural*, though not *mechanical* forces."[39] As to Leibniz's criticism of the vacuum and atoms, Clarke's opinion is clear. If one denies the existence of atoms and accepts the notion of infinite divisibility, it follows that bodies consist only of "pores and therefore should disappear completely. (Note that Clarke's argument is just as fragile as the one it purports to refute.)

But the argument continues. Leibniz's new essay raises a series of much more consistent objections to the atomistic concept. In nature, he remarks, there cannot be absolutely indiscernable entities, because "if there were some, God and nature would act without any reason in treating one differently from another."[40] But atoms are defined as absolutely indiscernable entities and, as such, they cannot really exist. The case of monads

---

[37] *Recueil de diverses pièces, sur la philosophie, la réligion naturelle, l'histoire, les mathématiques, etc. Par Messieurs Leibniz, Clarke, Newton, et autres auteurs célèbres* (1719), 2d ed. (Amsterdam, 1740), Vol. 1, p. 61.

[38] *Ibid.*, p. 82.

[39] *Ibid.*

[40] *Ibid.*, p. 94.

is different: "According to my demonstrations, every portion of matter is actually divided into differently moving parts and none of them is entirely like any other."[41] Other remarks treat the vacuum: "Space is but an order of the existence of things. Thus the image of a finite material universe immersed entirely in an infinite space cannot be admitted." Here follow valuable reflections on the nature of space, on the theological incompatibility of Newton's ideas about it, on the dubiousness of the proofs for the existence of a vacuum derived from the experiments of Otto von Guericke of Magdeburg and Torricelli, and so forth. Finally, Leibniz attacks the forces of attraction over distance. These are, he claims, absurdities similar to "the *occult qualities* of the Scholastics which are presented to us disguised under the specious name of *forces*, but which bring us back to the Reign of Darkness. It is *inventa fruge, glandibus vesci* [eating acorns, having found fruit]."[42] Indeed, if an action at a distance is contradictory, what about attraction through a vacuum? Clarke claims "that this means of communication is invisible, intangible, unmechanical. One can add with equal right that it is inexplicable, unintelligible, precarious, without foundation, without examples."[43] Under these conditions, to speak of the forces of attraction as natural but unmechanical would be to speak of a perpetual miracle, for "the natural forces are all subject to mechanical laws, just as the spiritual forces are subject to moral laws."[44] To sum up: the vacuum, atoms, and attraction over distance are typical of "*Philosophes à notions incomplettes*"[45] or of "those mathematicians who are concerned with nothing but their imaginative games."

It is not easy to reply to such cutting criticism on the level of great rational principles. Kant is right to place this sort of argument about the nature of things in itself among the antinomies of pure reason. Nonetheless, Clarke's reply is important and, from a scientific point of view, almost resolves the question. Defending Newton's *Optics* and the *Scholium generale* of the *Principia*, Clarke claims the right of science to set aside research into the "cause which makes the bodies tend towards one another" and to study only "the laws or the proportions according to which the bodies tend towards each other, in the way they are discovered by experiment, whatever their cause may be."[46] Neither are these notions "*incomplettes*," unlike the old theory of *parvum clinamen*. The theory of the attraction of bodies is something quite different. It is simply an exact description of that which has been confirmed by experiment and which awaits further explanation according to the laws of mechanics. Epicurus' doctrine ("*la déclinaison des*

[41] *Ibid.*, p. 95.
[42] *Ibid.*, p. 147.
[43] *Ibid.*, p. 150.
[44] *Ibid.*, p. 151.
[45] *Ibid.*, p. 99.
[46] *Ibid.*, p. 206.

atomes") is a renunciation *a priori*, intended only to introduce atheism into an older, sounder theory: atomism.

Clarke's answer is excellent, and suits a proper scientific epistemology. Science goes step by step, proceeding by degrees, addressing specific issues one at a time without claiming to have answers to ultimate questions. Its intention is to lighten the mule's load (remembering Galileo's metaphor), not to leap to a grand, all-embracing synthesis. This is no retreat towards "*notions incomplettes*," but the only rational instrument to fill in the lacunae and overcome the antinomies which permeate great metaphysical concepts.

6.6 Newton's "vis interna attrahens": Elasticity and Resistance

In the queries added to the second edition of the *Optics*,[47] the *Quaestio* 23 mentioned above reappears, expanded and renumbered, as query 31. In it, Newton proposed an idea which Rossetti had already expressed and which later became the basis of various hypotheses on the physical causes of elasticity: elasticity depends on the properties of attraction of the atoms which form the bodies. Newton claims that this attraction is also responsible for most natural phenomena such as chemical reactions, fermentations and explosive combustion.

The subject of query 31 was taken up by Pieter van Musschenbroek some ten years later in *Physicae experimentales et geometricae ... dissertationes* (1729) in a chapter entitled "Introductio ad cohaerentiam corporum firmorum," which had great influence up to the end of the eighteenth century. After making a few historical allusions to the subject and discussing various theories, the author criticizes the concept of elasticity of Bacon and other "metaphysical" authors, and supports Newton's hypothesis on the existence of internal forces, *vires internae*, which act the same way in every body as gravitation acts between bodies. "Such internal force," van Musschenbroek says, "was introduced by God into all bodies, and the omnipotent Creator wished them to operate in themselves according to that force: therefore its presence is a Law of Nature, like the other one which is called gravity."[48] The origin of elasticity is therefore an internal attractive force (*vis interna attrahens*).

At a certain level of generality, a scientific explanation can degenerate into inconclusive cosmological hypotheses, like the old Aristotelian *Physics*.

---

[47] I. Newton, *Optics, or a Treatise of the Reflections, Refractions and Colour of Light*, 2d ed., London, 1717.

[48] P. Musschenbroek, *op. cit.*, p. 451.

During the eighteenth century, Aristotelianism still flourished in a retrogressive official scholastic culture. In this milieu, Fr. Mazière, a priest of the Oratorium, published his essay *Les loix du choc des corps à ressort parfait ou imparfait, déduites d'une explication probable de la cause physique du ressort* (Paris, 1727) which earned him the prize of the Académie Royale des Sciences. In it, he developed concepts traceable to Descartes' *Principia*, in which elasticity is attributed to an *external* "subtle matter," wedged into the pores of the material. Mazière accepted this notion with a difference: he thought the "subtle matter" was not external but was enclosed in the body like air in a sponge. The thesis is less interesting than the ingenious and bombastic demonstration which "proves" it, and which the Académie seems to have appreciated. "A physical cause," Mazière writes, "is not an intelligence because that can only be God or the first cause; on the other hand, it cannot be a solid body, *ergo* the cause of elasticity must be a fluid." To this mysterious fluid, produced on such an unsteady metaphysical basis, Mazière applies the Cartesian theory of vortices, arriving at the following grand conclusion: "The subtle matter is the physical cause of elasticity by the effect of the centrifugal force of its small vortices."

In fact, the first steps to free elasticity from the bonds of a philosophy of nature may well have been made by two Italian scientists, Jacopo Belgrado and Jacopo Riccati.[49] Their theories are not altogether satisfactory, but at least they tried to find a solution in the strictly mechanical field at the phenomenological level—that is, in the motion of bodies. As a matter of fact, both scientists (especially Riccati) resort to the laws of elastic impact, in which both the quantity of motion and kinetic energy, or "living force," are conserved. Such laws explain the *vis elastica* as a reversible exchange of "living" and "dead" forces, using Leibniz's terms. Riccati's new hypothesis states that a deformation is always produced by a "living force" and is proportional to it, and that the force used to produce the deformation remains in the deformed body as a "dead force"—that is, as potential energy.

From a methodological point of view, this hypothesis is important. The idea that the action of a force can be translated into a sequence of infinitesimal impacts was to be taken up by Lazare Carnot,[50] and was central in all mechanistic interpretations of physical concepts. Saint-Venant himself

---

[49] J. Belgrado, *De corporibus elasticis, disquisitio physico-mathematica* (Parma, 1748); J. Riccati, "Verae et germanae virium elasticarum leges ex phaenomenis demonstratae," *De Bononiensi scientarium academia commentarii*, Vol. 1(1747), p. 523; J. Riccati, "Sistema dell' Universo," in *Opere del Conte Jacopo Riccati* (Lucca, 1761), Vol. 1, pp. 152–173.

[50] L. Carnot, *Principes généraux de l'équilibre et du mouvement* (Paris, 1803).

proposed it in 1851 with the aim of basing mechanics exclusively on motion and on the laws of communication of motion.[51] Hertz, getting to the heart of the epistemological discussion which shook the entire structure of science at the end of the nineteenth century, changed the hypothesis into a principle which organized his radical reform of mechanical laws.[52] The idea that elasticity is closely connected to the reversibility of transformations and the conservation of mechanical energy was visionary. It preceded a view of energy which, from George Green on, was to dominate the mechanical theory of elasticity. We shall return to this point later.

6.7 Boscovich's Reformation of the Old Atomism

In 1758, Fr. Giuseppe Ruggiero Boscovich, a Jesuit, published his major work *Theoria philosophiae naturalis redacta ad unicam legem virium in natura existentium.* This work, together with a number of essays and disquisitions presented at the Collegio Romano[53] made a considerable contribution to Newton's (and Rossetti's) idea. Boscovich avoids any reference to atoms, and dwells on a law of continuity which coordinates and dominates all natural phenomena. This law was, naturally enough, perfectly suited to the Roman Catholic church, which always emphasized continuity, although in quite a different sense (cf. the Apostolic Succession). *Natura non facit saltus*: nature never takes leaps; therefore "in any quantitative mutation from one magnitude to another, the transition cannot take place without going through all the intermediate magnitudes."[54] This is true even when it seems to be false, for example when two objects collide. At impact, each undergoes a rapid change of acceleration. The law of continuity cannot be obeyed unless

> before contact one of the two bodies accelerates and the other gradually slows down. Therefore this deceleration and acceleration occur before and after the contact respectively, and their cause must be called force. Since such force acts equally in the

---

[51] A. Barré de Saint-Venant, *Principes de mécanique fondés sur la cinématique* (Paris, 1851).

[52] H. Hertz, *Die Prinzipien der Mechanik in neuem Zusammenhang dargestellt* (Leipzig, 1894).

[53] G.R. Boscovich, *De viribus vivis, dissertatio, etc.* (Rome, 1745); G.R. Boscovich, *De continuitatis lege et ejus consectariis pertinentibus ad prima materiae elementa, eorumque vires, dissertatio* (Rome, 1754); G.R. Boscovich, *De lege virium in natura existentium dissertatio* (Rome, 1755) (hereafter cited as Boscovich, *De lege virium*); G.R. Boscovich, "De materiae divisibilitate et principia corporum, dissertatio conscripta jam ab anno 1748, et nunc primum edita," *Memorie sopra la fisica* (Lucca, 1757), Vol. 4.

[54] Boscovich, *De lege virium*, p. 4.

> opposite direction, in order to attain the equality of action and reaction, it tends to separate one body from the other, and must be called repulsive force.[55]

If we accept this hypothesis, we can understand the reason for the physical principle governing the image of the world which Boscovich suggests:

> In nature there are forces that are sometimes attractive and sometimes repulsive; more exactly, they are attractive at great distances such as those separating planets and comets; they are either attractive or repulsive at smaller distances; and only repulsive at minimum distances; they are also more repulsive the smaller the distances, and so on to infinity, so as to be able to annul even an arbitrarily great velocity.[56]

The existence of these forces in nature crowds the world with material bodies, since the prime element of matter is that point in space which is the center of the repulsive/attractive force whose variation depends on distance.

> The prime elements of matter are constituted by totally simple and unextended points; no doubt, because of a repulsive force increasing to the infinite for infinitesimal distances, no particle of matter can be joined to another without an interval, so as to form a solid and continuous part; [therefore]... prime elements of this kind must be without parts.[57]

Note the considerable difference between this hypothesis and the atomism previously discussed. According to the latter, we must accept the existence of particles of solid matter, endowed with poles (*vide* Rossetti) and definite form. Such particles are separate, interacting over empty space by attraction and repulsion over distance. According to Boscovich, however, there is no distinction between empty and full space. The notion of particles, as impenetrable substance, disappears completely. Boscovich's "particles" are not the solid ones accepted by Newton and others,[58] but simple geometrical points or positions in space, exerting force depending on their position and distance from each other. The presence of matter at point X, defined on three axes (x_1, x_2, x_3), is solely represented by the fact that a given function of X and another point ξ tends towards infinity when ξ tends towards X. Such a function expresses the attractive/repulsive force and has two arguments: point X and distance $d(X, \xi)$. Suppose that the

---

[55] *Ibid.*, p. 12.
[56] *Ibid.*, pp. 11–12.
[57] *Ibid.*, p. 13.
[58] *Ibid.*

force is described as $F = F[X, d(X, \xi)]$. Then the phenomenological condition of impenetrability no longer requires the presence of an impenetrable substance at X, but only the mathematical hypothesis that, as $d(X, \xi) \to 0$

$$\lim F[X, d(X, \xi)] \to \infty,$$

giving a positive sign to the repulsive force.

Boscovich is fully aware of the consequences of his theory. It elimimates the contradiction between traditional atomism and the apparent continuity of matter, by getting rid of the reason for the contradiction—the distinction between full and empty, matter and force. His solution has elements in common with the one proposed by the Anglican Bishop George Berkeley, who had also "solved" the difficulties of empiricism by suggesting the elimination of matter—though with a very different meaning. The internal contradiction engendered by *dualism* (*esse* and *percipi* in Berkeley, *matter* and *force* in Boscovich) has been disposed of in the simplest way, by denying that it exists. The distinction is artificial; thing and perception, matter and force are the same thing.

This solution is so daring as to be incredible. How can it possibly relate to the objects we perceive? Surely would any mass of matter tear itself apart if its particles repelled each other at short distances? Boscovich has the answer:

> This difficulty, which comes immediately to mind, was answered when we gave our verdict. In fact, we declare that by virtue of the condition imposed upon forces, according to which repulsion at shorter distances becomes attraction at longer ones, cohesion and the different types of body cohesion are perfectly well explained. And we reach this conclusion more effectively than by deducing it from the hypothesis of infinite attraction for infinitesimal distances. Let us consider two material points, placed at such a limit distance that, should they be forced to converge, they immediately repel each other, due to the repulsion acting with immediate force; and should they be forced to separate, the attractive force, acting at the greater distance, draws them together again. The chief property of cohesion, consisting of preserving the existing configuration, thus takes place. This happens if the two points are not separated by a stronger force than the one by which they attract each other, or if one of the two points is not acted upon by such a motion so fast that the relative forces cannot act in so short a time. In any case, this theory can explain various types of cohesion for the different nature of the conditions imposed upon the forces, and can clarify the differences between solid and fluid bodies, between elastic and soft bodies, and several other things of the kind, as we may find abundantly demonstrated in many places

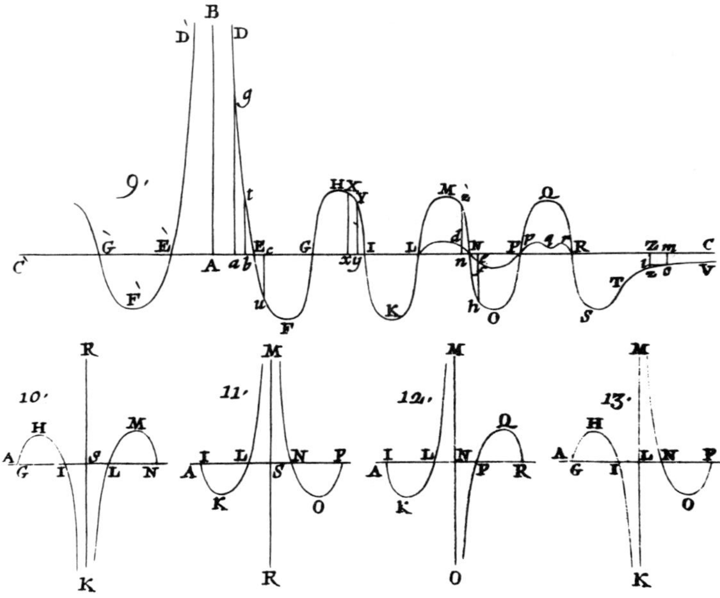

FIGURE 6.7. Boscovich's general law of natural forces. From *De lege virium...* (Rome, 1755).

> in the dissertations about them, and above all in the *Synopsis physicae generalis* mentioned above.[59]

The *Synopsis physicae generalis* was the work of Carlo Benvenuti, the Leghorn scientist who presented it to the Seminario Romano in 1754. It surveys the behavior of materials in terms of the variation of attractive and repulsive forces with distance. Several scholars were to use similar arguments, sometimes in the same spirit as Boscovich and Benvenuti, sometimes with the aim of finding a link between molecular and astronomical attraction, sometimes to prove that the two types of force had different natures, and so forth, right up to the mid-nineteenth century.

Boscovich conducts a similar survey at the end of the essay we are examining. But first he discusses at length the best algebraic form for the law of force. He expresses the relationship as a ratio of two suitable polynomials which have as their arguments the square of the distance $d(X, \xi)$. The law is represented in general terms in Figure 6.7, diagram 9 (derived from Boscovich's original), with diagrams 12 and 13 showing possible variants. Having established the general case, he adds that

> three classes of limit distances emerge, and in each class two genera. The first genus of the limits of the first class is the one

[59] *Ibid.*, p. 15.

> in which, by increasing the distance, a repulsive force changes into an attractive one, as represented by the points E, I, N, of fig. 9; the second genus is the one in which an attractive force changes to a repulsive one, as represented by the points G, L, P. In the first genus of the second class, the attractive curve is annulled, as IKL, between I, L, in which case, both before and after the contact there are repulsive forces. In the second genus, the repulsive curve becomes annulled, such as LMN, between L, N, in which case both before and after the contact there are attractive forces. In the first genus of the third class, a repulsive force changes to an attractive one, such as in fig. 12, at N. In the second, a change takes place from an attractive force to a repulsive one, such as in fig. 13, at L.[60]

These six cases, Boscovich claims, cover all possibilities. As always, such a closed categorization means that reality must be trimmed and squeezed to fit the model. Interpretation takes precedence over explanation. Classification becomes the goal of interpretation. Experimental criteria must give way to analogy. The sweep of the generalization, the elegance of the synthesis become themselves proof of the system's validity. Boscovich's system, if not positively Aristotelian, is in the spirit of the Scholastics. It is utterly alien to the strict hypothetical and deductive procedure of Galileian science and, for that reason, it is barren.

6.8 Developments of Boscovich's Theory: Early Nineteenth-Century Research on Elasticity

Newton's interpretation of elasticity as a consequence of molecular action is simple and appealing. It promises the possibility of describing the behavior of matter by summing the influence on any particle P of all the nearby particles when, by external force, they are forced to change position relative to P. The force/distance law defined by Boscovich can not only explain the phenomenon of cohesion and resistance but can also provide a basis for a phenomenological description of elastic behavior. We have arrived at a shift in the original aim of the First Day of the *Discorsi*: to begin to describe the behavior of matter under stress. The field was to be one of the richest and most interesting areas of research during the nineteenth century. We shall not examine these developments, but we should take a quick look at those early works which, while still concerned with Newton's and Boscovich's ideas and still retaining Galileian overtones, look towards an entirely new horizon.

---

[60] *Ibid.*, p. 37.

Boscovich's ideas had a considerable following. Laplace used similar concepts in his study of capillary action,[61] while Poisson applied the theory to the analysis of elastic surfaces.[62] But only Navier succeeded in transferring the vague conjectures of his predecessors into a powerful analytic deduction, finally connecting molecular action to observed phenomena. His paper, presented to the Académie des Sciences on March 14, 1821, marks the beginning of the modern mathematical theory of elasticity.[63] A few years later, Cauchy resumed and amplified Navier's work, examining every expressive capacity of molecular theory.[64] We no longer use this formulation, at least in the theory of structures, because of its complexity and because it can lead to false conclusions. Cauchy himself pointed to a clearer and more effective (if somewhat abstract) method.

But Navier's approach had its advantages. It corresponded to a challenging "program of research" that some of the most outstanding nineteenth-century scientists, such as Poisson, Clapeyron, de Saint-Venant and partly Cauchy, set forth: to create a "Physical Mechanics" (*Mécanique Physique*) as a new science parallel to Lagrange's "Analytical Mechanics," according to which any particular assumption concerning the mechanical behavior of materials and structures was to be interpreted in terms of molecular action.[65] Unfortunately, this program turned out to be useless and misleading: some inferences of the molecular theory of elasticity conflicted with experimental results and gave rise to the controversy over elastic constants which divided science for more than fifty years.

According to Navier and Cauchy, elasticity originates when an extremely large number of material particles $P, Q, \ldots$, originally sited at $P_0, Q_0, \ldots$, move after deformation to $P_1, Q_1, \ldots$. The relative distance between them changes, generating internal reciprocal attractive and repulsive forces.

Let us examine the behavior of two molecules P, Q, which are displaced from P_0, P_0 to P_1, Q_1 (Figure 6.8). The coordinates of P_0 are (x, y, z), while those of P_1 are $(x + u_x, y + u_y, z + u_z)$, where the quantities u_x, u_y,

---

[61] P.S. Laplace, *Annales de chemie et de physique*, Vol. 12 (1819).

[62] S.D. Poisson, "Mémoire sur les surfaces élastiques," *Mémoires de l'Académie des Sciences de l'Institut National* (1814), pp. 167–225; S.D. Poisson, *Traité de mécanique* (Paris, 1811), Vol. 1.

[63] L. Navier, "Mémoire sur les lois de l'équilibre et du mouvement des solides élastiques," *Mémoires de l'Académie Royale des Sciences de l'Institut National* Vol. 7 (1827), pp. 375–393.

[64] A.L. Cauchy, "Sur l'équilibre et le mouvement d'un système de points matériels sollicités par des forces d'attraction ou de repulsion mutuelle," *Exercices de mathématiques* (Paris, 1828)(hereafter cited as Cauchy, "Sur l'équilibre") Vol. 3, pp. 188–212; A.L. Cauchy, "De la pression ou tension dans un système de points matériels," *Exercices de mathématiques* (Paris, 1828), Vol. 3, pp. 213–236.

[65] S.D. Poisson, "Mémoire sur l'equilibre et le mouvement des corps élastiques," *Mémoires de l'Académie des Sciences de l'Institut National* (1829), pp. 357–627; *see* p. 361.

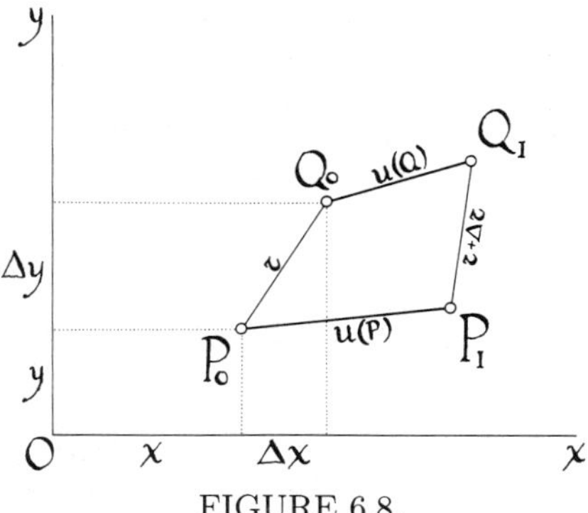

FIGURE 6.8.

u_z (assumed to be very small) define the vector $\boldsymbol{u} = (u_x, u_y, u_z)$, called the displacement of O. Then let $(x+\Delta x, y+\Delta y, z+\Delta z)$ be the coordinates of Q_0; those of Q_1 will be $(x+\Delta x+u_x+\Delta u_x, y+\Delta y+u_y+\Delta u_y, z+\Delta z+u_z+\Delta u_z)$ where Δu_x, Δu_y, Δu_z (even smaller than u_x, u_y, u_z) represent the variation $\Delta \boldsymbol{u} = (\Delta u_x, \Delta u_y, \Delta u_z)$ of the displacement of Q. Let r be the distance between P_0 and Q_0, and n_x, n_y, n_z the cosines of P_0Q_0. Then

$$\Delta x = rn_x, \qquad \Delta y = rn_y, \qquad \Delta z = rn_z. \tag{6.1}$$

The variation of the distance from P_0Q_0 to P_1Q_1 is

$$\Delta r = \Delta u_x n_x + \Delta u_y n_y + \Delta u_z n_z \tag{6.2}$$

and the direction of P_1Q_1 is given by

$$\frac{rn_x+\Delta u_x}{r+\Delta r}, \qquad \frac{rn_y+\Delta u_y}{r+\Delta r}, \qquad \frac{rn_z+\Delta u_z}{r+\Delta r}. \tag{6.3}$$

These are the "exact" expressions introduced by Cauchy.[66] Navier neglects the increments Δu_x, Δu_y, Δu_z in the numerator and increment Δr in the denominator, thereby identifying the direction of P_1Q_1 with that of P_0Q_0. We shall adopt this approximation. Furthermore, we shall assume that all the particles of the body have equal mass. At this point, according to the physical hypothesis, the displacement of the two particles gives rise to an attractive/repulsive force $\boldsymbol{R}^{(PQ)}$ in direction PQ, arising at P. This force is a function of the distance r and its change is Δr. Let the $f(r)$ be a function rapidly decreasing with r; then the X-component of $\boldsymbol{R}^{(PQ)}$ is

$$R_x^{(PQ)} = f(r)\Delta r \frac{rn_x+\Delta u_x}{r+\Delta r} \tag{6.4}$$

or, in Navier's approximation,

$$R_x^{(PQ)} = f(r)\Delta r n_x. \tag{6.5}$$

[66] Cauchy, "Sur l'équilibre," p. 190.

Furthermore, if we introduce the expressions

$$\Delta u_x = \frac{\partial u_x}{\partial x}\Delta x + \frac{\partial u_x}{\partial y}\Delta y + \frac{\partial u_x}{\partial z}\Delta z + \frac{1}{2}\left(\frac{\partial^2 u_x}{\partial x^2}\Delta x^2 + 2\frac{\partial^2 u_x}{\partial x \partial y}\Delta x \Delta y + \cdots + \frac{\partial^2 u_x}{\partial z^2}\Delta z^2\right) \tag{6.6}$$

and the analogous expressions for Δu_y, Δu_z, and if we recall equations (6.1) and (6.2), then equation (6.5) can be made more explicit:

$$R_x^{(PQ)} = rf(r)\left(\frac{\partial u_x}{\partial x}n_x^3 + \left(\frac{\partial u_x}{\partial y} + \frac{\partial u_y}{\partial x}\right)n_x^2 n_y + \cdots + \frac{\partial u_z}{\partial z}n_x n_z^2\right) + \frac{1}{2}r^2 f(r)\left(\frac{\partial^2 u_x}{\partial x^2}n_x^4 + 2\frac{\partial^2 u_x}{\partial x \partial y}n_x^3 n_y + \cdots + \frac{\partial^2 u_z}{\partial z^2}n_x n_z^3\right). \tag{6.7}$$

The same scheme can be repeated for any other pair of molecules. The resultant R_x acting at P is therefore given by the sum of these actions. Let us now consider the particular case in which, before deformation, the molecules around P are symmetrically distributed with respect to the three planes passing through P, planes that are parallel to the coordinate planes (x, y), (y, z) and (z, x). Given this hypothesis, we may state that the sum of the molecular actions on P will include only those terms of equation (6.7) that contain n_x, n_y, n_z raised to even powers. This means that R_x is represented by

$$2R_x = \frac{\partial^2 u_x}{\partial x^2}\left(S(r^2 f(r) n_x^4)\right) + \frac{\partial^2 u_x}{\partial y^2}\left(S(r^2 f(r) n_x^2 n_y^2)\right) + \frac{\partial^2 u_x}{\partial z^2}\left(S(r^2 f(r) n_x^2 n_z^2)\right) + \frac{\partial^2 u_x}{\partial x \partial y}\left(S(2r^2 f(r) n_x^2 n_y^2)\right) + \frac{\partial^2 u_x}{\partial x \partial z}\left(S(2r^2 f(r) n_x^2 n_z^2)\right) \tag{6.8}$$

where S stands for the sum extended to all the particles. Similar expressions apply to the other components R_y, R_z. By defining the "constants"

$$\mathcal{G}_{ij} = \mathcal{G}_{ji} = \frac{1}{2} S(r^2 f(r) n_i^2 n_j^2) \tag{6.9}$$

which depend on the nature of the material, and namely on the function $f(r)$, equation (6.8) and its analogues take the following form:

$$
\begin{aligned}
R_x &= \mathcal{G}_{xx}\frac{\partial^2 u_x}{\partial x^2} + \mathcal{G}_{xy}\frac{\partial^2 u_x}{\partial y^2} + \mathcal{G}_{xz}\frac{\partial^2 u_x}{\partial z^2} \\
&\quad + 2\mathcal{G}_{xy}\frac{\partial^2 u_y}{\partial x \partial y} + 2\mathcal{G}_{xz}\frac{\partial^2 u_z}{\partial x \partial z}, \\
R_y &= \mathcal{G}_{yx}\frac{\partial^2 u_y}{\partial x^2} + \mathcal{G}_{yy}\frac{\partial^2 u_y}{\partial y^2} + \mathcal{G}_{yz}\frac{\partial^2 u_y}{\partial z^2} \\
&\quad + 2\mathcal{G}_{yz}\frac{\partial^2 u_z}{\partial y \partial z} + 2\mathcal{G}_{yx}\frac{\partial^2 u_x}{\partial y \partial x}, \\
R_z &= \mathcal{G}_{zx}\frac{\partial^2 u_z}{\partial x^2} + \mathcal{G}_{zy}\frac{\partial^2 u_z}{\partial y^2} + \mathcal{G}_{zz}\frac{\partial^2 u_z}{\partial z^2} \\
&\quad + 2\mathcal{G}_{zx}\frac{\partial^2 u_x}{\partial z \partial x} + 2\mathcal{G}_{zy}\frac{\partial^2 u_y}{\partial z \partial y}.
\end{aligned}
\tag{6.10}
$$

In his memoir, Navier assumed that the sums appearing in equations (6.8) and (6.9) could be converted into integrals. But Cauchy and Poisson clearly showed that this hypothesis would lead to a paradox.[67] Later, Adhémar Jean-Claude Barré de Saint-Venant (a fervent and obstinate supporter of the molecular theory of elasticity) returned over and over to this subject throughout his long career. In a paper of 1844, he maintains that the impossibility of transforming the sums into integrals argues strongly in favor of Boscovich's theory, which he explicitly refers to, for it compels us to see atoms as unconnected, unextended points.[68] If, then, F_x, F_y, F_z denote the components of the external force per unit volume applied to particle P, the equilibrium equations at every point of the elastic body are

$$R_x + F_x = 0, \qquad R_y + F_y = 0, \qquad R_z + F_z = 0. \tag{6.11}$$

Now consider the simplest case, that of isotropy, in which the behavior of the material in no way depends on the orientation of axes x, y, z. In this case, the constants $\mathcal{G}_{xx}$, $\mathcal{G}_{yy}$, $\mathcal{G}_{zz}$ must be equal, and the same must hold true for distinct index constants. That is,

---

[67] *Ibid.*, pp. 203–205; S.D. Poisson, "Mémoire sur l'équilibre et le mouvement des corps élastiques," presented to the Académie Royale des Sciences, April 14, 1828, published in *Mémoires de l'Institut*, Vol. 8 (1829), pp. 357–570; see especially pp. 376–392.

[68] A. Barré de Saint-Venant, "Sur la question de savoir s'il existe des masses continues et sur la nature probable des dernières particules des corps," *Société philomatique de Paris* (1844), pp. 3–15. For a more detailed analysis of Saint-Venant's arguments in favor of Boscovich's theory, *see* E. Benvenuto, A. Becchi, "Sui principi di filosofia naturale che orientarono la ricerca di Saint-Venant," in *Omaggio a Giulio Ceradini* (Rome, 1988), pp. 125–138

$$\mathcal{G}_{xx} = \mathcal{G}_{yy} = \mathcal{G}_{zz}, \qquad \mathcal{G}_{xy} = \mathcal{G}_{yz} = \mathcal{G}_{zx}. \tag{6.12}$$

Next, we can demonstrate[69] that if $\mathcal{G}_{ij}$ is invariant upon a rotation of the reference, we necessarily have

$$\mathcal{G}_{xx} = 3\mathcal{G}_{xy}. \tag{6.13}$$

This means that, according to our model, all the constants in an isotropic body can be reduced to one, for instance $\mathcal{G}_{xy}$, which we will call $\mathcal{G}$. Then the local equilibrium equations (6.11) become

$$\mathcal{G}\left(3\frac{\partial^2 u_x}{\partial x^2} + \frac{\partial^2 u_x}{\partial y^2} + \frac{\partial^2 u_x}{\partial z^2} + 2\frac{\partial^2 u_y}{\partial x \partial y} + 2\frac{\partial^2 u_z}{\partial x \partial z}\right) + F_x = 0,$$

$$\mathcal{G}\left(\frac{\partial^2 u_y}{\partial x^2} + 3\frac{\partial^2 u_y}{\partial y^2} + \frac{\partial^2 u_y}{\partial z^2} + 2\frac{\partial^2 u_z}{\partial y \partial z} + 2\frac{\partial^2 u_x}{\partial y \partial x}\right) + F_y = 0,$$

$$\mathcal{G}\left(\frac{\partial^2 u_z}{\partial x^2} + \frac{\partial^2 u_z}{\partial y^2} + 3\frac{\partial^2 u_z}{\partial z^2} + 2\frac{\partial^2 u_x}{\partial z \partial x} + 2\frac{\partial^2 u_y}{\partial z \partial y}\right) + F_z = 0.$$

These are called Navier's equations.

Cauchy's two works on the same subject not only mostly confirm Navier's conclusions for the condition of isotropy, defined by a single constant, but also analyze what happens in the more general anisotropic case, "demonstrating" that there are at most fifteen constants. It was regarding these results that the controversy exploded. In further fundamental works, Cauchy took the mechanics of solids into a more formal and altogether more satisfying new direction. But the history of this interesting debate is, unfortunately, outside the scope of this book.

---

69 Cauchy, "Sur l'équilibre," p. 201.

7

The Initial Growth of Galileo's Problem

7.1 Introduction

In the preceding pages, we have tried to describe the development of the First Day of the *Discorsi* as a journey—albeit a digressive one—from a definite beginning to a less definite end. But the ideas presented in the Second Day had a less clear-cut fate. Their history is characterized by fragmentation; instead of a highway, we have dozens of footpaths, weaving all over the terrain, sometimes parallel, sometimes intersecting, never straightforward.

The topics presented in the Second Day vary, from the problem of solids of equal resistance, to questions concerning elasticity, deformation, the unstable equilibrium of an axially loaded beam, and so forth. Finally, there is the notable development of the methods of experimental analysis. We cannot set out a simple narrative of the evolution of these ideas because they are still the subject of lively debate.

All we can do is to look at how Galileo's immediate successors dealt with particular problems. A good many of our footpaths are side-tracks; others are dead ends. Instead of the great highway followed by such scientists as Leibniz, Bernoulli, Euler, Coulomb, Poisson, Saint-Venant and others we will explore the less well-known routes, for in this way we may learn more about the intellectual climate of the seventeenth and eighteenth centuries. But in terms of the intellectual climate of the seventeenth and eighteenth centuries, ours is no trivial exercise.[1]

As we know, the discussion of the resistance of solids jumps from *why* bodies resist fracture (First Day) to *how* their resistance varies under different conditions of constraint and load (Second Day). Galileo concentrates on the ways in which relative resistance changes for a given absolute resistance, not on the underlying physical reasons for the latter; these, he believes, do not affect the accuracy of a geometrical treatment. His analysis in the Second Day ignores causality in the interests of presenting a

[1] The reader can find an extensive account of the main arguments related to the "Galileo problem" in Clifford Truesdell's masterly book, *The Rational Mechanics of Flexible or Elastic Bodies: Introduction to Leonhardi Euleri opera omnia*, Vol. 11, 2d ser. of the critical edition of Euler's works (Basle, 1960) (hereafter cited as Truesdell, *Rational Mechanics*).

model whose advantage lies in simplicity and clarity rather than in phenomenological adequacy. Both the charm and the limitations of Galileo's whole approach are evinced by the very simple model to which the loaded beam is reduced. We know the difference between metal and wood and stone; we know that some materials can withstand very large deformations before breaking, and others that are almost undeformable until rupture. Some materials are resilient, some fragile. But the model ignores these differences. The cylinders and prisms of which Galileo speaks are very near the abstract solids of Euclidean geometry. To their form and dimensions he adds only the property of being able to support, within given limits, distributed or point loads. Any other criterion would be meaningless, just as it would be silly to specify the colour of the paint or the finish of the metal.

Galileo's model, then, is as simplified as possible. It can give rise only to the most elementary geometrical rules and the most obvious static laws. From one point of view, this is an advantage. Science has to progress by isolating the least number of meaningful parameters from the muddle of reality; only by doing this can we discover the most general, simple and certain laws. But what right do we have to do this? At what point does simplification become over-simplification to the point of distortion? The answer to this question is always difficult.

Science must not only discover the unifying rules; it also has to account for the differences. These objectives seem contradictory, and it takes enormous effort to find the harmony between them. But sometimes the history of science zigzags between the general and the specific—first reducing reality to such a level of abstraction that it ceases to be realistic and then trying to cope with all the anomalies. For example, Galileo reduces the cantilever from a real beam of wood, stone or metal, to an "ideal" cylinder or prism, to a lever. From this reduction he derives general laws, such as the fundamental rule establishing the equality of a resisting moment with an active moment. But then his successors build the model up, one cautious step at a time, until it looks rather more like a real beam again. The language used in description follows the same path; the vocabulary first narrows, and then expands and becomes richer and more descriptive. The initial linguistic reduction gives us terms and analogies that we carry over into the new, more elaborate model. For instance, Mariotte, Bernoulli and Leibniz substituted an elastic model for Galileo's rigid solid, but they still expressed the resistance of the beam at the joint with the wall in terms of force, and they carried out their calculations on the basis of Galileo's interpretation. The case has changed; the basic rules have not.

7.2 First Steps in the Controversy about Solids of Equal Resistance: Blondel's "Evidence"

Who were Galileo's first heirs? It is curiously difficult to decide; the first records indicate a heated quarrel among contenders for the honor. Whom shall we choose? Vincenzo Viviani? Alessandro Marchetti? François Blondel? What brings these three into conflict is not only the question of primacy, but, more importantly, the controversy on the subject of solids of equal resistance. This problem persisted throughout almost a century of quarrels and mutual accusations: "one of those scandalous conflicts," says P.S. Girard, "of which are unfortunately too many examples in the history of science."[2]

The specific object of the controversy was almost banal. Marchetti and Blondel (or perhaps Blondel and Marchetti) reached a conclusion different from Galileo's: the profile of a beam of equal resistance is not parabolic but elliptical. But what they thought was an error on Galileo's part was, in fact, just as correct as their new results. The two truths can coexist harmoniously if we accept two different definitions of "equal resistance." Having started with the divergence of the results, rather than with that of the hypotheses, the diatribe went on and on. This is common in philosophic and scientific disagreements; often the disputants carry on as though statement A necessarily invalidates statement B, when in fact the contradiction arises from a misunderstanding of the respective premisses.

The first debate provoked by Galileo's work on the resistance of structures concerned a typical architectural subject. How can we find new and unusual forms that, more effectively than those we inherit from tradition, can join beauty to rational perfection? This is the first point at which mechanics enters structural design. At the end of the eighteenth century, P.S. Girard, who concluded the long discussion, presented the optimization of structure as being the main aim of a science of building that conformed to the perfection and beauty of nature's works. For the moment, we shall confine ourselves to considering the events, from their beginnings. We shall hear the "evidence" of the authors who took part in the discussion; we shall call them one by one into the witness box, to let each tell his version of the truth, and we will let the reader form the verdict.

In the case in point, the heart of the question lay in the initial definition. According to Galileo and his supporters (e.g., Viviani and Grandi) a solid of equal resistance exists if, for a given load, every cross-section reaches the same level of maximum stress. For their opponents, on the other hand, such a solid is one that reaches the breaking point at whatever point a load of equal magnitude is placed along the axis of the beam. In other

[2] P.S. Girard, *Traité analytique de la résitance des solides et des solides d'égale résistance* (Paris, 1798), p. xx.

words, the Galileians dealt with the optimization of the structure for fixed loads, and their opponents instead regarded optimization for moving loads. Both treatments, obviously, are important in structural mechanics, though historically the first has been better developed.

Among those who claimed to be Galileo's heirs, Blondel has supporting testimony, especially that of Fr. Guido Grandi (a vituperatively prejudiced witness, we should note). François Blondel, architect, engineer and author, was born at Ribemont in Aisne in 1618. Most of his life was spent in travel. Between 1652 and 1655 he visited all Europe as the preceptor to the Count of Brienne. In 1657 he was invited on a diplomatic mission to Berlin and the year after to Constantinople. On his return to France in 1664, he was made responsible for restoration and engineering work in the city of Saintes. In 1668 he was elected to the Academy of Sciences. As a professor and (later) director of the Académie Royale d'Architecture, he set forth his views in a famous treatise *Cours d'architecture* (1675). He taught the Dauphin, and Louis XIV appointed him director of works of Paris, where he died in 1686.

One essay is of particular interest to us. The "*Quatrième problème résolu,*" in his long memoir "*Résolution des quatre principaux problèmes d'architecture,*"[3] contains his reflections on solids of equal resistance. He had apparently researched the topic, although when he had done so remains uncertain; he claims to have begun work on the subject in 1649. In a letter of 1661 he states

> perhaps there is nobody who has more esteem than I for all that comes from Galileo, among whose disciples I have had the honor of being one .... [I have] worked for so many years to extend this doctrine of the resistance of solids of which he is the inventor ... having composed on this subject the book that you saw ready for print more than twelve years ago, which I call Galilaeus Promotus de Resistentia Solidorum, and which, if some day it is published, will inform you of the respect I bear for this great man, whom our good friend Gassendi used to call the Plato of our century.[4]

Was *Galilaeus promotus* ever published? It seems unlikely. In the dispute between Guido Grandi and Alessandro Marchetti, the book keeps re-emerging like a jack-in-the-box. Grandi asserts its existence; Marchetti denies it. The matter was of some importance, since if the book existed, Marchetti may have been a plagiarist. Marchetti went so far as to doubt

---

[3] F. Blondel, "Résolution des quatre principaux problèmes d'architecture," in *Mémoires de l'Académie Royale des Sciences depuis 1666 jusqu'à 1699*, Vol. 5.

[4] F. Blondel, "Second discours, ou lettre au Sieur B., pour la résolution de ses doutes sur les propositions du premier discours," in "Résolution des quatre principaux problèmes d'architecture," *Mémoires de l'Académie Royale des Sciences depuis 1666 jusqu'à 1699*, Vol. 5 (1729), p. 529.

that the work existed even in Blondel's mind in 1649. In a long, pathetic letter to his Venetian patron, Bernardo Trevisano, in which Marchetti contests one by one Grandi's accusations, he writes:

> Not having published the said book, but only, as my antagonist admits, having claimed to have done so, Blondel may have desired to attribute to himself something he had not done .... Therefore, I feel much inclined to believe that either Blondel's said affirmation is a mere boast, or that, at most, besides those few propositions he printed about the matter ... he had found and demonstrated a few others of little importance which, when compared to mine, did not seem to him suitable for publication, and so he suppressed them spontaneously.[5]

Grandi strongly disagreed. As soon as he heard of Marchetti's letter to Trevisano, he set to work with a vengeance. He dedicated his reply to the Luccanese Vincenzo Nieri, his patron. It was published in 1712 under the title *Riposta apologetica del P. Maestro D. Guido Grandi Camaldolese ... alle opposizioni fattegli dal Signor Dottore A.M. nella sua dotta lettera diretta all' Eccellenza del Sig. B.T.* What, he demands, is Marchetti trying to propose?

> He wants to convince us that the book written as long ago as 1649 of which Blondel speaks in his second discourse of 1661 to Mr. Bout [*sic*; should be Buot] was *a mere boast.* I cannot imagine how such a suspicion could ever enter the mind of a sensible man. How could a mathematician so famous for so many celebrated works of architecture and of mechanics, as Blondel was, besides the other qualities he had of knowledge of languages, of antiquities, of *belles lettres*, ever be able to attribute to himself a non-existent work?[6]

Grandi is probably right. Marchetti's suspicion *was* rather rash, though so was Grandi's claim that Marchetti had plagiarized Blondel's discoveries.

According to Grandi's chronology, the contributions towards the theory of the resistance of solids, especially the question of solids of equal resistance, go as follows:

---

[5] A. Marchetti, *Lettera, nella quale si ribattono l'ingiuste accuse, date dal P.D.G.G. ... ad Alessandro Marchetti.Nella seconda edizione del suo libro "Della quadratura del cerchio, e dell' iperbola, ecc"....* (Lucca, 1711), p. 23.

[6] G. Grandi, *Riposta apologetica del P. Maestro D. Guido Grandi ... alle opposizioni fattegli dal Signor Dottore A.M. nella sua dotta lettera diretta all' Eccellenza del Sig. B.T.* (Lucca, 1712) (hereafter cited as Grandi, *Riposta*), p. 48. On the controversy between Marchetti and Grandi *see* also: G. Grandi, *Dialoghi ... circa la controversia eccitatagli contro dal Sig. Dottore Allessandro Marchetti ...* (Lucca,1712). These (unfinished) dialogues, however, do not concern our subject.

1649 Blondel composes the pamphlet *Galilaeus promotus de resistentia solidorum ....*

1657 On August 12, Blondel writes to Paolo Vulzio [Paul Würtz] in Sweden, the pamphlet in which he emends Galileo's mistake with this title: *F.B. epistola ad P.W. in qua famosa Galilaei propositio discutitur, circa naturam lineae, qua trabes secari debent, ut sint aequalis ubique resistentiae; et in qua lineam illam, non quidem parabolicam, ut ipse Galilaeus arbitratus est, sed ellipticam esse demonstratur, etc. Datum Farae Viromanduorum, pridie idus sextiles 1657.*

1659 Mr. M. [Marchetti] begins to work on his book, to which he initially wanted to give the title *Galilaeus ampliatus*, and which he later prints in 1669 with the title *De resistentia solidorum.*

1661 Blondel's letter written four years earlier ... is printed ... 1661. In the same year Blondel defends his pamphlet against a few objections made by Buot of the Académie Royale. This apology has as its title *Second discours, ou lettre au Sieur B. pour la résolution de ses doutes sur les propositions du premier discours.* The date of this second discourse is *À Paris le 18 Juillet 1661.* Whether it was printed or distributed as manuscript, I have not been able to find it; but it was published at least in 1673, after the first discourse which was reprinted in the same year.

1667 In *Antignome*, printed in this year by Donato Rossetti, mention is made of the work which Mr. M. [Marchetti] was preparing under the title *Galilaeus ampliatus.*

1668 The same work by Mr. M. under the same title is mentioned by Pietro Adriano Vandenbrecte, professor of humanities at Pisa, in his poem celebrating Mr. M.'s work on the comet.

1669 Marchetti publishes *De resistentia solidorum* in Florence after publishing his *Esercitazioni meccaniche* at Pisa in the same year.

1671 In the fifth Journal of Rome of May 29, there is a reference to the book *De resistentia* of Mr. M. by Francesco Nazari, who says that *twelve years earlier* he had demonstrated the propositions pertaining to the parabolic solid.

1673 Thanks to the Académie Royale, Blondel's pamphlet written in 1657 is reprinted ... *with the same date*, and with the second discussion written in 1661 in defense of the first; and together with other treatises by the same author and by other academicians, these essays are published in 1676.[7]

---

[7] *Ibid.*, pp. 46–47.

What meticulousness!

It would seem natural to think that the discoveries whose paternity aroused so much acrimony were of earth-shattering importance. But this was not so. Galileo's "error" resulted from a misunderstanding, and the proposed alternative solution is banal, the result of an elementary exercise. We can examine it by looking at Blondel's *Epistola ad P. W.* and his *Second discours*. In his letter to Würtz, part of a correspondence which has been lost, Blondel writes:

> You write to me that although you have carried out the cut of the beams according to Galileo's prescription, with a parabolic line according to the height, so that they should be of equal resistance everywhere, this has not answered your expectations at all .... Now, having thoroughly examined the question ... and as you are asking for my opinion on the matter, I affirm that I think—I do not hide it—that Galileo has made a mistake about this question, to the extent of believing that he could apply to beams supported on two sides those things that he had demonstrated with beams fixed on one side and jutting out freely on the other.[8]

The passage to which Blondel refers is the one we met earlier, in which Galileo points out that his fifteenth proposition applies to beams which "in big vessels" are used "in particular to support the decks." Blondel claims that Galileo's technical application is incorrect because transverse beams in ships are all supported at both ends, and these ends are strongly attached to the supports.[9] Galileo's error, according to Blondel, was to substitute a cantilever loaded at one end for a beam supported at both ends under different loading conditions. But Blondel's ideas on the loading conditions are themselves unclear. Let us look at the passage of the *Discorsi* preceding the fifteenth proposition, in which Sagredo discusses beams supported in two places, A and B:

> I am thinking that since the prism AB grows constantly stronger and more resistant to the pressure of its load at points which are more and more removed from the middle, we could in the case of large and heavy beams cut away a considerable part near the ends, which would notably lessen the weight, and which would be of no small advantage and utility in the rafters

---

[8] F. Blondel, *op. cit.*, p. 478. Leibniz called Würtz "an illustrious man" because of his military merits. He mentioned Würtz's experiments; carried out according to Galileo's hypotheses, these had not corresponded to his expectations. Würtz's experiments were known to Bülffinger; see Bülffinger, "De solidorum resistentia specimen," *Commentarii Academiae Petropolitanae* (1729), 2d ed. (Bologna, 1739), Vol. 4, pp. 164–181.

[9] F. Blondel, *op. cit.*, p. 479.

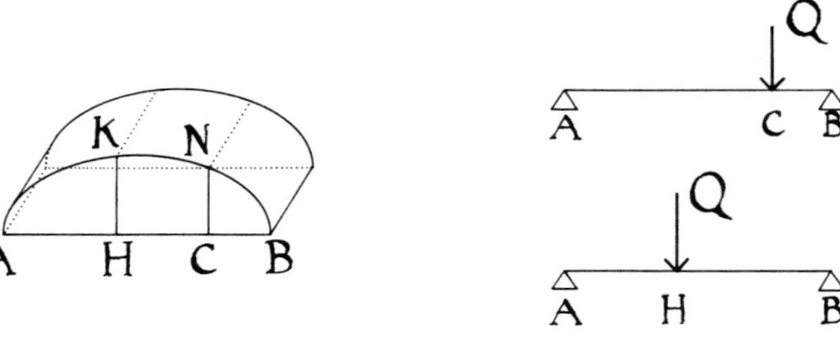

FIGURE 7.1.

> of great halls. It would be a fine thing if one could discover the proper shape to give a solid in order to make it equally resistant at every point, in which case a load placed at the middle would not produce fracture more easily than if it were applied at any other point.[10]

The last part of Sagredo's comment requires attention: the solid of equal resistance defined here does *not* correspond to the one which Galileo considered later. Sagredo's example is a beam, placed on two supports, which fractures when the same breaking weight is stationed anywhere along its axis. From this, Blondel derives the conclusion that Galileo was wrong, and he takes it upon himself to correct the master's error. His new proposition corresponds exactly to Sagredo's purpose.

If we accept the above definition of a solid of equal resistance and take the example of a beam on two supports, it follows that "what must be denied to a parabolic [or] ... hyperbolic profile ... can undoubtedly be allowed for a circular or elliptical profile; in fact, these profiles have all those properties that Galileo erroneously attributed to the parabola."[11] He reaches this result by observing that, for a solid with a circular or elliptical profile, the "moment of resistance" at C is to the moment of resistance at H both as $\overline{CN}^2$ is to $\overline{HK}^2$ and as $AC.CB$ is to $AH.HB$ (Figure 7.1). More explicitly, the bending moments in C and H resulting from load Q at C and H, respectively, are given by

$$M_C = Q\frac{AC.CB}{AB} \qquad M_H = Q\frac{AH.HB}{AB}.$$

On the other hand, the limit resisting moments in C and H are proportional to $\overline{CN}^2$ and $\overline{HK}^2$ respectively. The result follows from the geometry of a circular or elliptical profile. Of course, Blondel's results can also be derived from the Galileian definition of a solid of equal resistance, according to proposition 15 of the *Discorsi*, if we assume that beam AB is subject to

[10] Galileo, *Discorsi*, pp. 177–178.

[11] F. Blondel, *op. cit.*, p. 492.

a uniformly distributed load. In this case, we can demonstrate that equal resistance also obtains in a beam on two supports with a rectangular section of constant height and a base which varies according to a parabolic law.[12]

7.3 Marchetti's "Evidence" on Solids of Equal Resistance

Blondel's criticism—which enraged Galileo's pupils—can be reduced almost to a quibble: Galileo's approach, rather predictably, gives different results when applied to beams with two supports rather than cantilevers and to uniformly distributed loads rather than point loads. The criticism might be more significant if Blondel had specified what he meant by "load"; he maintained that the solid of equal resistance should be defined independently of the load condition. This conviction (partly justified in the case of a moving load) gave rather more significance to his results. Furthermore, his solution is elegant; he substituted one of the three curves of conic sections, the ellipse, for another, the parabola.

Probably this elegance did much to provoke Marchetti. He felt robbed. He claimed to have reached much the same conclusion at an early age (twenty-seven, according to him), and was probably banking on it to make his name in the scientific world. Instead, his discovery was pre-empted, and the results which were to be his triumph became instead a source of bitterness and disappointment which poisoned the rest of his life. We have heard Blondel's case; now let us listen to Marchetti's.

Alessandro Marchetti was born in Pontormo, near Empoli in Tuscany on March 17, 1633—that same spring that the Holy Office tried Galileo. He was a pupil of Alfonso Borelli, and later succeeded him at Pisa, where he taught philosophy and mathematics from 1660 until his death on September 6, 1714. He had literary as well as scientific interests, making an excellent translation of Anacreon (1707) and adapting Lucretius' *De rerum natura* into beautiful Italian verse. This latter was posthumously published in London in 1717. In 1669 he published two works, *Exercitationes mechanicae* and *De resistentia solidorum*. The former concerns, as we know, the law of the lever (see Part 1, section 2.4). The latter is of immediate interest in this discussion.

In his introduction, Marchetti mentions his change of projected title:

> You will be amazed, dear reader, because, in spite of what was promised in my name by two very famous gentlemen well known to me, Donato Rossetti ... and Pietro Adriano Vandenbroeke ... I have not in fact chosen the title *Galilaeus ampliatus*

[12] *Ibid.*, "Lettre à B.," p. 516.

> for my lucubration, but I have called it instead *De resistentia solidorum*.

About ten years ago, he writes, he was rereading the *Discorsi* when he came across Salviati's comments on the possibility of reducing the weight of beams in ships and buildings, and realized that "the application was incongruent to the solution." Exactly the same thought had occurred to Blondel. Did one of them learn of the problem from the other, or did each reach the same conclusion independently? However it happened, young Marchetti also found the same solution, the elliptical profile, and immediately wrote of his discovery to his teacher Borelli. "He approved my efforts, assented to the things I had begun to do, and exhorted me to advance with all my powers." But the task took time and care, he was distracted from it by the translation of Lucretius, and he was terrified of coming out in public with his criticism. One event in particular changed the course of his research; "while I was thinking of something more important and profound, to my help came the proposition that *the moments of the weights are in compound proportion to the weights and lengths*." He talked this over with his old schoolfriend Lorenzo Bellini. He and Bellini came up with different but related demonstrations, and both came to realize that everything could be deduced from this proposition.

> In fact, I saw, almost as through a cloud, that my results together with those of Galileo and many others, could, on this principle, be demonstrated much more briefly and easily, not to say more surely. I tackled the matter with a light heart, comforted by the praise and exhortations of Bellini, and in the end put all in order in the hoped-for manner. Because of this, not entirely unreasonably, I substituted the title *De resistentia solidorum* for *Galilaeus ampliatus*, as I had the right to do.

This is Marchetti's version of events, given in 1669. Note that the question of the elliptical solid is put to one side. Marchetti's objective is much more important: he wants to reformulate Galileo's treatment from the start, creating a new, more natural and powerful synthesis. He intends to replace Galileo's brilliant but sometimes fragmentary and fortuitous solutions with a uniform, straightforward method, deduced *a priori* from the proper definitions and hypotheses. If he did "borrow" Blondel's results, the sin is trivial, since it concerns only one proposition (book 2, proposition 39), in which Marchetti asserts that "elliptical and circular solids resting on two extremities are of equal resistance in every part"[13]

Why, then, should Guido Grandi accuse Marchetti of plagiarism 40 years later? In the second edition of Grandi's *Quadratura circuli, et hyperbolae*

---

[13] A. Marchetti, *De resistentia solidorum* (Florence, 1669) (hereafter Marchetti, *Resistentia*) p. 124.

per infinitas hyperbolas, parabolas geometricè exhibita... (Pisa, 1710), the author adds an incongruous preface which sings Blondel's praises and attacks Marchetti. When Marchetti complained in a letter to Bernardo Trevisano (Grandi "could have left out that long and completely unnecessary preface in which anybody who is not blind can see ... that he has no other intention than that of discrediting my poor labors"), Grandi responded by accusing Marchetti of stealing others' ideas. Why?

We will never know for certain. The motives are buried with the men. What Grandi may have had in mind was to deflate Marchetti's reputation in order to favor the work, not yet published, of a scientist dear to Grandi's heart, Vincenzo Viviani. Viviani had planned to reformulate Galileo's work for years, possibly since soon after the master's death; Grandi claims that Viviani's intentions went back to 1644. Marchetti's synthesis was in direct competition with Viviani's. This rivalry may have been a symptom of a violent schism among the post-Galileians. For example, a letter from Viviani to Carlo Dati mentions the suspicion that Borelli had pushed Marchetti into publishing his treatise prematurely, taking Viviani's ground out from under him, as a sort of pre-emptive strike. Borelli in fact claimed that Viviani had tried to steal credit for Borelli's mechanical explanation of the action of muscles (*De motu animalium*) by encouraging his own pupil, Schenone, to publish his treatise *Myologia* in 1667.

Cardinal Leopoldo de' Medici intervened to try to soothe wounded feelings. He made Marchetti promise to delay publication in order to let Viviani complete his work. This was a customary courtesy of the time, and scientists were expected not to transgress it. Galileo had done the same for Luca Valerio, and Viviani for Antonio Oliva. But Marchetti burned with impatience; he feared, as he wrote Carlo Dati in February 1668, "the manifest peril of being preceded by someone else and wasting ... seven years of almost continuous work." At least the fundamental "mechanical theorem" upon which the theory of *Resistentia solidorum* was based could be published without offending anyone. In fact, he gave his *Exercitationes mechanicae* to the printers in 1668. But Grandi, in his *Riposta apologetica* of 1712 insisted that this was a diabolical plot against Viviani: "The publication of that book was instrumental in diverting Viviani's attention, and to say the truth it was a well-studied diversion."[14] Whether Marchetti published with malice aforethought is, of course, impossible to determine. But as his letter to Blondel in February 1668 shows, Viviani did indeed slacken off, leaving the completion of his own work to the indefinite future.

---

[14] Grandi, *Riposta*, p. 83.

7.4 Marchetti's Axiomatic Approach to the Resistance of Solids

It is none of our business to decide if Marchetti were trying to undermine Viviani, as Grandi claimed. From the scientific (and more interesting) point of view, the publication of both the *Exercitationes mechanicae* and *De resistentia solidorum* in the same year is hardly coincidental. The axiomatic/deductive method which Marchetti used makes the first of these treatises almost a preface to the second, as though the *Exercitationes* were a preliminary exercise on a simpler conceptual model. As we mentioned above, in dealing with the law of the lever Marchetti desired principally to define axiomatically the relationship between three concepts, moment, weight and distance, demonstrating that the fundamental proposition on the moment derives from the relation among the three. In *De resistentia solidorum* the method is the same; here, however, the basic "lexicon" consists in four concepts: resistance, weight, the moment of resistance, and the moment of weight. The axioms (or "*Definitiones*") establish their mutual relations, while the relation between moment and weight is established in the first proposition, with an implicit reference to the more specific treatment in the *Exercitationes*.

Here are the definitions:

> I. A solid whose moment of weight exceeds its moment of resistance we call a solid of lesser resistance; II. The opposite is true for a solid of greater resistance; III. A solid whose moments are equal is called a solid of equal resistance; IV. The name "maximum weight" for a solid of greater resistance is given to the weight, whose moment added to the moment of the solid's own weight is equal to the moment of resistance of that solid; V. The words "cylinder," "cone," "prism," "pyramid" are expressed by the single word "solid" of any type; if no further distinction is added, it means that the solids are geometrically similar.[15]

To be honest, these definitions are not "sufficient" as they were in the *Exercitationes*. We have to read between the lines to understand some other hypotheses, which Marchetti thought so obvious that they could be taken for granted. For example, he omits from his list of axioms the assertion that resistance is proportional to the area of cross-section at the joint. This lapse may be explained by means of a curious reasoning that was to be made more explicit in Viviani's treatise. In fact, according to both Marchetti and Viviani, the limiting resistance of the fixed section can be represented by a weight that balances the breaking load. This weight can be pictured as

[15] Marchetti, *Resistentia*, p. 1.

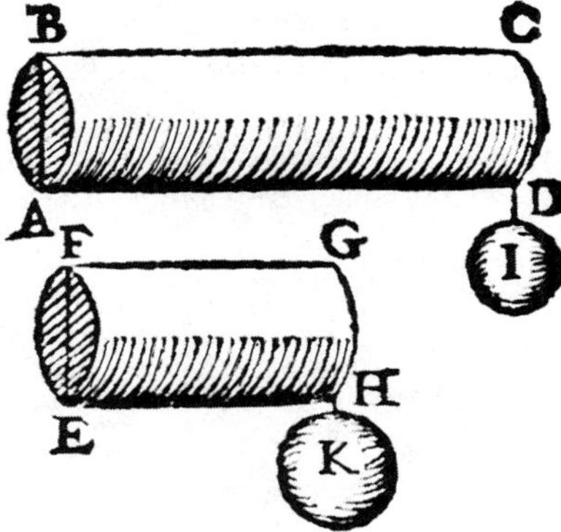

FIGURE 7.2. Figure used by Marchetti to state his basic propositions on the resistance of solids. From *De resistentia solidorum* (Florence, 1669).

a heavy plane surface (in the sense of Archimedes' *Equilibrium of Planes*), whose form and size are the same as those of the cross-section. In this way, the rupture stress takes the role of the specific gravity of the surface, and therefore the resistance is proportional to the area of this surface—that is, of the cross-section.

From the first three definitions, and the use to which Marchetti puts them, we can conclude that resistance R is proportional to the ratio between the moment of resistance M_R and the moment of weight M_Q. That is:

$$R \propto \frac{M_R}{M_Q}. \tag{7.1}$$

This result underlies the numerous and sometimes obscure propositions which make up the rest of the treatise. For example, proposition 2 affirms that "the moments of resistance of two similar solids are to each other as the cubes of the homologous sides; that is, as the cubes of the diameters of the bases."[16] The rather sketchy demonstration presupposes that resistance is measured from the surface of the base, and that the moment of resistance is therefore proportional to that surface and to a distance proportional to the diameter. Proposition 3 implicitly uses equation (7.1) to show that

$$R \propto \frac{\overline{AB}^3}{l} \tag{7.2}$$

(Figure 7.2), where AB is the diameter of the base and l the length of the solid (cf. definition 5). "The relationship between the resistances of similar solids of any kind is in compound proportion to the ratio between the cube of the diameter of the base and the inverse of the length."[17]

---

[16] *Ibid.*, p. 2.
[17] *Ibid.*, p. 3.

Many of the results given in the Second Day of the *Discorsi* derive from relations such as equations (7.1) or (7.2). The change of emphasis is reasonable, since Marchetti is more concerned with premisses than demonstrations. He frequently fails to keep his promise of greater simplicity and conciseness, as Grandi nastily pointed out. (Grandi actually counted the lines in Marchetti's demonstration of proposition 31 and found them to be twice the number in Galileo's.)[18] As often happens, Grandi's detestation was so great that he failed to recognize Marchetti's real contribution towards unifying the method to enrich the field of applications.

Marchetti's method generally amounts to determining the conditions that ensure the existence of a given ratio between moment of resistance and moment of weight for different forms of solids. To this end, he examines cylinders and prisms in book 1 (propositions 19–36) and parabolic conoids (37–81), parabolic solids (82–109), hyperbolic conoids and solids (110–113), hemispheres and hemispheroids (114–116) and finally semicircular and elliptic solids (117), all in terms of the cantilever. In book 2, he develops a similar articulation for the resistance of a beam with two supports. The material is so dense that inevitably errors creep in. Finally, Marchetti states as a corollary the results obtained by Blondel on solids of equal resistance.

7.5 Viviani's "Evidence"

We have one more witness: the most authoritative, the "last pupil of Galileo," the master's most loyal heir, Vincenzo Viviani. He was born in Florence in 1622, and went to study with Galileo in 1639, acting as auditor and amanuensis to the master until the latter's death. He went on to continue his studies with Torricelli. By then he had established a considerable reputation; Casimir of Poland and Louis XIV both invited him to accept high office, but he desired to remain in Florence. He was a member of the Accademia del Cimento, of the Royal Society (from 1696) and of the Académie Royale des Sciences (from 1699). He died in Florence in 1703.

Unfortunately, his version of events is inextricably mixed with that of Guido Grandi, whose fervent advocacy we have already encountered. Moreover, Viviani's treatment of the resistance of solids arrived on the scene rather late—seventy-six years after the master's death and fifteen years after his pupil's, to be precise. Grandi included it in the third volume (pages 195–305) of the works of Galileo, published in Florence in 1718, giving it the title "*Trattato delle resistenze principiato da Vincenzo Viviani per illustrare le opere di Galileo*," and adding his own comments. By 1718, progress in the field of resistance—especially the mathematical revolution

[18] Grandi, *Riposta*, p. 39.

of Leibniz and Newton—made it necessary for Grandi to try to revise Viviani's treatise in the light of new developments, just as Viviani had tried to modernize Galileo.

It is precisely this pattern of improvement (all done with an air of apologetic and defensive deference) that makes the transition from Galileo to Viviani and Viviani to Grandi so fascinating, even if we cannot assign exact dates to the advancement of an idea. There can be no doubt that Marchetti's synthesis at least partly determined Grandi's arrangement and interpretation of Viviani's material, even though Viviani may have composed his work first. The axiomatic/deductive structure of Marchetti's and Viviani's treatises is strikingly similar, although there are many variations and improvements in the latter. But while Marchetti forges ahead with only five definitions—as though all he had to do was establish his vocabulary—Viviani takes the time to present nine hypotheses which govern his system. Furthermore, he adds five "primary" and four "secondary" definitions.

The first three hypotheses concern equilibrium; the next five deal with the specifics of resistance.

> IV. Any resistance can be exceeded by a weight or a force or a moment which is greater than that resistance; V. ... the flexibility of bodies which yield must be disregarded since these can change the proportions examined; in the same way one must leave ... the tempering and hardening of metals out of consideration. In fact, the yielding qualities of solid materials alter the proportion of resistance to the point that the same iron will be now more, now less resistant according to differences in tempering and to its flexibility ...; therefore, if the absolute resistance of a section of a solid is given, in order to find the relative resistance ... it is more convenient to assume that the material does not yield, since the more it yields, the greater is the weight required to break it; VI. The diversity of the material alters resistance, since two equal solids or similar solids of different material, for example a glass one and a steel one, resist unequally; VII. If a solid splits crosswise, the fracture occurs in all points [of the fractured section] simultaneously, both at points far from or near the support and closer to the middle; moreover the fracture occurs with the regular motion of one of the two surfaces which moves in relation to the other surface, which remains still. VIII. Pulling directly on a solid, two resistances have to be considered; one, that of the attachment of the solid's filaments, which is as different as the diversity of the

> material, ... and the other, that of the vacuum, which is always the same for all materials.[19]

These, then, are Viviani's hypotheses. More correctly, they are Grandi's and Viviani's, since Grandi reconstructed them with considerable critical acumen. Hypothesis 8 has clearly been inserted in order to strengthen the case that Viviani had written his treatise before 1644, when the *horror vacui* had not yet been discredited. The others try to define Galileo's model solid, an almost Euclidean abstraction with finite resistance, and to establish this model's superiority to more realistic ones. Viviani defends Galileo's simplification by two arguments. First, eliminating such variables as flexibility allows us to derive a general theory using only one experimental datum, that of absolute resistance. Second, the model represents *minimum* resistance; deformable bodies can be expected to be more resistant. By opting for the lower limit of resistance, we can (at least partly) ignore the differences between materials. Galileo's initial outline of resistance, "geometrically demonstrated" without regard to the "imperfections of matter," is therefore confirmed, although Viviani and Grandi are perfectly aware of its limitations. In his commentary on hypothesis 7, Grandi refers to attempts by Mariotte, Leibniz, Varignon and Bernoulli to introduce elasticity into Galileo's model. He concludes that "these differences of opinion show ... how difficult it is to determine the true and natural hypothesis." The rule might vary with the material; it is therefore better, at least for the moment, to disregard the latter "to illustrate the material which we have at hand theoretically, as Galileo did together with our author [Viviani], leaving philosophers and the skilled observers of nature to take account of the differences."[20]

The ninth hypothesis is given as follows: "One can consider sections of solids as *heavy*, and, like Archimedes, imagine that planes have weight and then that these planes [are] placed on the lever's support ..."[21] The text is incomplete but the idea is clear; we have seen this notion in Marchetti's work, but here it is fully formulated. Just as Stevin had transformed a geometrical constraint into reactive force, so Viviani tries a similar reduction of the phenomenon of resistance. Although debatable, this attempt is highly imaginative. It justifies Galileo's hypothesis that resistance can be seen as force applied to the center of gravity of the section. Grandi's comment takes this position. But when Viviani's work was published, Galileo's hypothesis was not, in general, much trusted. For this reason, Grandi includes this hypothesis with the four secondary definitions. These distinguish the planes of "equal gravity *in species*" (uniform superficial specific weight) from planes

---

[19] V. Viviani, "Trattato delle resistenze ...." in G. Galilei, *Opere* (Florence, 1718), Vol. 3 (hereafter cited as Viviani, "Trattato"), pp. 201–202.

[20] *Ibid.*, p. 202.

[21] *Ibid.*

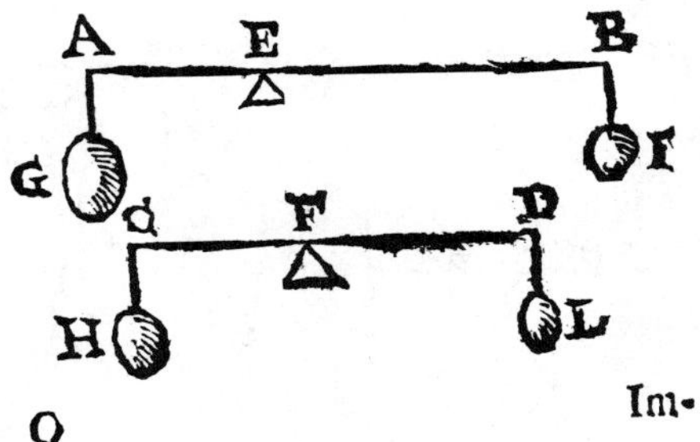

FIGURE 7.3. Viviani's figure for demonstrating his fundamental theorem 9. From the third volume of Galileo's *Opere* (Florence, 1718).

of "different gravity *in species*" (non-uniform superficial specific weight). "These different manners of gravity," Grandi writes, "according to our author, I think are equivalent to different resistances, either uniform or not, as defined above."[22]

From these premisses, Viviani works slowly and surely with precise geometrical steps. For example, in dealing with Galileo's first proposition, he prepares the ground with four theorems which examine the effect on the moment of resistance of variations in section area, base, height and diameter. But what most strongly characterizes Viviani's reformulation of the *Discorsi* is the central position given the law of the lever. In this, Marchetti and Viviani are at one. Marchetti had intended to mark the close connection between resistance and the law of the lever in the formal analogy between the *Exercitationes* and *De resistentia solidorum*. Viviani's *Trattato* states the relationship more specifically. For example, theorem 9 states: "if the resistances of two levers [i.e., the weights acting on two levers] are to each other as the lengths of the counterlevers, raised to the same power m, are to each other, and if the distances of the weights from the fulcrum are equal, then the counterweights are to each other as the lengths of the counterlevers raised to the power $m+1$ are to each other."[23]

If $AE = CF$ (Figure 7.3), and $I : L = \overline{EB}^m : \overline{FD}^m$, where m is an integer, then the law of the lever lets us obtain

$$I = G\frac{AE}{EB} \qquad L = H\frac{CF}{FD}$$

from which we derive

$$G : H = \overline{EB}^{m+1} : \overline{FD}^{m+1}$$

Q.E.D. In fact, *all* of Galileo's theorems can be derived from theorem 9 by using Viviani's hypotheses, especially the ninth. Marchetti tried, not

[22] *Ibid.*, p. 203.
[23] *Ibid.*, p. 209.

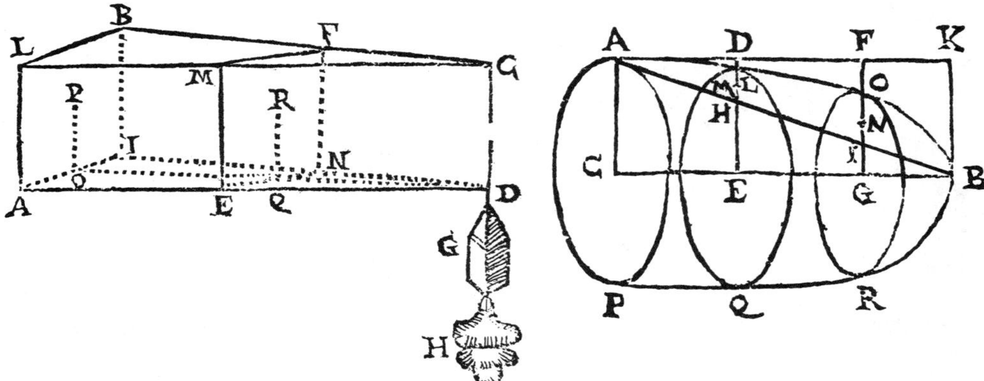

FIGURE 7.4. Viviani's solids of equal resistance: a cantilever subject to a load at its end. From the third volume of Galileo's *Opere* (Florence, 1718).

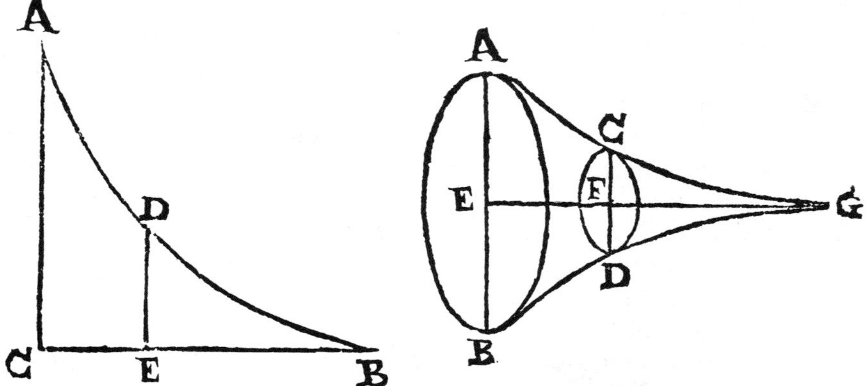

FIGURE 7.5. Viviani's solids of equal resistance: a cantilever subject to its own weight. From the third volume of Galileo's *Opere* (Florence, 1718).

altogether successfully, to make Galileo's work simple and concise. Viviani and Grandi succeeded where Marchetti failed.

Proposition 48 and its successors deal with solids of equal resistance. First they consider loads concentrated on the end of a cantilever (Figure 7.4), implicitly confirming Galileo's results. Then they treat the cantilever under its own weight (Figure 7.5), bringing new forms into being and clarifying the complicated question which troubled Blondel and Marchetti. By now we can no longer distinguish Viviani's work from Grandi's; the problems and intuitions of the former elicit answers and demonstrations from the latter. This is true of proposition 59, question 20, which examines the fifth solid of equal resistance: "To seek what that solid would be ... which held vertically has equal resistance in every section; that is, whose sections have the same relation to each other as the solids placed under them have to each other." The answer is: "Such is the solid formed by the logarithmic curve *AHB* [Figure 7.6] revolved around its asymptote *DO*."[24] Grandi quotes

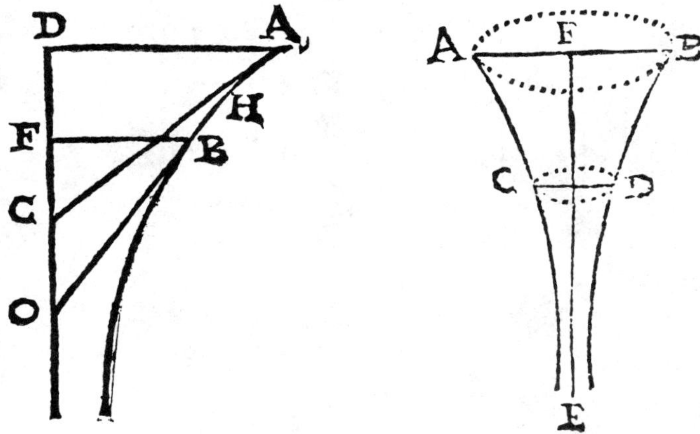

FIGURE 7.6. Viviani's and Grandi's contribution to the theory of solids of equal resistance: a heavy solid held vertically. From the third volume of Galileo's *Opere* (Florence, 1718).

his own theorems for the curve of the logarithm and its properties.[25] But Viviani does more than simply pose questions; in theorem 40 he demonstrates that "the plane figure revolving around its own axis, whose surfaces cut at different applicates are to each other as the same applicates, is not a figure of a proportional increase." By "figure of proportional increase" he means a figure like a triangle or parabola (Figure 7.7) in which area EAF is to area BAG as the area of the rectangle EG is to the area of rectangle BD.

The last part of the treatise considers the solids of equal resistance discussed by Blondel and Marchetti. It fully clarifies the misunderstanding of Galileo's work, distinguishing the two types of equal resistance discussed above and "rehabilitating" the master.

The Viviani–Grandi treatise is a masterly achievement in the old manner. The delay in its publication robbed it of some of its freshness but also allowed it to be perfected. In it, the authors conclude the Galileian project of establishing the science of resistance on an abstract level, leaving out any reference to the real behavior of material (*"le imperfezioni della materia"*) under stress. Viviani and Grandi exhaust the Galileian model; it cannot be taken any further than they take it.

But matter *is* imperfect, and we must consider its imperfections as well. These flaws may be worked into an exact mathematical treatment that improves Galileo's model and brings theory into better conformation with

---

[24] *Ibid.*, p. 245.

[25] G. Grandi, *Geometrica demonstratio theorematum Hugenianorum circa logisticam, seu logarithmicam lineam ....* (Florence, 1701).

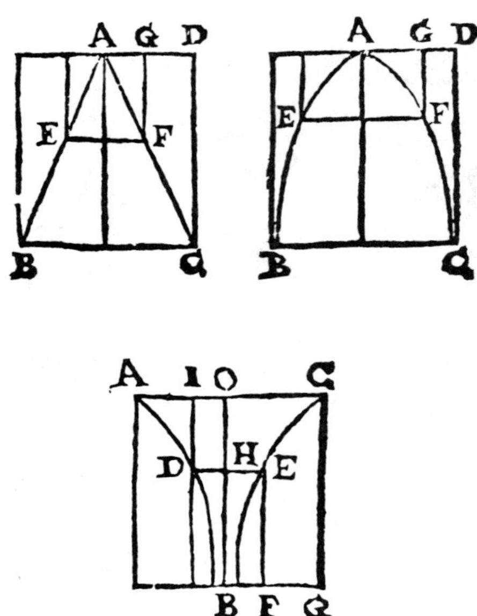

FIGURE 7.7. Viviani's drawings for explaining his concept of "figure of proportional (or non-proportional) increase." From the third volume of Galileo's *Opere* (Florence, 1718).

reality. The legacy of the Second Day does not end with the work of Viviani and Grandi; it also inspires new and stimulating strands of research which we will briefly inspect.

7.6 Antony Terill and Solids of Ultimate Dimensions

It may seem odd, given his problems with the Church, that Galileo's heirs included the Jesuits. But during and after the mid-seventeenth century, a number of scientific Jesuit schools accepted Galileo's resistance of solids with attention, often with enthusiasm. They were engaged in a general renewal of sciences along traditional Aristotelian guidelines, creating a new Scholasticism.

Two scientists in particular, Honoré Fabri and Ignace Gaston Pardies, were deeply concerned with the question of the resistance of solids, and offered ingenious contributions. Before we consider their work, we should remember that hostility towards Galileo was by no means completely extinguished. As we know, some Jesuits took offense at his distinction between geometry and logic—a distinction which effectively separated mechanics from the physical system of Aristotelianism.

Antony Terill tended to favor logic. Terill was born in Dorsetshire in 1623 but lived most of his life in Italy. He joined the Society of Jesus in 1647 and became a professor of philosophy and theology at Parma. Later he taught mathematics and theology at the English College in Liège. In 1660 he published a curious mathematical/philosophical treatise, his *Problema mathematico-philosophicum tripartitum, de termino magnitudinis, ac virium in animalibus*, which approaches resistance and dimensional limits in a philosophical frame of mind. In it, Terill goes to Simplicio's defence. Simplicio, he implies, could reach the same conclusions as Salviati, since Peripatetic philosophy lets him prove that geometrically similar bodies have different resistances.

To this end, Terill tries to link the resistance of solids to a general principle concerning nature's tendency to preserve continuity and contiguity between parts of a body. Cardano had introduced a similar concept, and G.B. Porta had used the notion to explain the action of siphoning. It was also employed by opponents of Torricelli's and Otto von Guericke's new views on the vacuum.[26]

Terill assumes that this natural tendency creates a force on continuity and contiguity (*virtus continuativa, vel contiguativa*) on which the strength of materials depends. Furthermore, he emphasizes the distinction between resistance to fracture and the "agglutinating force" or "adhesion" (*vis agglutinativa, adhaerescentia*) by which the particles of a body are held together. The former derives from the nature of continuity and depends on the size of the surfaces resulting from fracture, while the latter is a "specific continuity force" (*virtus continuativa in specie*) and depends only on the material. This dependence on surfaces is enough to establish *a priori* (that is, without any other mechanical proof) that the resistance of homogeneous similar bodies decreases with their increase in volume and mass. The weight responsible for fracture increases with the cube of the length, and the "continuity force" with its square. Therefore every natural heavy body has an intrinsic dimensional limit. Things can only get so big, no bigger.

We could perhaps see Terill's distinction between "specific continuity force" and resistance to fracture as an early version of the concept of tension as a contact force acting on each particle of a cross-section. But this is about all we can do with Terill's treatise. The rest of his work is of no particular importance.

---

[26] G. Cardano, *De subtilitate* (Lyon, 1580), p. 18; G.B. Porta, *Pneumaticorum ibri tres ....* (Naples, 1601), p. 10; P. Casati, *Vacuum prescriptum. Disputatio physica ....* (Genoa, 1649); G. Schott, *Mechanica hydraulica-pneumatica* (Würzburg, 1657).

7.7 Fabri: Elasticity as an "Intermediate Force"

Honoré Fabri's contribution is another matter. He published a monumental work in ten volumes, his *Physica, id est scientia rerum corporearum*, dedicated to Cardinal Leopoldo de' Medici, at Lyon from 1669 onwards. It is a comprehensive work, starting with elementary properties of bodies and ending with the corporeality of man and with philosophical anthropology. Our interest lies in book 2 of the first treatise, which concerns "tension and pressure," and in book 5 of the second treatise, in which Fabri presents an extended paraphrase of the *Discorsi*. Fabri had considerable respect for Galileo.

> Galileo, the famous mathematician, was the first and the last (as far as I know) who considered such resistance of bodies. With few, it is true, but with very accurate and therefore even more ingenious propositions, he brought the whole theory to an end. But as he published it only in Italian, which is rather little known to those who are dedicated to research, and as he omits many other things that pertain to the above-mentioned resistance, I thought it worth while to consider the resistance of bodies in this particular book.[27]

Fabri's work contains definitions, hypotheses and axioms, in which he hopes to demonstrate every proposition as accurately as possible. The vocabulary is clearly Aristotelian, with plenty of traditional distinctions—power *vs* act, substance *vs* accident, and so forth—but he means to adapt this vocabulary to the new science. This creates notable differences between his and the traditional Peripatetic treatment of physics. Some of his chapters have a strictly physico-mathematical character; for example, the first book, *De compresso et tenso*, of the first treatise. "In this book," Fabri says in the introduction,

> you will not find anything that does not pertain to physics. In fact, we are going to speak of the pressure and tension of bodies. We shall explain in what manner they manifest themselves. Furthermore, we will introduce a new motive power, a generator of tension and pressure whose really admirable effects we are going to explain. I say it is new, since nobody up to now has ever examined it, as far as I know, and in order that it should not lack a name, I will call it "intermediate force"[28]

The "intermediate force" he proposes is defined later as that force "because of which a compressed or taut body returns to the natural state of its

---

[27] H. Fabri, *Physica, id est scientia rerum corporearum* (Lyon, 1669) (hereafter cited as Fabri, *Physica*), Vol. 2, p. 514.

[28] *Ibid.*, Vol. 1, p. 42.

extension, or tends to return to it."[29] Between the composition of Fabri's work and its publication, the author comments, other scientists had begun to study this force, calling it an elastic one. But he prefers his term because it is more expressive. The intermediate (or elastic) force is halfway between an external force (*ab extrinseco*) such as gravity, and an internal one (*ab intrinseco*) such as the vital force of animals.

Fabri's book contains 298 propositions, in laborious and pedantic Latin, many of which are erroneous or futile. For this reason historians have largely ignored him, and for this reason we shall pass over his obscure line of reasoning. But embedded in this morass of material are results of some interest and importance. For example, proposition 56 affirms that "the same cord can be more or less taut, and between the tensions there is the same ratio as between the stretching forces." Proposition 59 adds that "the tensions of the same cord, whether more taut or less, are to each other as are the respective excesses of tension." The meaning of the different expressions can be gleaned from the context. The stretching force (*vis tendens*) is external—it might be a weight, for example. The tension (*tensio*), corresponding to the force of tension (*vis tensionis*), is the physical state produced by the stretching force and is proportional to it.[30] Finally, the "excess of tension" is a geometrical quantity. More precisely, it is the increase in the cord's length after stretching. Therefore propositions 56 and 59 affirm that elongation is proportional to the stretching force. But this, surely, is Hooke's law, *ut tensio sic vis*. True, but this does not discredit Hooke, who based his law on experiment. For both Hooke and Fabri, the proportionality between force and elongation was a universal property of bodies, reflecting the proportion which must exist between cause and effect. It is, in effect, more important as a measure of force than as a phenomenological law.

Fabri's very real insights are, unfortunately, rather weakly demonstrated; they are supported by analogies, not by real proofs. For example, in proposition 63 Fabri uses the analogy between the motion of a pendulum and the vibrations of a plucked string to affirm that

> from here we can deduce a consequence, more important than any other; and that is that all the reductions [i.e., reversions to original configuration] of the same taut cord are isochronous, that is, they are acted on at the same time; and this has never been demonstrated up to now, as far as I know.[31]

The author dwells at length on the subject of vibrating cords, presenting results which are occasionally correct. He takes into account such factors as tension and density; for instance, proposition 139 states that "the periods of vibration are proportional to the square root of the densities of matter."

---

[29] *Ibid.*, p. 43.
[30] *Ibid.*, pp. 61–63.
[31] *Ibid.*, p. 64.

But he fails to escape some serious traditional prejudices. In particular, he claims tension and pressure are analogous but essentially different phenomena; proposition 224 states that tension can be increased to infinity, but pressure cannot. In fact, if pressure is proportional to "decrement of extension," there must be a limit.[32]

This distinction between tension and pressure often gets Fabri into trouble, but the analogy has even worse results. If for no other reason, Fabri should be remembered for founding a long-standing error which trapped scientists from Mariotte to Leibniz to Jakob Bernoulli: "In the bending of a bow, there is some compression; however, from a physical point of view, all is the same as if there were only tension" (proposition 256).[33] The demonstration of this proposition is both obscure and inconsistent; essentially Fabri claims that, if we take the most-pressed fiber as a reference, all the others are taut in relation to it. His successors used much the same argument.

This proposition is followed by others that deal—imperfectly, but for the first time—with the question of the deformation of an inflected beam. He considers three cases of inflection. In the first, a "very long rod" bends of its own weight; in the second, a bow is fixed at its central point and bent by a cord tied to its ends; and the third concerns a beam supported at the ends with a point load on its middle. In the first case, he states, the rod curves in the same way as a heavy cord; like Galileo, Fabri believes that this curve is a parabola.[34] In the other two cases, the curve is not parabolic, and the question of its shape remains unresolved.

In book 5 of the second treatise, *De resistentia corporum*, Fabri "translates" Galileo's *Discorsi*. His adaptation resembles Viviani's, but it is less rich and well-developed. Like Viviani, Fabri aims to give Galileo's theories a deductive axiomatic form, using the law of the lever systematically. When he comes to solids of equal resistance, Fabri sees a "fault," as Blondel and Marchetti did, but the fault is a different one. Fabri demonstrates that a "prism having triangular bases parallel to the horizon" (proposition 72) is a solid of equal resistance. Viviani dealt with the same solid in his treatise, and both he and Fabri are surprised that Galileo went to the trouble of using a parabolic solid when there was such a simple and convenient alternative. In proposition 74, he considers a parabolic solid "complementary" to Galileo's (i.e., solid AML; see Figure 7.8) inserted in the mortise ML, stating (deceptively) that this solid is of equal resistance in relation to its own weight.[35] The treatise ends with some propositions about arches and vaults; we will consider these later.

---

[32] *Ibid.*, pp. 132–133.
[33] *Ibid.*, p. 164.
[34] Galileo, *Opere*, Vol. 8, pp. 369–370.
[35] Fabri, *Physica*, pp. 594–595.

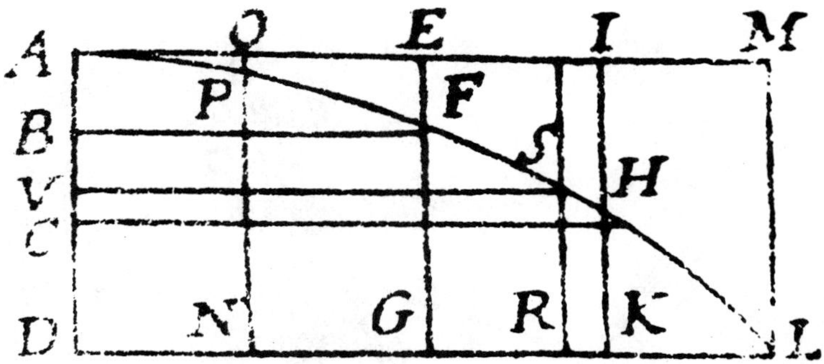

FIGURE 7.8. Fabri's figure for his propositions on Galileo's solid of equal resistance and its complementary. From the second volume of *Physica* (Lyon, 1669).

7.8 Pardies' Statics

Fabri's contribution is by no means negligible, but it has attracted surprisingly little attention. Fabri's friend and fellow Jesuit, Ignace Gaston Pardies, has been rather more fortunate. His works were reprinted several times in the early eighteenth century. Truesdell's historical essay discusses their merits in some detail. Interested readers should consult Truesdell; we shall limit our discussion to a few of the results which Pardies presented in his *De la statique, ou de la science des forces mouvantes* (1673).

The author is forthright about the dismal past, and clearly prefers his own era: "it has only been in recent times," he writes in his preface, "times so happy for those who desire to make new discoveries, that a few people have appeared who dedicated themselves to the cultivation of this science [mechanics], or rather who have made a new science of it."[36] He alludes not only to Galileo but also to Torricelli and Wallis. Pardies proposes to summarize this new science, "following the fine idea that Pappus has given us in order to gather together all that the different authors have found on the subject, adding to it what I myself will discover, if I am lucky."[37] Pardies intended *Statique* to be the first of six discourses which would complete his work. The short treatise, though not divided into formal parts, deals with three themes: statics in general (by which he means "the science of forces necessary to move bodies notwithstanding the resistance of contrary forces"),[38] the equilibrium of cords, and the resistance of solids.

The second part is the most original. In it, Pardies formulates a remarkable theorem on the catenary (Figure 7.9), asserting that "the continuous

---

[36] G.D. Pardies, *Œuvres*, new ed. (Lyon, 1725), p. 213.
[37] *Ibid.*, p. 216.
[38] *Ibid.*, pp. 226–227.

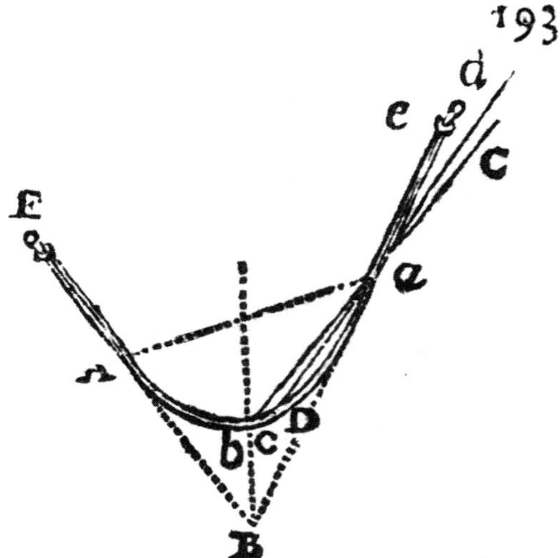

FIGURE 7.9. Pardies' theorem on catenary (1673). From Pardies' *Œuvres* (Lyon, 1725).

tangents cross in the prolonged line bB, so that by raising the vertical straight line from the common point B, we find the center of gravity of the cord Aba."[39] That is, whatever curve a heavy rope assumes, the point of intersection B of two lines tangent to A and a lies on the vertical line passing through the center of gravity of the part of the rope below cord Aa.

The truth of this proposition follows directly from the hypothesis that the tension in every point of the rope has the direction of the tangent at that point. Pardies' demonstration is rather more tortuous, and he makes use of the application of the "principle of solidification." He goes on to refute Galileo's thesis (adopted by Fabri) that the curve of a rope is parabolic. "As a matter of fact," Pardies writes,

> if there were a chain composed of very tiny rings (Figure 7.10) ... drawing the tangents at b (i.e., bD) and a (i.e., aD), these two tangents would intersect in D in the line of direction DC of chain aCb ... and the center of gravity of chain ab should be in C. Now, if figure aCb were parabolic, line DCE would divide aF into two equal parts; but the part aC of the parabola is longer than Cb; and it is quite easy to show that the center of gravity of parabola ab cannot be in C.[40]

Using the same argument, we can also see (Figure 7.10) that "if we conceive

[39] *Ibid.*, p. 280.
[40] *Ibid.*, p. 281.

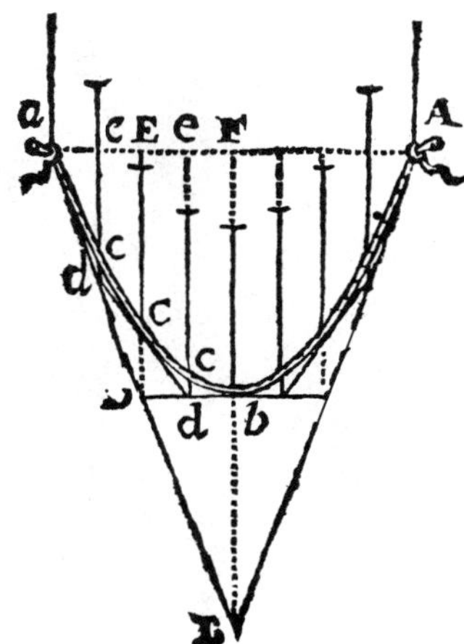

FIGURE 7.10. Pardies' drawing for the problem of the suspension bridge (1673). From Pardies' *Œuvres* (Lyon, 1725).

a thread without weight on which are resting an infinite number of equally heavy lines EC, ec, parallel and equidistant from each other, then thread $aCbA$ is perfectly parabolic." In fact, the center of gravity of the load acting on aCb lies on the vertical line DE which bisects aF. "Now, this is a property of the parabola, and the geometricians know that there does not exist another curve where these conditions are met."[41] He goes on to discuss cases of loads producing elliptical and hyperbolic curves.

Pardies next considers the subject of the resistance of solids. He professes the principle that "no body breaks except by traction."[42] The fracture caused by bending or inflection is due to the traction of the fibers in the taut part. This, he adds, is why an egg cannot be broken by squeezing the ends between one's fingers—the "problem of the egg," which had been discussed for centuries (Figure 7.11). For the egg to break, its "meridians" ACB would have to bend, but the consequent traction is hindered by the parallels CD. Pardies considers the example of a column of wood planks (Figure 7.12), but the reader might think of a cooper building a barrel with wooden staves, reinforced by iron hoops.

The discussion now turns to the study of a rod (*bâton*) which is pulled lengthwise or bent. Pardies introduces a rather odd but interesting model; he pictures the rod as a series of short cylinders on a cord, like a necklace composed of cylindrical beads (Figure 7.13). His explanation of the difference between the effects of traction and those of flection differs sharply from Galileo's. It is obvious, he claims, that the cord running through the

---

[41] *Ibid.*, p. 282.
[42] *Ibid.*, p. 288.

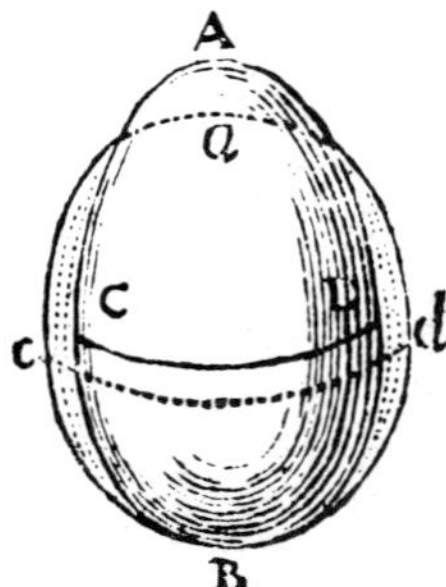

FIGURE 7.11. Pardies' explanation of the "problem of the egg" (1673). From Pardies' *Œuvres* (Lyon, 1725).

FIGURE 7.12. The example of a column of wood planks used by Pardies to maintain that no body breaks except by traction (1673). From Pardies' *Œuvres* (Lyon, 1725).

cylinders is "incomparably more taut when the rod is horizontal; and those who have considered this matter seem not to have reflected enough."[43] Pardies knows that his model is too schematic, using only one central cord through the cylinders when there should be many. We can imagine them distributed throughout, and notice that, because of bending, the upper cords are lengthened, while the lower ones are shortened. But Pardies makes the same mistake as Fabri, by assuming that this opposite behavior of the upper and the lower cords confirms his model: "what is gained with the upper cords that are getting farther from the supporting point e, is lost in the lower cords that get nearer the same supporting point"[44] and therefore we can get by with taking only the central cord.

---

[43] *Ibid.*, p. 292.
[44] *Ibid.*, p. 296.

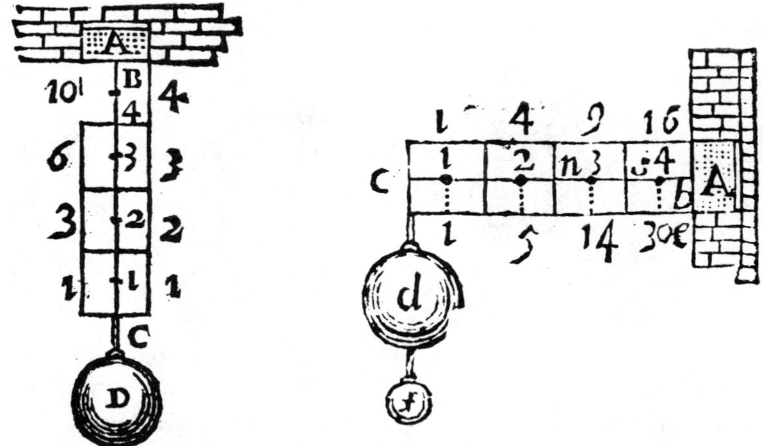

FIGURE 7.13. Pardies' model for explaining the behavior of a bent rod (1673). From Pardies' *Œuvres* (Lyon, 1725).

8

Early Theories of the Strength of Materials

8.1 Elasticity Enters the Theory of Resistance

The contributions of Fabri and Pardies have led us towards new points of view. Fabri's "intermediate force" opens up questions on the deformation of a beam, and points towards an explanation of resistance itself. Pardies's study of the equilibrium of ropes, and his peculiar model for a beam, a "string carrying beads," change the context in which Galileo's problem must be placed. For the first time, the concept of lengthening and shortening longitudinal fibers (the cords in the model) is connected to resistance.

We come close to one of the major highways in the history of mechanics, one taken by the great scientists at the end of the seventeenth and the beginning of the eighteenth century. It has been well mapped by historians of science, from Todhunter and Pearson to Timoshenko. We cannot follow their steps; we refer the reader to Truesdell's cited work for the details.[1] We shall limit our attention to a short chronological summary of the introduction of the concept of elasticity into the study of the resistance of solids.

As early as 1620, Isaac Beeckman had noticed that the fibers on the convex edge of a bent beam are taut, while those on the concave edge are pressed. But he failed to elaborate on this fact.[2] On the other hand, in a letter to Mersenne in 1630, Beeckman gives the first quantitative estimate of elastic deformation. He notes that if a weight is attached to a spring, the longer the spring is, the lower the weight descends.[3] Obvious as this may seem, it is an important insight because it gives rise to a new concept. It can be formalized as follows: if weight F is constant, then the lengthening Δl of the spring is proportional to the initial length l of the spring, that is

$$\text{for } F = \text{constant}, \quad \varepsilon = \frac{\Delta l}{l} = \text{constant}$$

where ε indicates elongation per unit length and usually is called (linear) *strain*.

---

[1] C. Truesdell, *Rational Mechanics*, *cit*; *see* also *Essays in the History of Mechanics* (New York, 1968) (hereafter cited as Truesdell, *Essays*).

[2] I. Beeckman, *Journal tenu par Isaac Beeckman de 1604 à 1634*, facsimile ed. C. De Waard (The Hague, 1939–1953), fols. 137 bis v, 139 bis v.

[3] *Ibid.*, fol. 362r.

The English took an early interest in elasticity. As usual, the initial impulse was to look for underlying causes; we can, for example, cite William Petty's memoir, *The Discourse made before the Royal Society concerning the Use of Duplicate Proportion, together with a New Hypothesis of Springing or Elastique Motion*, which was published in 1674. Todhunter's comments are caustic:

> Although absolutely without *scientific* value, this little work throws a flood of light on the state of scientific investigation at the time .... There is an appendix (page 121) on the new hypothesis as to elasticity. The writer explains it by a complicated system of atoms, to which he gives not only polar properties, but also *sexual* characteristics, remarking in justification that the statement of Genesis 1:27:—"male and female he created them"—must be taken to refer to the very ultimate parts of nature, or, to atoms as well as to mankind![4]

Fortunately, the next English writer on the subject had more rational ideas. Robert Hooke was not only a mathematician, mechanician and physician, but after the Great Fire of 1666 he proposed plans for the rebuilding of the city of London. In 1678 he published his fundamental work on elasticity, *Lectures de potentia restitutiva, or of Spring, Explaining the Power of Springing Bodies.* He may not have known of Fabri's work; at the beginning of the treatise he writes, "The Theory of Springs, though attempted by divers eminent Mathematicians of this Age has hitherto not been Published by any."[5] Hooke claims to have discovered the theory of springs eighteen years before his book appeared, but he had not wanted to publish it because he meant to apply his theory to a particular use: "About two years [ago] ... I printed this Theory in an anagram at the end of my Book on the Description of Helioscopes, viz., *ceiiinosssttuu*, id est, *ut tensio sic vis.* That is, the Power of any Spring is in the same proportion with the Tension thereof." He then describes his experiments (Figure 8.1). Hooke notes that the same linear relation holds in the case of the flection of a cantilever loaded with a weight at its free end. The same results seem to follow, even when he varies the structure and material ("Metal, Wood, Stones, baked Earths, Hair, Horns, Silk, Bones, Sinews, Glass, and the like"). He naturally comes to the conclusion that

---

[4] I. Todhunter and K. Pearson, *op. cit.*, Vol. 1, pp. 4–5.

[5] Quoted in R.T. Gunther, *Early Science in Oxford* (Oxford, 1931), Vol. 8, pp. 331ff.

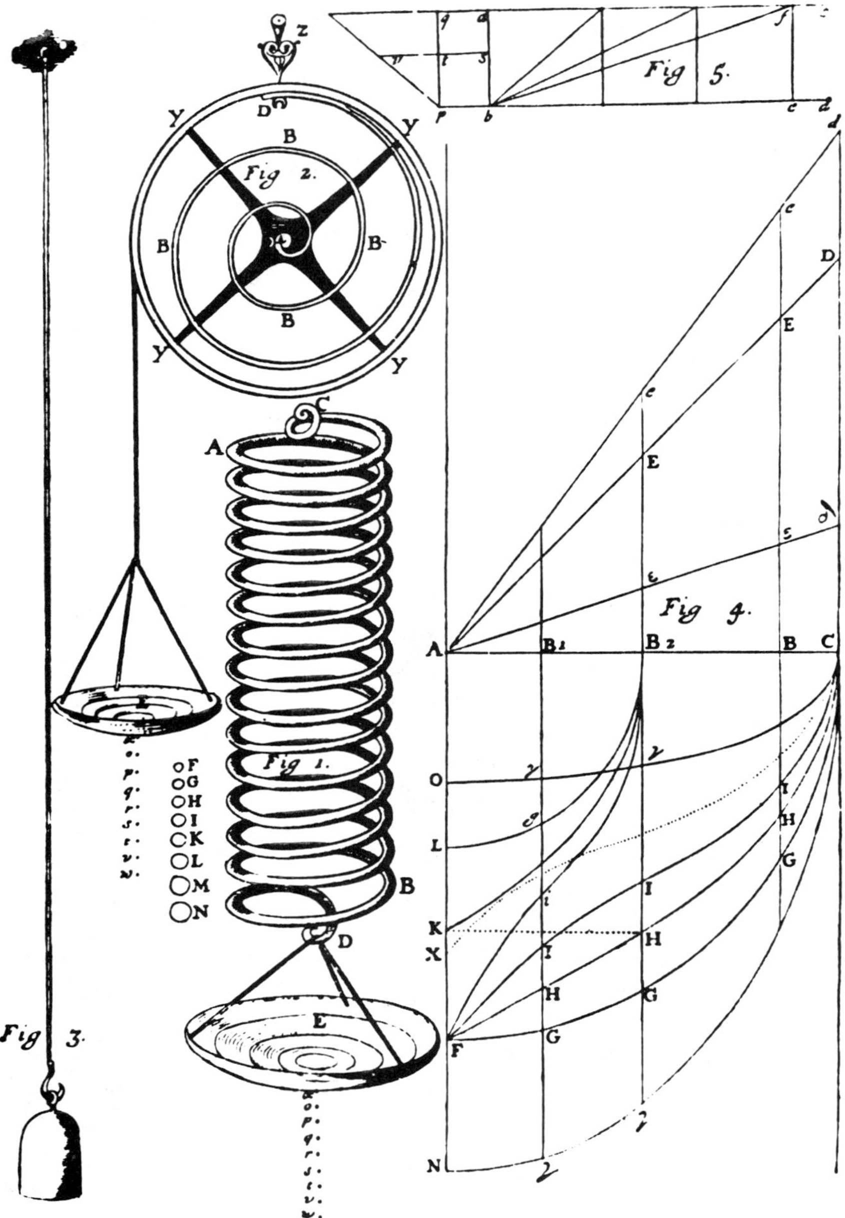

FIGURE 8.1. Hooke's experiments on the power of springing bodies (1678). From Vol. 8 of Gunther's *Early science in Oxford* (Oxford, 1931).

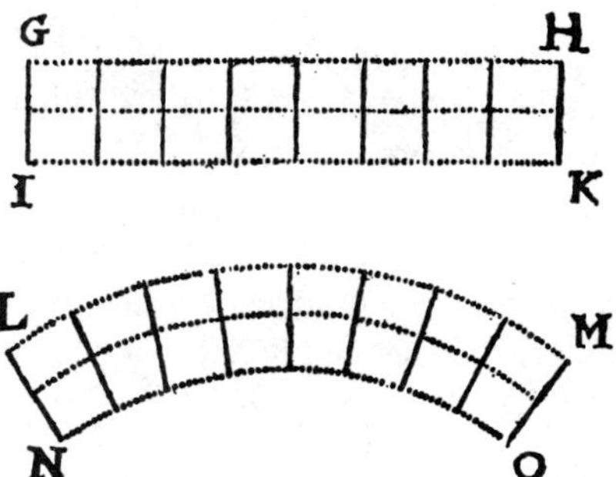

FIGURE 8.2. Hooke's figure showing the extension and the contraction of the fibers of a bent beam (1678). From Vol. 8 of Gunther's *Early science in Oxford* (Oxford, 1931).

> it is very evident that the Rule or Law of Nature in every springing body is, that the force or power thereof to restore it self to its natural position is always proportionate to the Distance or space it is removed therefrom, whether it be by rarefaction or separation of its parts one from the other, or by a Condensation, or crowding of these parts nearer together.[6]

Hooke is an acute observer of the behavior of a bent beam (Figure 8.2). To explain its "compound way of springing" he considers "two lines joyned together as at $GHIK$, which being ... bended into the form $LMNO$, LM will be extended, and NO will be diminished in proportion to the flexure, and consequently the same proportions and Rules for its endeavour or restoring it self will hold."[7] As Truesdell observes, this "compound way of springing" will remain the principal problem of elasticity for the next century. Hooke presents no ideas on correlating the curvature of a fiber with a bending moment, nor does he take into account the reciprocal reaction between two fibers.[8]

8.2 Mariotte's Contribution

Edmé Mariotte, physician of Dijon and prior of St. Martin sous Beaune, was the first to introduce elasticity into the study of Galileo's problem. He was born in 1620 and died in Paris in 1684. His reputation rests on other, more important contributions, set forth in his four *Essais de physique* (1676-1679). In the second one, *De la nature de l'air*, he presents his ideas on the behavior of an ideal gas which usually go under the name of Boyle's

---

[6] *Ibid.*, p. 336.
[7] *Ibid.*, p. 337.
[8] Truesdell, *Rational Mechanics* p. 55.

Law. In his *Traité du mouvement des eaux et des autres corps fluides*, he deals with the movement of water in pipes, establishing a simple formula for constructing pipes of the proper size.

In the second *discours* of the fifth part of this *Traité* (published posthumously in 1686), Mariotte criticizes Galileo's hypothesis for the fracture of a cantilever. Like Viviani, he recognizes the fact that Galileo's hypothesis implies the simultaneous fracture of all parts of the beam at its intersection CD with its support (Figure 8.3, sketch 106).

> This contrasts with many experiments I have made with wood and glass solids, where I have found that it was necessary to take the ratio between FD and segment less than DI, as a quarter of DC or a third, etc., and not between FD and half of DC. To find this ratio and reject Galileo's, I reason as follows: first of all, I suppose that wood, iron, and the other solids have fibers and particles which are mutually interwoven so that they cannot be separated except by a certain force, and all their resistance in breaking manifests itself when they are pulled ... lengthwise. I further suppose that these parts can be lengthened more or less because of different weights; and finally, that there is a degree of elongation which they cannot tolerate without breaking.[9]

This last hypothesis contains a criterion of fracture and resistance which differs from Galileo's concept. No longer is fracture the effect of exceeding the maximum supportable weight or tension; now it results from exceeding the maximum elongation. In the nineteenth century, Saint-Venant would take up a general criterion of fracture with reference to deformation. For now, Mariotte formulates his own version of Hooke's Law: "if a weight of 500 pounds makes a wooden solid two *lines* longer, a weight of 125 pounds will lengthen it by half a *line*, and one of 250 pounds about one."[10]

When he comes to the case of a bent beam, Mariotte constructs a curious model. (It seems that the development of Galileo's problem involves a good deal of model-making, from Pardies's to Mariotte's to Leibniz's.) It is still a lever (Figure 8.3, sketch 107) but one in which the joint between beam and wall is replaced by elastic elements placed at different distances from fulcrum C. When the beam is loaded, these elements are subject to elongation. The further they are from C, the greater is their elongation. Perhaps this image persuaded Mariotte that a bent cantilever was similar

---

[9] E. Mariotte, "Traité du mouvement des eaux et des autres corps fluides ...," in *Œuvres de M. Mariotte, de l'Académie Royale des Sciences* ..., new ed. (The Hague, 1740), p. 461.

[10] *Ibid.*, p. 462.

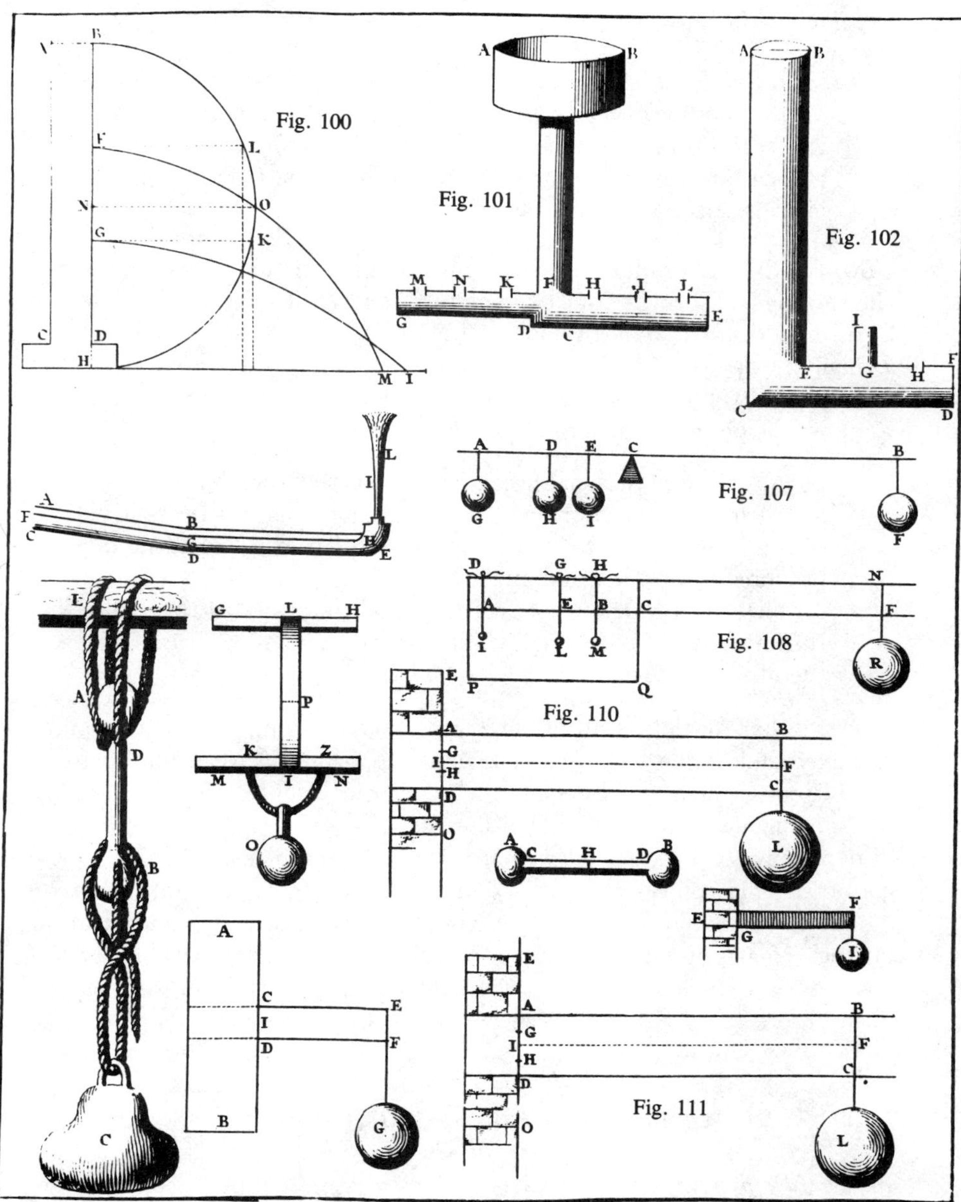

FIGURE 8.3. Drawings used by Mariotte to explain his new hypothesis (1684). From Mariotte's *Œuvres* (The Hague, 1740).

to a structure whose longitudinal fibers lengthen in proportion to their distance from the intrados, and so he fell into Fabri's error. But Mariotte's reasoning is far more subtle—and more insidious.

> If $ABCD$ is a square staff fixed in the wall [Figure 8.3, sketch 111] it can be conceived that between D and I, I being at half AD, the fibers are pressed by weight L and the ones near D more than the ones near I; and that between D and A they lengthen, as has been explained.[11]

Now, the distribution of the lengthening and shortening of the fibers over the thickness AD singles out two points, G and H. The former is placed at a distance from A equal to one-third of AI, the latter at a distance from D equal to one-third of ID. They are the points on which are applied the resultants of the traction and pressure resistance, respectively. So far, the argument is perfect. But now we get to the point:

> As it is very reasonable to assume that these pressures resist as much as the tractions and that the same weight is necessary to produce them, these tractions and these pressures will share the force of weight L: adding a third of thickness IA to a third of thickness ID, the whole will be equal to a third of the total thickness AD; from which it follows that everything is as if all the fibers lengthen.

It seems that Mariotte had no intention of establishing, for the tension of the fixed end of the beam, the triangular distribution with point zero at the intrados which was later attributed to him. He sees the distribution of the elongations as exactly the same as that which was to be the point of departure for Parent's treatment in 1713, a treatment which led to the correct elastic solution of the problem. His wrong interpretation, "that all the fibers would lengthen," derives from his (incorrect) concentration on the "length of the counterlever," without establishing a value for the resistance—that is for the total moment of the stresses operating on the fixed base.

8.3 Leibniz's *New Demonstrations*

In his "*Demonstrationes novae de resistentia solidorum*," Leibniz gave physical significance to Mariotte's formal equivalence. Leibniz knew about Mariotte's research, although it is not quite clear to what part of it he really wished to refer when he wrote the following passage:

[11] *Ibid.*, p. 465.

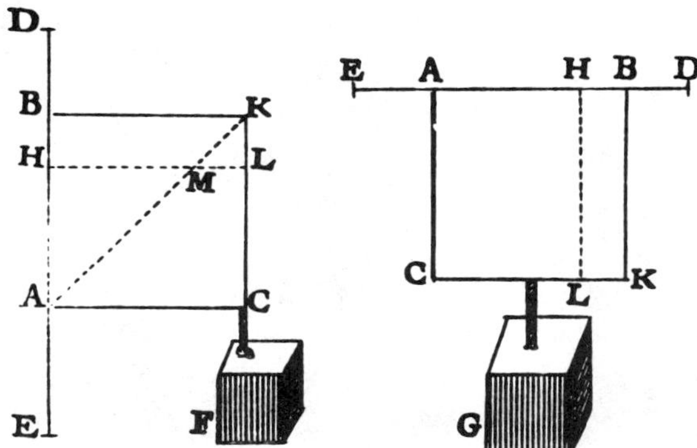

FIGURE 8.4. Leibniz's figure for comparing the breaking load of a cantilever to that of a taut beam. From the *Acta Eruditorum* of 1684.

> The illustrious Mariotte ... by means of ingenious calculation obtained that weight F is about the fourth part of weight G [G and F are respectively the breaking load for the beam subject to traction and for the cantilever, when the length and height of the beam are equal (Figure 8.4)]. But as I have had an occasion to consider the thing more deeply in order to reduce it to the laws of Geometry, I found, at last, the true proportions and demonstrated, among other things, that weight F must be the third part of weight G, and thus the robustness of bodies resisting breakage is in sequialter proportion [i.e., 3:2] less than the one indicated by Galileo.[12]

The model which Leibniz proposes is shown in Figure 8.5. Every fiber is likened to a spring connecting the beam to the wall. Having hung weight F from the beam, Leibniz supposes that the beam rotates around fulcrum A.

> Point B of the beam moving away from the wall in ${}_1B$ goes to ${}_2B$, drawing with it the fiber that connects it with the wall and stretching it like a cord, lengthening it with respect to its natural state of ${}_1B\,{}_2B$; in the same manner, point H stretches its fiber in ${}_1H\,{}_2H$; furthermore, fiber ${}_1H\,{}_2H$ resists the weight that stretches it less than fiber ${}_1B\,{}_2B$, and this is in proportion to the square of the distance from A.[13]

According to Leibniz, this is due to the fact that the elongation of the fiber at distance y from A is proportional to y; the same goes for the tension ("for

[12] G.W. Leibniz, "Demonstrationes novae de resistentia solidorum," *Acta eruditorum Lipsiae* (July, 1684), pp. 319–325.

[13] *Ibid.*, p. 321.

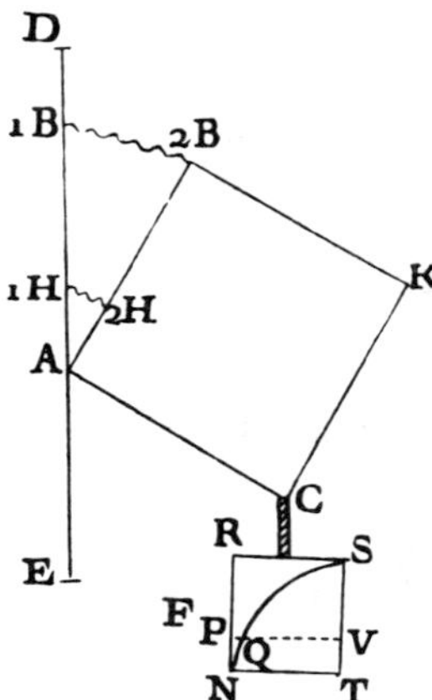

FIGURE 8.5. Leibniz's mechanical model for calculating the bending moment on the cross-section of the cantilever. From the *Acta Eruditorum* of 1684.

the hypothesis elsewhere confirmed that the elongations are proportional to the stretching forces"[14]). Therefore the moment about A is proportional to y^2.

From this it follows that "in the simultaneous tension of all the fibers existing at every point, the resistances at every point are among themselves as the square of the distances from the lowest point." By "resistance" he means the moment about A of the resisting force, or stress. To sum this up: the resistance in H is to the resistance in B as the square of AH is to the square of AB.

> Let us suppose that the weight F ... [in Figure 8.5] be the parabolic body $NRSQN$, freely hanging from C, in which height NR is equal to base RS (just as AB is equal to AC), and in which ... PQ is to RS as the square of NP is to the square of NR. Thus, given that base RS represents the resistance in B, the ordinate PQ will represent the resistance in H, if the heights NP, NR are proportional to the heights corresponding to AH, AB. Therefore, the entire concave parabolic trilinear line $NRSQN$ will represent the resistance of the whole line AB.... On the other hand, the square $RNTS$ circumscribing this parabolic trilinear represents the resistance [this time, he means the resultant force of resistance or stress] that would be exerted by line AB if the beam were subjected to traction [Figure 8.4, sketch 2].... In fact, if the beam is subjected to traction, the resistance of all of its points is equal .... Now, since square RT is the treble of the inscribed concave parabolic

[14] *Ibid.*, p. p. 322.

> trilinear, i.e., of $NRSQN$, the direct resistance of a line such as AB will equally be three times the transversal resistance.[15]

This passage is really significant. Although it is based on a particular example, the argument contains all the factors which, coordinated by the new differential and integral calculus, would produce the general theory. We find the concept of tension as a contact force affecting each particle of the transverse section; we find the notion of summing the contributions of the tensions relative to each point; we find at least a hint of the proportionality between bending moment and inertial moment of the section. But what this passage lacks is the language itself. If Leibniz had used the integral calculus, the statement that the tension σ is linearly related to the distance y of H from A, and that the relative moment depends on y^2, would have led to a formula of the type

$$\sigma = \sigma_{\max}\frac{y}{h}, \qquad M = \int_S \sigma_{\max}\frac{y^2\,dS}{h}, \tag{8.1}$$

where $\sigma_{\max}$ is the tension at B, h is the height AB, M is the reactive moment, and S is the section. From equation (8.1) it follows that $M = \sigma_{\max}J_A/h$, where J_A is the moment of inertia of the section with respect to the horizontal axis passing through A. We have to wait another twenty-odd years before the calculus manages to get a firm grip on *resistentia solidorum*, an event which gives the new technique an excellent chance to show its resources.

8.4 New Problems: Catenaries and Elastic Curves

Leibniz's memoir offers other valuable contributions. The author confirms and extends the Galileian results on solids of equal resistance, and affirms the connection between elastic and acoustic properties of bodies. All this, according to Truesdell, lets us consider Leibniz "the father of the mathematical theory of elasticity."[16]

From 1684 to the first years of the next century, the history of mathematics and mechanics was preparing for great events. During those years, the differential and integral calculus took their first effective forms; during those years, this powerful new technique allowed scientists to tackle problems unsolved in human history.

In 1690, one of these questions gave rise of a major debate, when Jakob Bernoulli brought up the problem of the catenary in a memoir on another subject in the *Acta eruditorum* of Leipzig. Leibniz replied at once, announcing that "his key" (the infinitesimal calculus) could be used to reach

[15] *Ibid.*, pp. 322–325.

[16] Truesdell, *Rational Mechanics*, p. 63.

a solution. In the same year, Huygens demanded information from Leibniz on the new algorithm; at the same time, he studied the problem on his own and obtained some (rather limited and involuted) results which he sent Leibniz for publication by the Leipzig *Acta*.

The June 1691 issue of the *Acta* was memorable. In his introduction the editor wrote:

> The benevolent reader will certainly remember the problem proposed by the illustrious professor Jakob Bernoulli of Basel .... The famous Wilhelm Gottfried Leibniz had promised to make known a solution obtained with his method, if no one else had found it by the end of the year [i.e., 1690]. Actually, the challenger's brother, Mr. Johann Bernoulli, a scholar of medicine and very proficient in this research, found the solution and sent it to us last December; furthermore ... he asked us very kindly to add that of Leibniz to his. For this reason, we urged the famous gentleman ... to publish his result .... Also the noble Christiaan Huygens deigned to embellish our journal with his solution to the problem. Thus we offer you, benevolent reader, the two solutions of these illustrious noblemen together with that of Bernoulli, but in the order in which we received them.

Truesdell's comment is apt: "these three solutions ... exhibit the mathematics of the future, the present, and the past."[17] Huygens' solution represents the past; he approaches the problem with a complex, though skilful, geometrical method. Leibniz, using his new technique, reaches a correct analytical formula for the catenary (Figure 8.6). In the plane (x, y), $y/a = (b^{x/a} + b^{-x/a})/2$, where a is segment ΘA and b is a pure number which, in fact, corresponds to base e of the natural logarithm. We should note that Leibniz's mechanical explanation is inadequate, but his reasoning was based on Pardies' theorem, as he himself explained in a letter to Huygens in 1694, in response to a "long sequence of requests, complaints and accusations."[18] Finally, Johann Bernoulli's work (entitled "*Solutio problematis funicularii*") supplied two correct constructions of the catenary and listed thirteen properties, but without complete proofs. His "*Lectiones mathematicae de methodo integralium aliisque, conscriptae ... cum auctor Parisiis ageret, annis 1691 et 1692*" presents valid statical arguments, and brings to light a new and important concept for continuous mechanics: the need and the possibility of expressing the equations of equilibrium in differential form.[19]

[17] *Ibid.*, p. 66.
[18] *Ibid.*, p. 71.
[19] J. Bernoulli, *Opera omnia tam antea sparsim, quam hactenus inedita* (Lausanne and Geneva, 1742) Vol. 3, pp. 385–558.

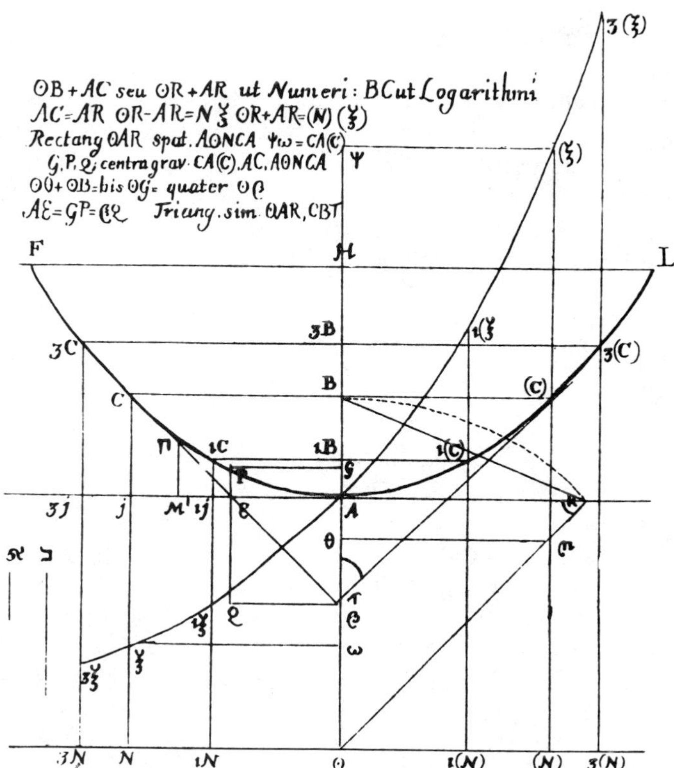

FIGURE 8.6. Leibniz's figure for the catenary. From the *Acta Eruditorum* of 1690.

From this, it was a short step to a complete explanation of the catenary. In 1697–1698, Jakob Bernoulli was the first to derive the general equations that not only solved the problem, but also permitted the treatment of the more general theme of the equilibrium of a flexible rope, subject to any distribution of tangential (f_t) and normal (f_n) forces. Bernoulli's equations are, as we know,

$$\frac{dT}{ds} + f_t = 0, \qquad \frac{T}{r} = f_n,$$

where T is the tension, s the curvilinear abscissa, and r the radius of curvature.

Around the end of the seventeenth century, discussion about the catenary was disrupted by a rude and fundamentally silly quarrel involving Leibniz, Johann Bernoulli and David Gregory, who had published an essay on the subject in the *Transactions* of the Royal Society (August 1697). Gregory's essay intended to show how Newton's method could be used to determine "the fluxions by a relation given with the fluents."[20] (Gregory's contribution exerted remarkable influence, mainly because he recognized a connection between a catenary and a simply compressed arch, as we shall see somewhat later.) The dispute smoldered on for some years in the *Acta eruditorum* with an acrimony that was really disproportionate to the importance of the subject. But the catenary was only an excuse; the real source of the argument was the more convoluted quarrel between Newton and Leibniz about the new calculus.

8.5 Jakob Bernoulli's Fundamental Work

The problem of the cord led directly to another, "equally interesting problem," according to Jakob Bernoulli, that of "the inflection or of the curving of beams, of bows, of elastic elements of every kind, because of their own weight or of a weight applied to them or of any other acting force."[21] Bernoulli pointed out the difficulty of the problem in 1691, declaring that it could be solved in a few months. In fact, it took him three years. In 1694, he revealed his results in a memoir in the *Acta eruditorum* entitled "*Curvatura laminae elasticae. Ejus identitas cum curvatura lintei a pondere inclusi fluidi expansi. Radii circulorum osculantium in terminis simplicissimis exhibiti, una cum novis quibusdam theorematis huc pertinentibus, etc.*" ("Curvature of an elastic plate [rod]. Its identity with the curvature of a linen [cloth] due to the pressure of an internally held fluid. Very simple determination of the radii of osculatory circles, with some new theorems pertaining to this subject") Bernoulli was to return to the same topic, with greater maturity and clarity, in a letter of March 12, 1705, to the Académie Royale, published in its *Mémoires* in 1706 under the title "*Véritable hypothèse de la résistance des solides, avec la démonstration de la courbure des corps qui font ressort.*"

This memoir deserves attention. In it, Bernoulli developed and completed the complex arguments laid out in 1694 (see the discouraging figure [Figure 8.7] which illustrates his proposed geometrical construction), deriving the formula

$$\frac{1}{r} = f(Q)$$

---

[20] D. Gregory, "Catenaria," *Philosophical Transactions*, Vol. 19, No. 231 (August 1697), pp. 637–652, reprinted in *Acta eruditorum Lipsiae* (July 1698), pp. 303–321; *see* p. 306.

[21] J. Bernoulli, "Specimen alterum calculi differentialis in dimetienda spirali logarithmica, loxodromiis nautarum, et areis triangulorum sphaericorum: una cum additamento quodam ad problema funicularium, aliisque," *Acta eruditorum Lipsiae* (June 1691), pp. 282–290. Reprinted in *Opera Omnia*, Vol. 1, pp. 442–453 and in *Leibnizens mathematische Schriften*, Vol. 5, pp. 252–254.

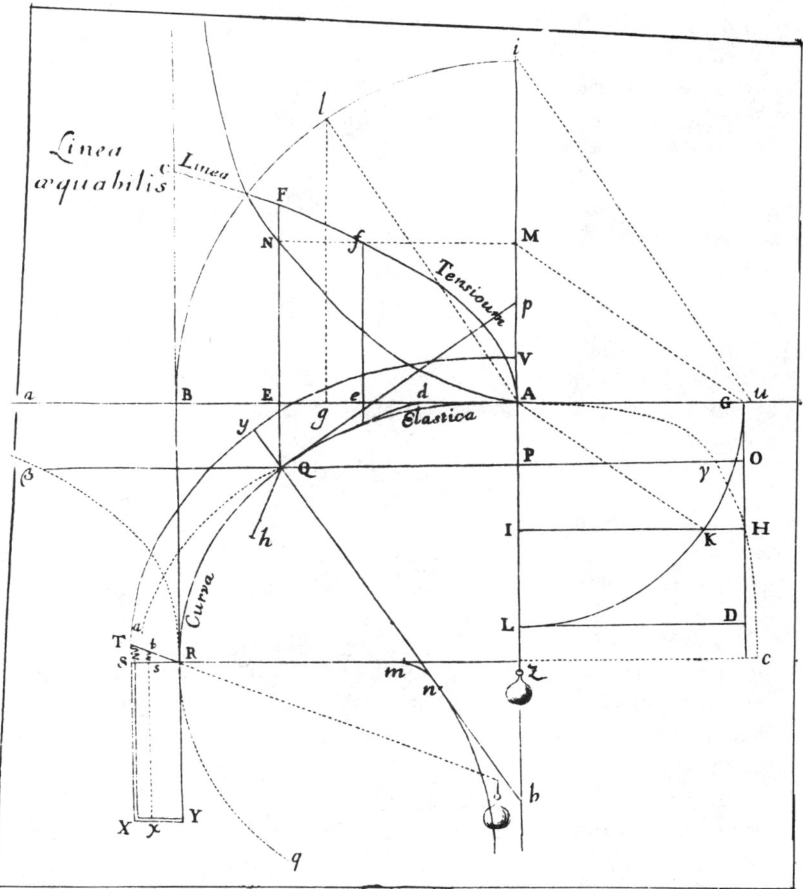

FIGURE 8.7. Jakob Bernoulli's figure for determining the elastic curve. From the *Acta Eruditorum* of 1694.

where r is the radius of curvature of the elastic rod and $f(Q)$ is a function of load Q. But above all, the memoir presents four lemmas which allow us to introduce elasticity with some exactness into the study of the resistance of solids. Furthermore, the first three lemmas introduce the notion of a *local* stress–strain relationship:

I. Fibers of the same material and of the same width, or thickness, extended or compressed by the same force, lengthen or shorten proportionally to their length.

II. Homogeneous fibers having the same length, but a different width or thickness, equally lengthen or shorten, if they are acted upon by forces proportional to their width.

III. Homogeneous fibers of the same length and thickness, but loaded with different weights, neither lengthen nor shorten proportionally to these weights; but the lengthening or the shortening caused by the large weight is to the lengthening or shortening caused by the small weight less than the ratio that the first weight has to the second.

As we can see, the first two lemmas imply the possibility of rewriting Hooke's law (elongation proportional to force) in the form

$$\Delta l \propto \frac{l}{S} F \tag{8.2}$$

where Δl is the elongation, l is the length of the "fiber," S is the cross-sectional area, and F is the force. Equation (8.2) is equivalent to the stress–strain relation

$$\varepsilon \propto \sigma$$

if we set $\sigma = F/S$ and $\varepsilon = \Delta l/l$. The third lemma refers to non-linear behavior.

Truesdell comments,

> Thus Bernoulli is the first to introduce a *stress–strain relation* as distinct from a formula such as Hooke's for elongation Δl as a function of applied force F. This, too, must not be exaggerated, since still more than a century ahead lie the *local* concepts of stress and strain used in modern theories of materials. Bernoulli refers here only to simple push or pull, and his explanations indicate also that he regards the phenomenon as occurring *homogeneously* over the length and cross-section of the specimen. But Lemmas I and II together assert that *there is an elastic law ... which is common to all specimens of a given material*, be their lengths and areas what they may. It is the first time since Galileo's formula for rupture that a *material property* appears in rational mechanics.[22]

Finally, there is the fourth lemma (Figure 8.8):

> The same force that makes a beam or a rod $ABCD$ bend from AB to GF, stretching a part of its fibers to the measure of triangle BSF and pressing the other part in the measure of triangle ASG, will be able to stretch the whole of the fibers on support A in the measure of triangle ABF, or to press this assemblage around support B or F, in the measure of triangle BAG or FAG.

---

[22] C. Truesdell, *op. cit.*, p. 106.

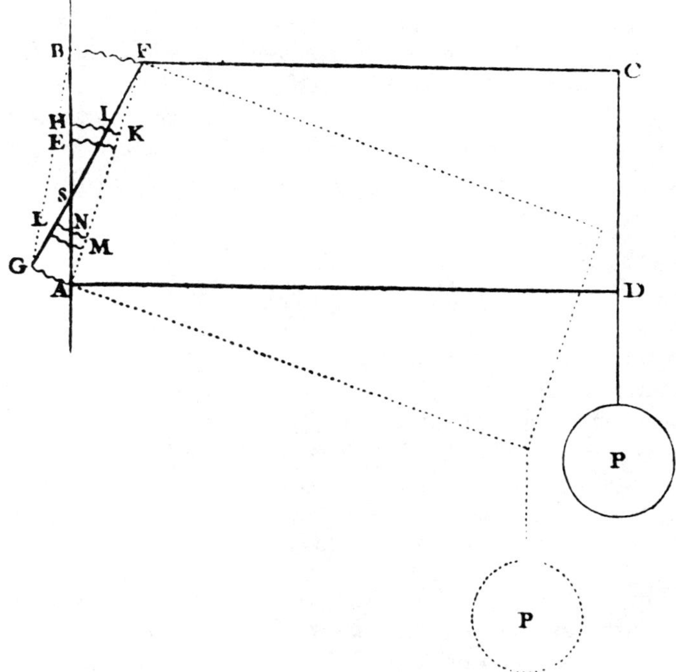

FIGURE 8.8. Jakob Bernoulli's figure for stating his "véritable hypothèse." From the *Mémoires de l'Académie Royale* of Paris (1706).

Bernoulli inherited this hoary thesis which, since Fr. Fabri's time, continued to be upheld and his demonstration is exactly like Fabri's: "the fiber in H being stretched on support A to the length HK and pressed at the same time on support F to the length KI, behaves as if it were only stretched to the length $HI = HK - KI$."[23]

8.6 Varignon and the Galileo–Mariotte Dichotomy

Why was Bernoulli impelled to formulate a thesis that was not merely wrong, but superfluous as well? Probably Varignon's work has something to do with it. It was based on a "dichotomy" between Galileo and Mariotte, as if their systems could share the field. Three years before Bernoulli's letter, in 1702, Varignon presented a memoir to the Académie Royale. This

[23] J. Bernoulli, "Véritable hypothèse de la résistance des solides, avec la démonstration de la courbure des corps qui font ressort," *Mémoires de l'Académie Royale des Sciences* (1706), pp. 176–186; *see* p. 181.

memoir was conceptually close to Leibniz's work of 1684, but brought in the powerful new integral calculus.[24] It expressed and generalized the second equation of (8.1), which had been implied by Leibniz's treatment.

Varignon's "fundamental rule of the resistance of solids" for the cantilever establishes the ratio between the breaking load at flexure P and that at traction Q, avoiding any hypotheses about the law $\sigma = \sigma(y)$ by using the following formula (Figure 8.9):

$$P = \frac{Q \int \sigma(y) y x \, dy}{\sigma_{\lim} l \int x \, dy} \tag{8.3}$$

where $l = DT$, $y = DH$, $\sigma(y) = HK$, $\sigma_{\lim} = BC$ and $x = x(y) = EF$. Applying equation (8.3) to two hypotheses ($\sigma =$ constant, or σ proportional to y), Varignon easily derives Galileo's and Mariotte's results, putting them in a particularly expressive form. "Galileo's hypothesis" is that the resultant of stresses passes through the center of gravity of the breaking section; Mariotte's is that the resultant passes through the center of percussion. The Académie's commentator was enthusiastic:

> The comparison between the centers of gravity of Galileo's hypothesis and the centers of percussion of Mariotte's hypothesis throws new and brilliant light on the whole of this matter, and produces a harmony and order that satisfy the spirit.[25]

A number of scientists defended the distinction between Galileo's and Mariotte's hypotheses for a long time, clinging to a distinction between equally persuasive systems which divided material behavior into two different classes. The center of gravity and the center of percussion—derived from antiquity and recently reformulated by Huygens—typified the theory itself. They seemed quite worthy to share the field. Perhaps for this reason, the treatises that taught eighteenth-century engineers and architects cite only Galileo's and Mariotte's solutions. Bernard Forest de Belidor's *Science des ingénieurs* (1729), Giovan Battista Borra's *Trattato delle resistenze* (1748) and Francesco Milizia's *Principii di architettura civile* (1781) are examples of this trend.

Of course, the Galileo–Mariotte "dichotomy" is somewhat misleading, since the two solutions provide an answer—albeit deceptively—to different problems. Galileo's hypothesis deals with the ultimate load that a cantilever can support without collapsing, whereas Mariotte's treatment applies to the elastic behavior of the structure.

---

[24] P. Varignon, "De la résistance des solides en générale pour tout ce qu'on peut faire d'hypothèses touchant la force ou la ténacité des fibres des corps à rompre; et en particulier pour les hypothèses de Galilée et de M. Mariotte," *Mémoires de l'Académie Royale des Sciences* (1702), pp. 66–100.

[25] In *Histoire de l'Académie Royale des Sciences, année 1702* (Paris, 1720), p. 108.

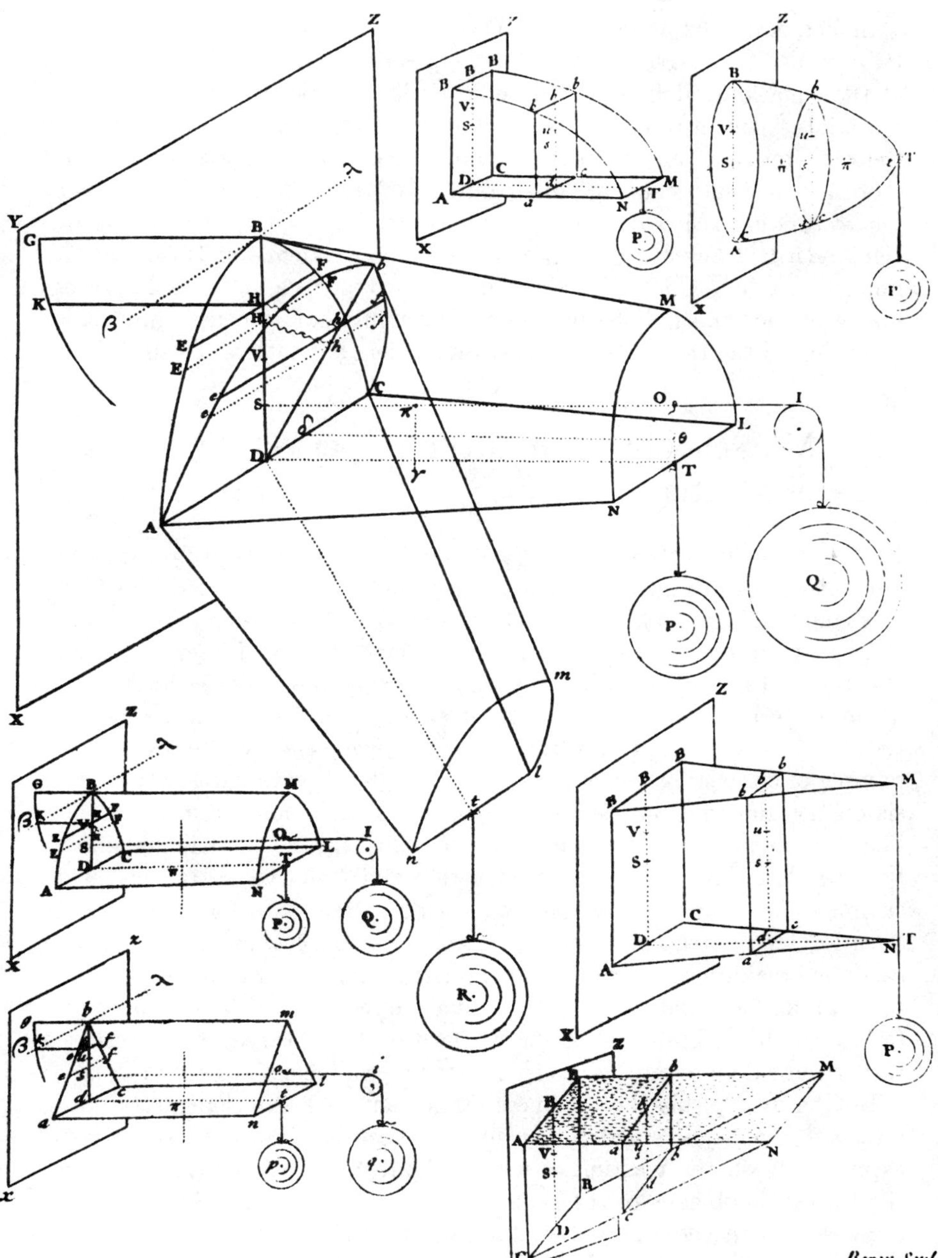

FIGURE 8.9. Varignon's drawings for his memoir on the resistance of solids. From the *Mémoires de l'Académie Royale* of Paris (1702).

In fact, Antoine Parent disproved the Fabri–Bernoulli hypothesis in 1713. He obtained the correct solution for the position of a neutral fiber, arriving at the equations which can still be found in textbooks on the strength of materials.[26] But it took until the end of the eighteenth century to dissipate the old prejudice, and to persuade engineers to rely on less elegant, but more useful treatments. In this regard, Coulomb's famous *Essai sur une application des règles de maximis et minimis à quelques problèmes de statique relatifs à l'architecture*, presented to the Académie in 1773, plays an important role.[27] Coulomb rediscovers and extends Parent's results, recognizing in the "rules of the maximum and minimum" a new argument for returning to the theory "a harmony and order that satisfy the spirit."

8.7 Musschenbroek and the Imperfections of Matter

During the eighteenth century, a good many scientists began to consider the "imperfections of matter" which Galileo deliberately excluded and Mariotte collated. To do this was to call into question the over-simplified models that we mentioned above. Among these scientists was Pieter van Musschenbroek. His fundamental work *Physicae experimentales et geometricae demonstrationes* ..., published at Leiden in 1729, contains a dense chapter ("*Introductio ad cohaerentiam corporum firmorum*") which deals with a number of interesting topics. Musschenbroek dwells with particular emphasis on the difference in the behavior of solids when subjected to traction, bending or compression, distinguishing as usual (in the Galileian tradition) between "absolute resistance" for traction and "relative or transversal resistance" for bending. The author does not attempt to formulate a constitutive law, but he presents a good many examples, measuring in each case the breaking load for traction and the elongation or contraction of the resistant surface, and carefully recording the results. His data induced him to look at the differences, not the similarities. For this reason, his research rather casts doubts on the received formulae than confirms them.

But because of his thoroughness, Musschenbroek can demonstrate and interpret new aspects which had hitherto been ignored. For example, his experiments on the traction of wires (Figure 8.10) led him to contradict the thesis "so obvious to the geometers" that the ratio between force and cross-section reaches a constant limit for rupture, whether one measures

[26] A. Parent, "De la véritable mécanique des résistances relatives des solides, et reflexions sur le système de M. Bernoulli de Bâle," *Recherches de mathématique et de physique* (Paris, 1713), Vol. 3, pp. 187–201.

[27] C.A. Coulomb, "Essai sur une application des règles de maximis et minimis à quelques problèmes de statique relatifs à l'architecture," *Mémoires de mathématique et de physique, par divers savants*, Vol. 7 (1776), pp. 343–382.

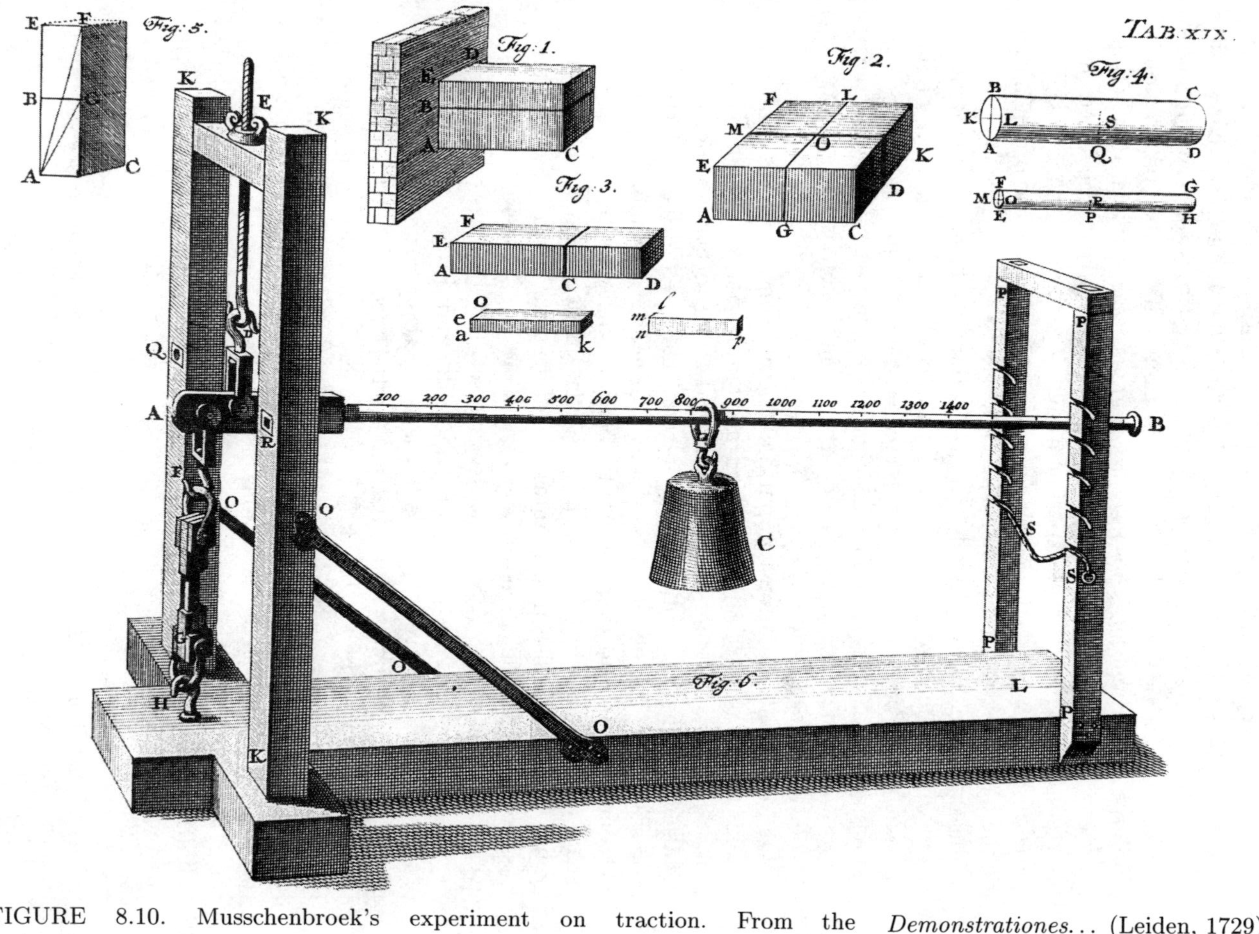

FIGURE 8.10. Musschenbroek's experiment on traction. From the *Demonstrationes*. . . (Leiden, 1729).

the cross-section of the wire before or after the experiment. Moreover, these experiments revealed the production of heat during deformation or breaking. Musschenbroek was the first to note this non-conservative aspect of material behavior. He attributed it to "internal friction" —that is, to the "friction of the parts among one another and to strong lateral compression" which occurs "when the metal thins out because of traction." Similarly, Musschenbroek proposes to eliminate the artificial distinction between Galileo's and Mariotte's formulae for flexure. He is convinced that the results must be derived experimentally, by using the apparatus illustrated in Figure 8.11. "The doctrine of resistance is always too problematic for the insight of mathematicians," he observes.[28]

Musschenbroek also critically revises Bernoulli's formula. His experiments with wooden beams showed him the considerable influence of the non-elastic reaction, or rather the increased time required to deform the beam under a constant load. To Musschenbroek, this suggested the extreme proposition that "there are as many types of inflection as there are types of wood under examination."

But Musschenbroek's most important original contribution is that concerning the relation of the limit load at compression to the length of a thin rod. Others had noted that a thin rod tends to bend before breaking (see Section 6.2), but the phenomenon had never been subjected to proper experimental observation. Musschenbroek uses the apparatus shown in Figure 8.12 to find that the limit load is inversely proportional to the square of the length of the rod. He goes even further, formulating the following law, which also incorporates the dimensions of the (rectangular) cross-section: "parallelepipeds of the same wood ... compressed along their length, exert *resistance forces* that vary inversely to the square of the length, directly to the thickness of the uncurved side, and directly to the square of the side that is curved."[29] This law is not entirely correct, as Truesdell points out; it holds for fracture but not for the preceding instability, that is, for the bend before the break.[30] Nevertheless, as far as the dependence between limit load and length is concerned, Musschenbroek's experimental law is of considerable historical interest. Chronologically, it is second only to Leonardo's imperfect observations on the same subject (see Figure 8.13, taken from Leonardo's manuscripts). Musschenbroek's law precedes Euler's definitive result by about fifteen years. "This fact alone," Girard remarks, "deserves to be recorded in this history of science, by the number of people,

---

[28] P. Musschenbroek, *Demonstrationes physicae, experimentales et geomtricae* (Leiden, 1729), p. 622.

[29] *Ibid.*, p. 660.

[30] Truesdell, *Essays*, p. 125.

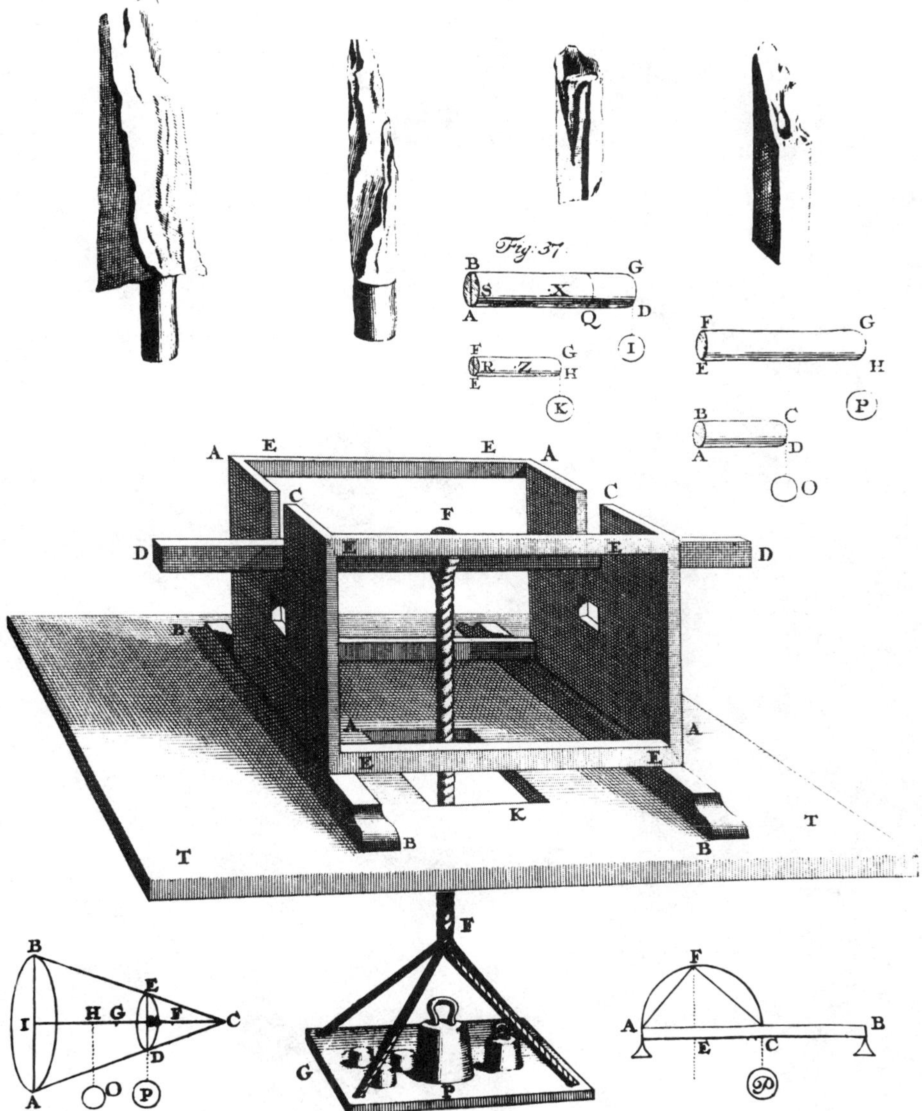

FIGURE 8.11. Musschenbroek's experiment on bending. From the *Demonstrationes...* (Leiden, 1729).

all too scant, who, in the field of physical research, are in pursuit of the fundamental point common both to theory and experiment."[31]

[31] P.S. Girard, *Traité analitique de la résistance des solides et des solides d'égale résistance* (Paris, 1798) p. xliii.

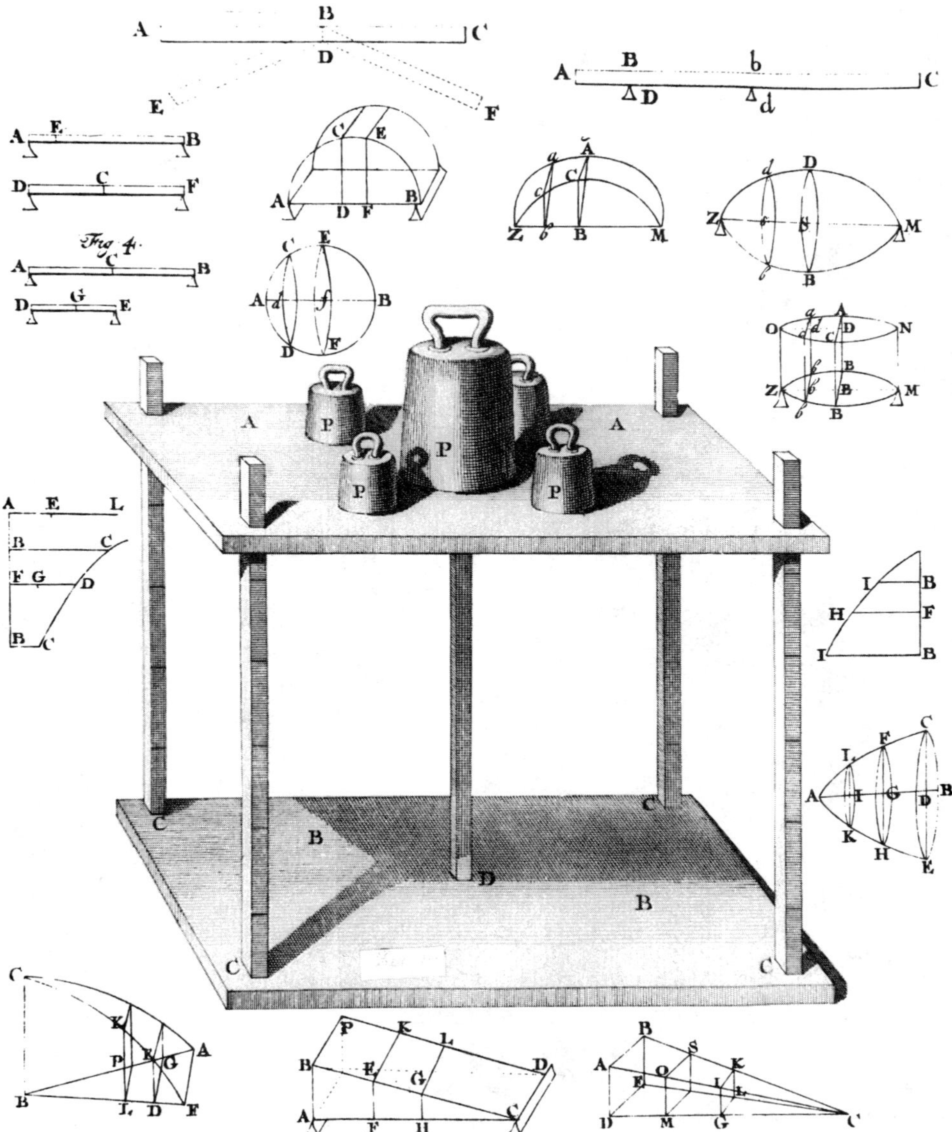

FIGURE 8.12. Musschenbroek's experiment on bending before breaking. From the *Demonstrationes...* (Leiden, 1729).

8.8 The Last of the Eighteenth-Century Treatises on Resistance

We should conclude this incomplete historical survey by mentioning the *Traité analytique de la résistance des solides et des solides d'égale résistance*

FIGURE 8.13. Leonardo's sketches on the buckling of a compressed beam.

by the engineer "*des ponts et chaussées*," Pierre Simon Girard. Girard's treatise was published in Paris in 1798, at the ebb of the eighteenth century, when Napoleon was rising toward his apogee and the French Revolution, in one way, and the Industrial Revolution, in another, had shattered the old traditions and ideas.

The title page of Girard's work includes a quotation from d'Alembert's *Élémens de philosophie*, which neatly expresses Girard's experimental dedication: "Experimentation will no longer be useful solely for the confirmation of theory but, being distinguished from theory without being superior to it, it will lead to new truths which theory alone could never have been able to reach." This is particularly true, Girard observes in his introduction, for all that concerns the analysis of the resistance and elasticity of materials. The author's explanation shows his remarkable perception of the "epistemological" difference between the general equations for equilibrium and the "constitutive" equations linking force to deformation. His words anticipate by more than 150 years a critical clarification of the laws of mechanics, which today's scientists consider a modern development:

> If in the theory of statics it is correct to consider levers a means by which the mobile bodies act one upon another as if they were endowed with perfect inflexibility, such a hypothesis ceases to be admissible in the application of this science to the calculation of machines, insofar as Nature has created no substance whose parts cannot be separated one from another by the action of a certain force. Thus there are two kinds of equilibrium that must be considered in the levers, and in machines that embody them; the one concerns the opposing forces that are counterbalanced, and the other one concerns a certain function of these forces and the consistency of the parts of which such machines are composed. It is possible to state accurately the conditions of the first type, but those of the second type are only approximately expressible. In fact, in order to arrive at their determination, it would be necessary to know the form and disposition of the elementary molecules of the bodies, what constitutes the force whereby they resist their separation, and since this knowledge is beyond our means, it is possible to apply the calculation to the resistance of solids only by resorting to various hypotheses, among which the most likely is the one whose results are closest to experimental data.[32]

The introduction from which this passage comes is a fine historical synthesis of the subject, from Galileo to the author's own time. Girard recognizes Galileo as the founder not only of the theory of structural resistance,

[32] *Ibid.*, pp. ix–x.

but also (quite correctly) of the theory of solids of equal resistance, which Girard examines at length. These themes, the author asserts, are fascinating and well worth consideration:

> by making bodies of *equal resistance*, we make them of *minimum volume*, and in this way we conform to the laws of Nature whose operations seem to be constantly carried out with the greatest possible economy of means. What distinguishes Nature's productions from those of human industry, is that cause and effects essentially harmonize. The first cannot undergo any modification without the others being changed; i.e., from a new cause a new effect always results. On the contrary, in human productions, there is no necessary proportion between the effect and the cause that produces it. If, for example, we must support a given weight, we can use without distinction either a thread that has just enough resistance, or an immense cable .... The seat of perfection is in that single determination at which Nature spontaneously arrives, while we must work experimentally through the immense distance that separates us from Nature.[33]

This perfectly expresses the essence of illuministic science. It refers implicitly to the teleological view of natural laws that animated the debate about Maupertuis' principle of least action. Nature's works are perfect, elegant and economical; man's works are imperfect, disorderly and spendthrift. This was an ingredient of the philosophy of the time. We have only to remember Rousseau's proposal that we should return to a state of nature, that civilization is corrupt and benighted. Therefore, to Girard, research into solids of equal resistance has additional importance if it can produce a more "natural" use of materials. It might allow man to be as elegantly parsimonious as nature.

> The theory of solids of equal resistance is, in fact, none other than an application of this same principle, since between an infinite number of bodies of a given resistance they [such solids] contain the minimum quantity of matter. Thus to reduce to such a form, so far as circumstances permit, all the bodies that are present in the construction of machines that we use to increase our strength, means to bring ourselves as near as we can to the perfection that so strikingly characterizes all the works of nature ... in short, to conform to her views.[34]

Can we not see in a bird's wing a beam mortised at one end, narrowing continually from the fixed end to the free end?

---

[33] *Ibid.*, p. xiv.
[34] *Ibid.*, pp. xvi–xvii.

The introduction traces the history of Galileo's problem and Bernoulli's theory of elastic curves to Euler and Coulomb. Girard quotes the latter's essay of 1773, applying the maximum and minimum rule to such technically important applications as the dimensions of retaining walls. Finally, Girard discusses the experimental studies of resistance, referring to Parent, Musschenbroek, Perronet, and Buffon (who studied timber, including the effects of humidity, seasoning, etc.) All studies of wooden beams, Girard observes, have failed to take into account the longitudinal adhesion of wood fibers.

> It is evident that this cohesion must render their inflexion more difficult. In fact, we have already seen in our first experiments that absolute elasticity which, according to the hypotheses of the geometers about the organization of fibrous bodies, should be due only to the dimensions of their bases of rupture, still depends on the length of their constituent fibers. Consequently, we have investigated the function of this length, which represents the longitudinal consistence in different types of wood, in order to obtain their "specific absolute elasticity."[35]

With similar observations (not always valid or well thought out) Girard approaches the recognition of the effects of shearing force, which generally goes with bending.

But there is more: the author had done his homework. In the course of his work on materials, he illuminates some aspects of structural behavior which go beyond simple elasticity. "We observed that resistance diminishes to the extent to which the load continues to act."[36] This is a typical anelastic phenomenon and concerns the "viscous component of the material." It lies outside the terms of elasticity, where stress is connected only to actual deformation. "This observation," Girard says, "opening up a new field of research, unfortunately allows us to foresee the extreme difficulty of obtaining strictly exact results."

We have to admire the clarity with which Girard recognizes the need to include time among his parameters. But not all his work is equally effective. The last of the four sections comprising his *Traité* quickly runs up against nearly insurmountable analytical complications and does not correspond well to the nature of the physical problem. The study of anelastic phenomena had to wait for almost a century.

Figures 8.14 and 8.15 reproduce Girard's original tables. They show his carefully designed apparatus (described at great length in the text) and a large number of wooden beams on which experiments were carried out, both to determine the coefficient of elasticity and to evaluate breaking strength.

---

[35] *Ibid.*, pp. lii–liii.
[36] *Ibid.*, p. liii.

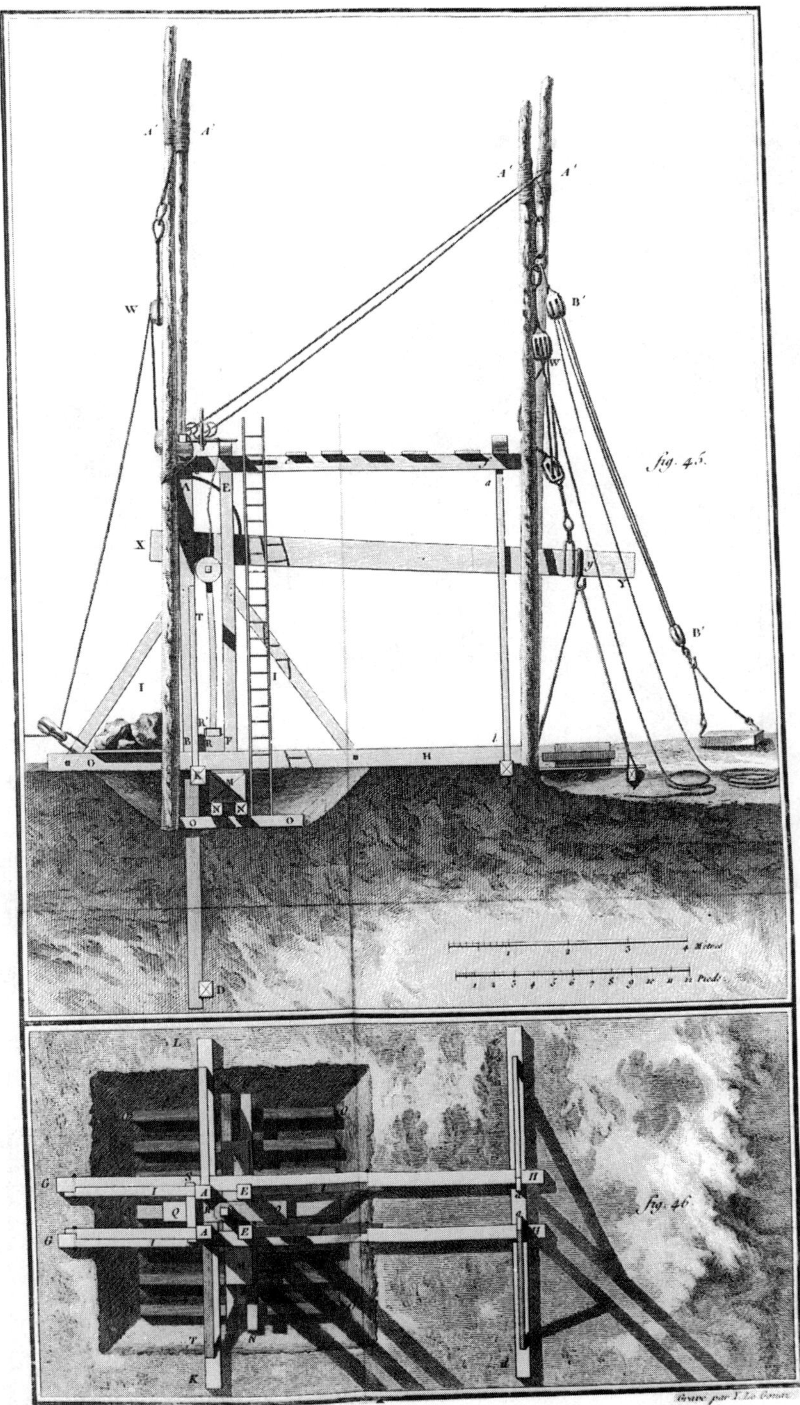

FIGURE 8.14. Girard's experimental apparatus. From the *Traité...* (Paris, 1798).

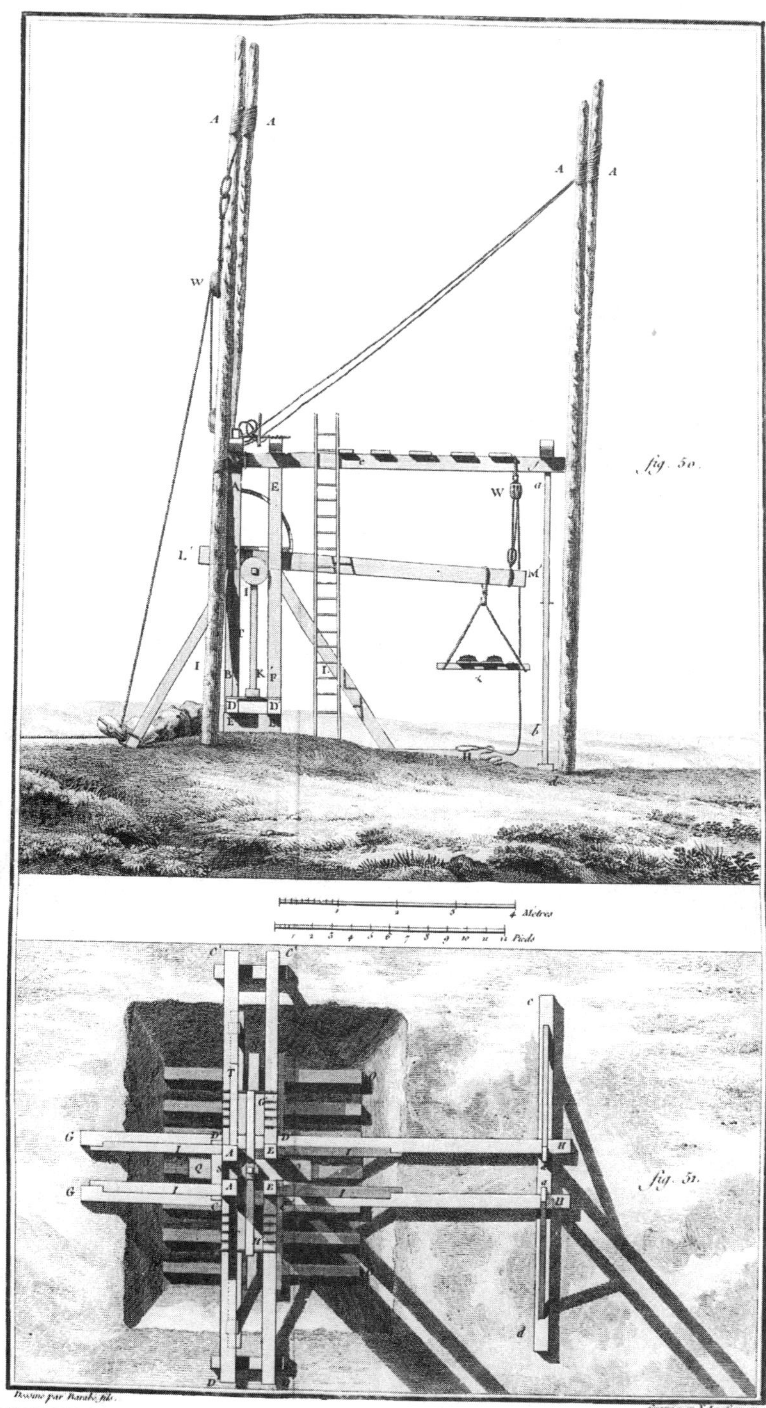

FIGURE 8.14 continued.

(5)

II. BOIS DE CHÊNE.

TABLE des expériences faites pour déterminer l'élasticité absolue des bois qui ont été chargés debout, et qui ont supporté les charges rapportées dans la Table précédente.

Numéros des expériences et des pièces qui y ont été soumises.	INDICATIONS DES DIMENSIONS et du poids des pièces.		FLECHE de courbure au milieu de la longueur.	CHARGE.	ÉLASTICITÉ ABSOLUE: à l'instant où la charge commence à agir.	après que la charge a agi pendant un certain tems.	Expression théorique du poids sous lequel la pièce commencerait à fléchir étant chargée debout, calculée pour le 1er instant de la charge.	calculée pour un instant où la charge agit depuis quelque tems.	TEMS écoulé depuis le commencement de l'expérience jusqu'au moment de l'observation.
	Mètres.	*Kilogr.*	*Mètres.*	*Kilogr.*			*Kilogr.*	*Kilogr.*	*H. déc.*
No 1. N° 1 de la 1re Table.	Longueur de la pièce...2,5978	....	0,0180	1884,67	....	38250	....	13985	10,00
	Haut. de l'équarrissage.0,1579	....	0,0238	2379,85	....	36524	....	13354	15,00
	Larg. de l'équarrissage.0,1285	....	0,0283	2932,81	....	37876	....	13848	20,00
	Poids de la pièce......	54,7843	0,0373	3467,55	....	33954	....	12414	30,00
				Moyennes........		36651	....	13400	
No 2. N° 1 de la 1re Table	Longueur de la pièce...2,2731	....	0,0158	1884,67	....	29174	....	13929	17,50
	Haut. de l'équarrissage.0,1579	....	0,0215	2395,34	....	27266	....	13017	26,66
	Larg. de l'équarrissage..0,1285	....	0,0248	2930,47	....	28909	....	13806	36,66
	Poids de la pièce......	54,7844	0,0300	3470,49	....	28304	....	13517	46,66
				Moyennes........		28413	....	13567	
No 3. N° 3 de la 1re Table.	Longueur de la pièce...1,6237	....	0,0045	1879,33	....	37238	....	34885	10,00
	Haut. de l'équarrissage .0,1330	....	0,0056	2390,00	....	38054	....	35620	15,00
	Larg. de l'équarrissage..0,0992	....	0,0074	2928,06	....	35280	....	33023	20,00
	Poids de la pièce......	42,0666	0,0101	3466,61	....	30599	....	28641	25,00
				Moyennes........		35293	....	33035	
No 4. N° 7 de la 1re Table.	Longueur de la pièce...1,9484	....	0,0045	1882,47	....	64461	....	41897	3,33
	Haut. de l'équarrissage .0,1579	....	0,0056	2393,63	....	65864	....	42809	8,33
	Larg. de l'équarrissage .0,1285	....	0,0079	2932,18	....	57186	....	37170	19,58
	Poids de la pièce.....	45,9747	0,0095	3468,43	....	56251	....	36561	21,25
			0,0113	4007,08	....	54634	....	35510	23,75
			0,0153	4542,69	....	45744	....	29732	41,66
				Moyennes........		57356	....	37279	
No 5. N° 7 de la 1re Table.	Longueur de la pièce...1,6237	....	0,0056	1877,42	....	29893	....	27980	22,91
	Haut. de l'équarrissage.0,1579	....	0,0068	2388,09	....	31312	....	29309	81,66
	Larg. de l'équarrissage.0,1285	....	0,0079	2926,64	....	33032	....	30918	90,00
	Poids de la pièce......	38,3164	0,0090	3465,19	....	34331	....	32134	93,50
			0,0101	4003,25	....	35341	....	33080	93,75
			0,0113	4559,41	....	35977	....	33675	109,58
			0,0119	4976,43	....	37288	....	39901	112,91
			0,0141	5512,05	....	34844	....	32615	129,58
				Moyennes........		34002	....	32451	

FIGURE 8.15. Girard's experiments on wooden beams. From the *Traité*... (Paris, 1798).

Note that the method used for to determine flexional rigidity (and therefore the elastic coefficient) is the same as that which Daniel Bernoulli and Euler had proposed. Girard uses a formula to express the deflection of the free end of a cantilever when a load is set on it. In the linear case, this takes the following form:

$$f = \frac{Ql^3}{3EJ}$$

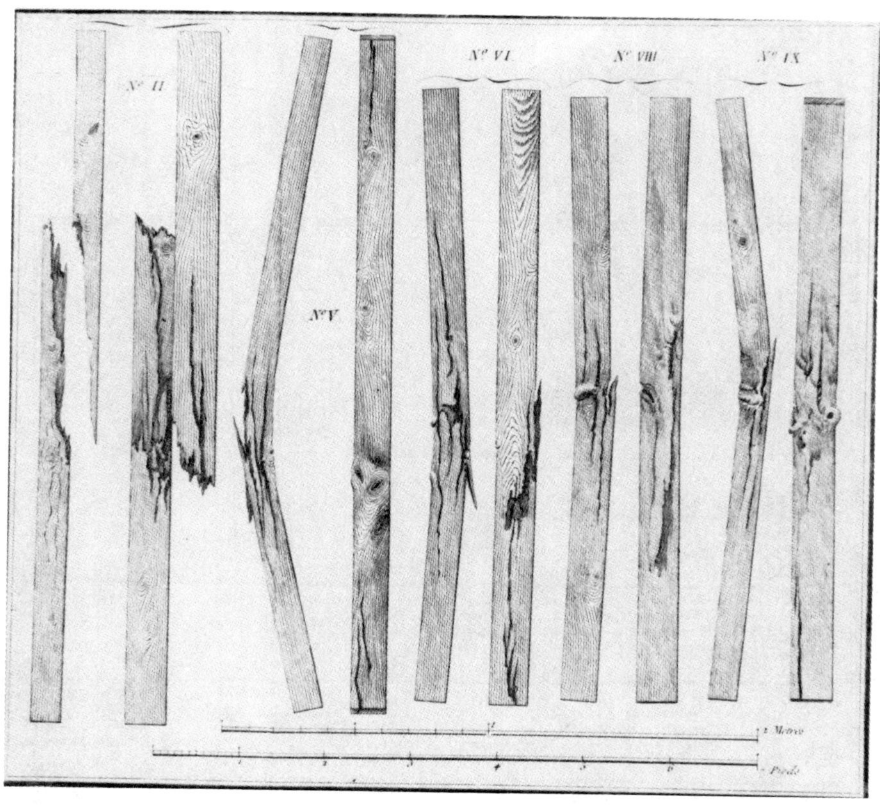

FIGURE 8.15 continued.

so that

$$EJ = \frac{Ql^3}{3f}.$$

Measuring the weight Q, length l and displacement f, he reaches his goal.

Girard's *Traité* is principally concerned with the "imperfections of materials," but it has no shadow of the doubts about the legitimacy of mathematical treatment which we find in Musschenbroek. Indeed, Girard has such faith in calculation that he includes it in the title: *Traité analytique* .... This is most obvious in Chapter 2, which is dedicated to the study of solids of equal resistance. Girard finally perfects the original theme of the post-Galileian debate, achieving admirable simplicity. The basic formulae needed to construct the entire theory, he claims, are those relating the maximum (or minimum) stress to the bending moment of an inflected beam, or to the axial force of a pulled or compressed beam. If we take X and Y as the dimensions of the cross-section (rectangular, as in Figure 8.16) and abscissa z as the axis of the beam, these formulae yield:

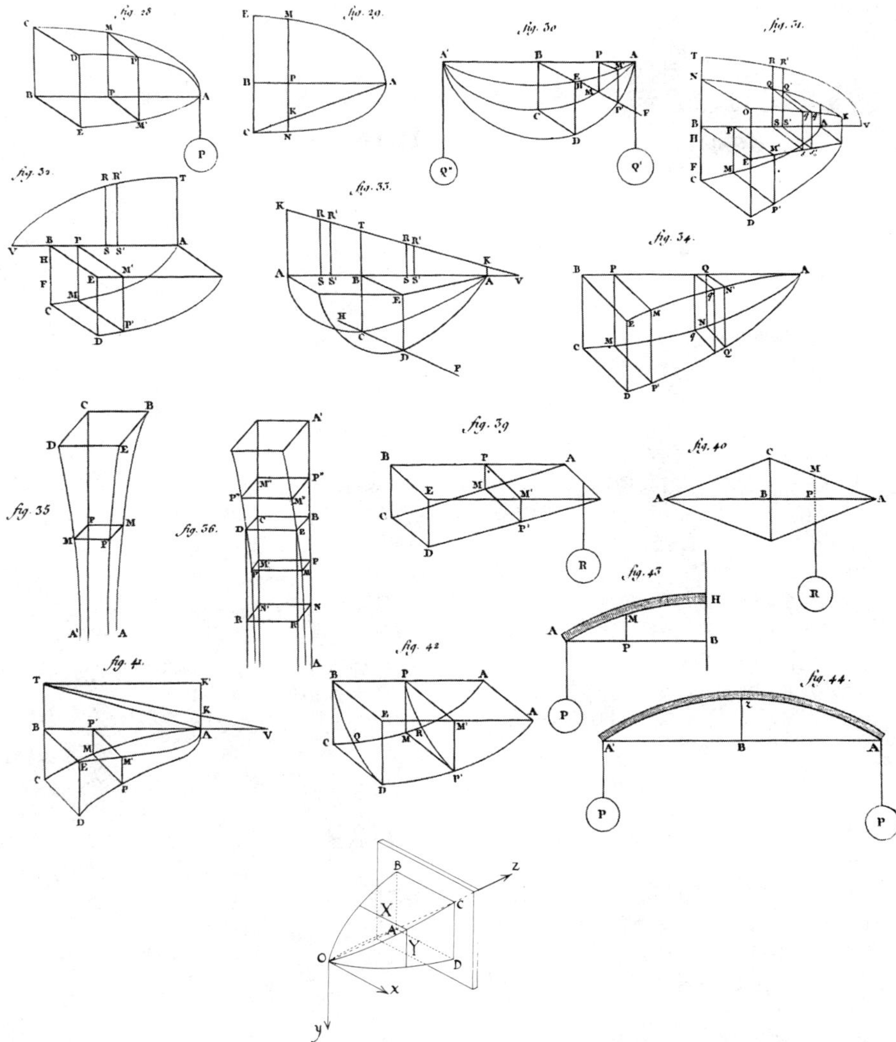

FIGURE 8.16. Girard's figures on the subject of solids of equal resistance.

$$XY^2 = F(z), \qquad XY = G(z),$$

where $f(z)$ and $G(z)$ are appropriate functions of z, corresponding to given distributions of loads (transverse or axial, respectively). Girard develops several examples, both in hypothesizing that stress is produced only by external loads and in taking into account the beam's own weight (cf. Figure 8.16).

We shall not dwell on the subject. But it is pleasant to conclude this discussion of Galileo's problem with these last, fine developments. They splendidly express the master's original intention: "we should come as near as we can to the perfection which so strikingly characterizes the works of nature: ... in short, to conform to her views."

Author Index

Abel, N.H., 134, 141
Aczél, J., 16
Aggiunti, N., 190
Albertus Magnus, 17
d'Alembert, J. Lerond, 14, 63, 66, 127–129, 132, 134–137, 284
Amaldi, U., 141
Ammannati, G., 147
Ampère, A.M., 108–113, 115
Anacreon, 241
Andrade, J., 8, 126fn
Antiphon, xv
Archimedes, 17, 43–56, 61, 63, 66–68, 79, 101, 146, 175, 176, 180, 244
Aristotle, xvii, 17, 20–29, 32, 39, 40, 43, 61, 68, 71, 145, 146, 174, 175, 194, 206, 211
Arrighetti, A., 190
Arrighetti, N., 190

Bacon, R., 175
Baliani, G.B., 170, 180, 200
Banfi, A., 156fn
Barberini, M. card., (later Urban VIII), 150
Bartoli, D., 167
Basnage, H., 41
Beaugrand, J. de, 73fn
Beeckman, I., 201, 262
Belgrado, J., 222
Belidor, B. Forest de, 35, 279
Bellarmino, R. card., 148–150, 156
Bellini, L., 242
Benedetti, G.B., 54, 176
Benvenuti, C., 226
Benvenuto, E., 34fn, 128fn
Berkeley, G., 225
Bernardo, G., 123
Bernoulli, D., 14, 31, 32, 116–121, 124, 126fn, 127, 137, 141, 288
Bernoulli, Jakob, 233, 234, 248, 256, 271, 272, 274–277, 279, 283, 287
Bernoulli, Johann, 4, 31, 77, 88–91, 272
Berti, D., 151fn
Berti, G., 200
Bessel, F.W., 149
Bettino, 68
Blondel, F., 197, 235–242, 246, 250, 251, 256
Bohr, N.H.D., 214
Boltzmann, L., 10, 115
Bolyai, J., 126, 134
Bonaventura de Bagnoregio, 17
Bonaventura, T., 199
Bonfigli, G.V., 211
Bonola, R., 126fn
Borelli, A., 241–243
Borelli, G.A., 193
Borgia, G. card., 150
Borra, G.B., 279
Boscovich, G.R., 223–227, 231
Bossut, C., 35
Boswell, 86
Bourbaki, N., 140
Boyle, R., 265
Bradwardine, T., 175
Brelot, M., 10
Bressan, A., 10
Bridgman, P.W., 6fn, 179
Brienne, Count of, 236
Brunelleschi, F., xx
Buffon, G.L. Leclerc de, 287
Bülffinger, G.B., 31, 122–124, 239fn
Buot, 237, 238
Buridan, J., 27
Burley, W., 171

Cajetan, T. de Vio, 20
Caramuel, J. Lobkovitz, 68
Cardano, G., 253

Carnot, L.M.N., 3, 7, 14, 32, 109, 116–117, 223
Carteron, H., 21fn
Caruso, A., 169, 170fn, 175, 187, 190fn
Casati, P., 68, 156
Castelli, B., 148, 173, 200
Cauchy, A.L., 62, 120, 130, 134, 137–140, 228–232
Caverni, R., 188
Cavina, V., 123
Ceredi, G., 18
Ceva, G., 140
Clarke, S., 218–221
Clifford, W.K., 8
Colli, G., 9fn
Colonna, Egidio, 171
Colonne, L. delle, 148
Copernicus, N., 148
Coriolis, G.G., 32
Cornford, F.M., 23fn
Costanzi, E., 150fn
Coulomb, C., 35, 233, 279, 280, 287
Cournot, A., 105
Crew, H., 44fn
Cristina di Lorena, Grand Duchess, 148
Crombie, A.C., 172
Culmann, K., 141
Curtze, M., 56

D'Addio, M., 149fn
Darboux, G., 139–141
Dati, C., 243
Delambre, J.J., 126
Democritus, 145, 152, 173, 175, 204
Descartes, R., 6, 26, 30, 67, 73fn, 77, 79, 85–88, 119, 122, 145, 157, 158, 201, 202, 206–208, 222
Deussen, P., xviifn
Dieulamant, M. de, 41
Dini, P., 148
Dorna, A., 138–139
Dubarle, A., 149fn
Du Bois-Reymond, E., 8
Dugas, R., 9, 88
Duhem, P., 14, 28, 37, 40, 56, 77fn, 79, 149, 201
Duns Scotus, 171, 175

Einstein, A., 26
Elea, 21
Eleatic school, 21, 23
Elzevier, L., 151
Epicurus, 152, 202, 221
Eschinardo, F., 156, 157
Euclid, 43, 56, 70, 77, 125
Euler, L., 14, 27, 127, 205, 233, 282, 287, 288
Eutochius, 17

Fabri, H., 156, 157, 252, 254–256, 262, 276, 279
Faraday, M., 26
Favaro, G., 150fn
Ferdinand II, Grand Duke, 150, 209, 210
Fermat, P., 73fn, 202
Ferrari, M., 130
Finetti, O., 210fn
Foncenex, F. Daviet de, 63, 126–133, 136, 137
Foscarini, P.A. fr., 148
Fossombroni, V., 98–101, 115
Foucault, J.B.L., 149
Fourier, J.B., 66–67, 98, 101–105, 115
Fracassati, C., 210fn
Fracastoro, Girolamo, 41
Francis da Paola, 201
Friedman, Y., 192
Frost, R., 158
Fuss, P.H., 32fn

Galilei, G., 5, 6, 16, 18, 27, 30, 35, 40, 41, 43, 53, 55, 61, 80, 82–84, 87, 96, 104, 107, 118, 145–152, 206, 208, 209, 211, 221, 227, 233–236, 239–243, 246, 247, 251, 252, 256, 257, 262, 265, 266, 276, 277, 280, 286, 287, 291
Galilei, V., 147
Galluzzi, P., 16–18
Gassendi, P., 202, 236
Gauss, C.F., 105, 134
Genocchi, A., 126
Georges, E., 142
Geymonat, L., 150, 169, 170fn, 175fn, 187, 190fn

Girard, P.S., 235, 283–291
Gonzaga, L., 123
Grandi, G., 197, 198, 236–252
Grassi, O., 150
Grazia, V. di, 173
Green, G., 32, 223
Gregory, D., 273
Grioli, G., 11
Guarini, G., xx
Guericke, Otto von, 220, 253
Guerrini, B., 151
Guidobaldo del Monte, 4, 52, 54, 67, 80–82, 95, 187
Guiducci, M., 190
Guldin, P., 61
Gunther, R.T., 263fn
Guthrie, W.K.C., 28fn

Hamel, G., 10, 141
Hankel, H., 141
Hegel, G.W.F., xx, 7, 12
Heiberg, J.L., 34, 44fn, 56
Heidegger, M., 19
Hermann, J., 31, 122
Hero, 45, 50, 51, 77
Herschel, F.W., 149
Hertz, H., 8, 10, 26, 223
Hesse, Mary B., 26
Hett, W.S., 34fn
Hevelius, J., 41
Hilbert, D., 9–12, 142
Hill, T.E., 119fn
Hire, P. de la, 58, 64–66, 125
Hobbes, T., 202
Hooke, R., 6, 180, 217, 255, 263–266, 276
Høyrup, J., 77
Huygens, Christiaan, 6, 30, 31, 56, 58–60, 63, 86, 202, 271, 272, 279
Huygens, Constantijn, 202

Inchofer, M., 150
Isozaki, A., 192

Jammer, M., 8, 9
Jensen, J.L., 141
Jonas, W., 192
Jordanus de Nemore, 14, 77, 79

Kant, I., 141, 220
Kenzo, Tange, 192
Kirchhoff, G., 8
Koyré, A., xx, 169

Lagrange, L., 4, 13, 14, 18, 31, 55, 56, 61, 66, 68, 79, 85, 95–98, 100–101, 105, 109, 113–115, 126–127
Lamy, B., 41
Lana-Terzi, F., 203
Laplace, P.S., 14, 109, 228
Leibniz, G.W., 6, 26, 30, 119, 122, 145–146, 218–221, 233, 234, 239fn, 247, 248, 256, 266, 268–273, 277
Lenoble, R., 201
Leo XIII, pope, 149
Leonardo da Vinci, 41, 78, 187, 283
Leopoldo de'Medici, card., 243, 254
Leucippus, 175
Libri, G., 187
Liceti, F., 154
Lobachevski, N.I., 126, 134
Loria, G., 84fn
Louis XIV, 236
Lucretius, 152, 204, 241, 242
Luther, M., 148

Mach, E., 8, 10, 49–50, 55, 83, 97, 117, 125fn, 149
Maculano, V., 150
Magnus, L.J., 16
Mansart, J., xx
Marchetti, A., 61–64, 181, 197, 235–246, 248–251, 256
Marcolongo, R., 48
Marini, L., 125
Mariotte, E., 97, 217, 234, 248, 256, 265–268, 277, 280
Maritain, J., 20
Maupertuis, P. Moreau de, 74–76, 287
Maurolico, F., 17, 54
Maxwell, J.C., 26
Maymont, P., 192
Mazière, 222
Mendoza, D. Hurtado de, 78
Mercato, V., 54

Mersenne, M., 73fn, 87, 157, 201–206, 262
Micanzio, F., 151
Michelangelo (Buonarroti), xix
Michelini, F., 190
Milizia, F., 279
Miller, V.R. and R.R., 30fn
Möbius, A.F., 142
Moerbeke, W. of, 17
Monge, G., 125
Montanari, G., 210, 212
Montucla, J.É., 126
Morpurgo-Tagliabue, G., 149fn, 150
Musschenbroek, P. van, 203, 221, 280–283, 287, 290
Mûzâ ibn Shâkir, 56

Napolitani, P.D., 55
Navier, L., 228–232
Nazari, F., 238
Neumann, C., 115
Newton, I., 6, 11, 26, 27, 40, 145, 218–221, 223, 224, 227, 247, 274
Nieri, V., 237
Noailles, Duc de, 202
Noll, W., 10–12, 142

Oliva, A., 243
Oreggi, A., 150
Ostrogradski, M.W., 105

Painlevé, P., 10
Pappus, 45, 50–52, 257
Pardies, I.G., 156, 252, 257–260, 262, 266, 272
Parent, A., 197, 268, 279, 280, 287
Parmenides, 21
Pascal, B., 203
Pascal, É., 73fn, 203
Pasqualigo, Z., 150
Pearson, K., 8, 197fn, 262, 263
Peiresc, N. Fabri de, 202
Pelacani, B., 14
Peri, D., 190
Peripatetic Author of the *Mechanical Problems,* 34–40, 67, 77, 105
Perronet, J.R., 287
Petty, W., 263
Piccolomini, A., 18
Piccolomini, A., archbishop, 151
Pieroni, G., 151
Pincherle, S., 141
Piola, G., 101
Plato, 173, 174, 236
Poincaré, H., 8, 32–34, 149
Poinsot, L., 14, 105–108, 115, 125
Poisson, N., 85, 86fn
Poisson, S.D., 32, 129, 137, 228, 231, 233
Poncelet, J.V., 32
Popper, K., 155
Porta, G.B., 253
Poupard, P., 149fn
pseudo-Euclid, 56–58

Ramsey, F.P., 136
Rankine, W.J.M., 32
Redi, F., 209, 212
Redondi, P., 153, 174
Reech, F., 8
Rey, J., 201, 203
Riccati, J., 222
Riccati, V., 31, 91–95, 98, 122–125
Ricci, M., 199, 200
Ricci, O., 147
Richard of Middleton, 171
Riemann, G.F.B., 126
Riso, Domenico de, 68
Roberval, G. Personne de, 4, 41, 42, 68, 73fn, 202, 203
Rossetti, D., 152, 154, 209–217, 221, 223, 224, 238, 241
Rosso, A., 175
Rousseau, J.J., 287
Ruarus, M., 203
Russo, F., 149fn
Rutherford, E., 214

Saccheri, G., 69–74, 76
Saint-Venant, A. Barré de, 8, 223, 231, 233, 266
Salvio, A., 44fn
Schenone, 243
Schiapparelli, G.V., 134
Schnuse, 141
Schopenhauer, A., xvii–xx

Siacci, F., 130, 141, 142
Simplicius, 40
Sinno, de, 130
Sizzi, F., 148
Sneed, J.D., 19
Soufflot, J.G., xx
St. Augustine, 149
Stevin, S., 4, 41, 55, 68, 81–82, 101, 122, 248
Stokes, G., 134
Sturm, C., 115

Tait, P.G., 8, 32
Tartaglia, N., 14, 78–79, 147
Tchebichef, P., 141
Tempier, É., 152, 171
Terill, A., 156, 252–253
Thomas, I., 46, 48fn
Thomson, W. (Lord Kelvin), 32
Timoshenko, S., 262
Todhunter, I., 197, 262, 263
Tonti, E., 142fn
Torricelli, E., 74, 84, 96, 104, 198–200, 206, 220, 246, 253, 257
Toupin, R., 10–13
Trevisano, B., 243
Trojano, C., 79
Truesdell, C., 9–13, 19fn, 142, 233fn, 257, 262, 265, 271–272, 276, 276fn, 283

Urban VIII, pope, 149, 150

Vailati, G., 37, 44, 45, 50–53, 56, 66, 77fn, 79
Valerio, L., 52, 54, 61, 68, 176
Valerio, V., 243
Valéry, P., xviii, 9, 13
Vandenbrecte (or Vandenbroeke), P.A., 238, 241
Varignon, P., 4, 31fn, 40, 67, 68, 82, 88–91, 103, 118, 197, 248, 277
Vassura, G., 84fn
Vaux, Carra de, 50, 52fn
Venturi, G.B., 78
Ville, A. de, 169
Vitruvius, Pollio, xvii, 17
Viviani, V., 175, 179, 181, 188, 197, 198, 235, 243, 246–251, 256, 266
Volkmann, P., 10
Volpi, R., 141

Waard, C. de, 200, 201fn
Wallis, J., 6, 30, 52, 53, 67, 257
Weierstrass, K., 134
Westphal, A.H., 142
Wickstead, P.H., 23fn
William of Ockham, 171
Woepcke, F., 56
Wohlwill, E., 34
Woolf, L., xxi
Wren, C., xx, 6, 30
Wright, F.L., 192
Würtz, P., 197, 238, 239

Young, T., 32

Zanotti, M., 141
Zeller, E., 34
Zeno, 21

Subject Index

Absolute resistance, 178, 181–182, 188, 204, 280

Act (Greek: ἐντελέχεια or ἐνέργεια; Latin: Actus)
  Aristotelian concept of–, 21–22; Aristotelian explanation of motion in terms of– and power, 22–23; Aristotelian principle of conservation concerning a necessary proportion between– and power, 27–29; metamorphoses of– in modern science, 30–32

Action
  principle of– and reaction, 6, 11, 164, 224; – at a distance, 25–26, 218–221, 224–225; contact –, 25–27; V. Riccati's concept of "– of power", 91–94, 123–124

Actual energy, 32

Actus, *see* Act

d'Alembert's
  synthesis of mechanical principles, 127–128; functional equations, 132–133, 134–135, 136–137; demonstration of the parallelogram of forces, 134–137

Ampère's
  argument against the principle of superimposition of equilibria, 109–110; demonstration of the principle of virtual velocities, 110–113

Archimedes'
  concept of weight, 43; seven postulates for the foundation of his statics, 44–45; first demonstration of the law of the lever, 46–48; second demonstration of the law of the lever, 53–54, 177; four axioms for the center of gravity, according to Vailati's reconstruction, 51; lemma for the demonstration of the law of the lever, according to Vailati's reconstruction, 53; proverb, 55, 68–69; mirrors, 175

Aristotle's
  philosophy of nature, 20–24; concept of motion, 22–23; demonstration of the existence of one or more Immobile Movers, 23–24; physical principles, 25–30; postulate of contact action, 25–27; law of motion ("–axiom"), 27–29; 37, 40; principle of conservation, 23–24, 27–29; refutation of the existence of a vacuum, 166–167; wheel, 174

Aristotelian *Mechanical Problems,* 34–42, 105, 194;
  demonstration of the parallelogram rule for the composition of movements, 35–36; explanation of the law of the lever, 36–39

Atmospheric pressure, 170, 199–203

Atomism
  Galileo's–, 152–154; Lucretius' atomistic explanation of the mechanical behavior of bodies, 165; Galileo's first intimations of an atomistic theory of resistance, 169–173; ambiguity of Galileo's atomism, 173–175; Mersenne's opinion, 204–205; Rossetti's atomistic explanation of resistance and elasticity, 209–217; Newton's– and his explanation of resistance in

Atomism (cont.)
terms of molecular attractions, 218, 221; Leibniz's criticism of– 219–220; Clarke's reply, 220–221; Boscovich's reformation of– 223–227; the molecular theory of elasticity in the 19th century, 227–232
Axiomatic method, 9–14;
Hilbert's application of– to geometry, 9–10; Noll's application of– to mechanics, 11–12

Baliani's argument against the power of the vacuum, 170, 200; criticism of Galileo's solution of his "problem", 180
Batavian drops, 209–212
Beeckman's first quantitative estimate of elastic deformation, 262
Bent beam, *see* Galileo's fundamental problem, Deformation of a bent beam
Bernoulli, D.'s
concept of mechanics, 117–118; definition of force, 118–119; hypotheses on forces, 119; demonstration of the parallelogram of forces, 119–121, 136, 137
Bernoulli, Jakob's
equations for flexible ropes, 272–273; solution to the problem of the bent beam, 274–276
Bernoulli, Johann's
(and Varignon's) definition of virtual velocity and energy, 89–90; general statement of the principle of virtual velocities, 89
Bisymmetry property of the composition of forces, 131, 134
Bohr's and Rutherford's atomic model, 214
Borelli's mechanical explanation of the action of muscles, 193, 243
Boscovich's
law of continuity, 223; concept of atoms as unextended points, 224, 231; concept of matter in terms of repulsive and attracting forces, 224–226; law of forces, 225–227
Boscovich's and Benvenuti's explanation of the behavior of materials under stress, 226–227
Bright atoms, 212
Buckling of a thin rod, 205, 282

Carnot, L.'s criticism of force, 7–8, 116–117
Catenary
Pardies' theorem, 257–258; Huygen's, Leibniz's and Johann Bernoulli's contribution, 271–272; Jakob Bernoulli's solution, 272–273; Gregory's approach, 273–274
Cauchy's
functional equation, 62, 130, 138–139, 140; demonstration of the parallelogram of forces, 137; contribution to the molecular theory of elasticity, 228–232
Center of gravity
Archimedes' postulates on the –, 44–45; Hero's and Pappus' axioms on the–, 51; Vailati's reconstruction of the Archimedean theory of the–, 51–53; Wallis' demonstration of the existence of the–, 53
Central plane, 51–52
Central straight line, 51–52
Ceva's theorem concerning the trihedral angle, 140
Collapse load, 181–182, 186, 277–279, 283
Commutative property of the composition of forces, 131

Composition and decomposition of forces, *see* Parallelogram of forces
Composition and decomposition of movements, 35–36, 40–41, 116–117, 127
Conservation
  Aristotelian principle of–, 22–24; the Immobile Mover, 23, 25; modern metamorphoses of the Immobile Mover, 30–34; – of movement, 23; according to Descartes, 30; – of live/dead forces: Leibniz, Huygens, Lagrange, 30–31; V. Riccati's approach in terms of "actions", 31, 91–94; – of energy, 31–32; Poincaré's remark, 33–34
Conspiring forces, 119
Constitutive equations, 163, 180, 286
Constraints, 105–115
Contact, action, forces, 25–27
Continuum mechanics, the foundation of–, 10–12
Controversy about solids of equal resistance
  Blondel's criticism of Galileo's theory, 235, 239–241; Marchetti's approach, 244–246; Viviani's solution, 246–252; Fabri's remark, 256; Girard's treatise, 286–292
Controversy on elastic constants, 228, 232
Controversy on the nature of mechanical principles, 14, 116–118, 127–128
Copernican system, 147–151
Corpi quanti, 173–175

Darboux's
  axioms for the composition of forces (or lines), 139–140; demonstration of the parallelogram of forces, 139–141
De la Hire's
  formulation and demonstration of Archimedes' lemma for the law of the lever, 64–65; demonstration of the law of the lever, 65–66
De Maupertuis'
  law of rest, 74; criticism of the law of the lever, 75–76
Deformation of a bent beam
  Fabri's deceptive theorem, 256; Pardies's model of a necklace composed of cylindrical beads, 259–260; Mariotte's contribution and his confirmation of Fabri's error, 266–268; Leibniz's new approach, 268–271; Jakob Bernoulli's fundamental memoir, 274–276; Musschenbroek's experiments, 282
Descartes'
  principle of conservation of motion, 30; "only one principle of all machines" (intimation of the principle of virtual work), 30, 79, 85–88; concept of "two-dimensional" force (or action), 86–87; explanation of the resistance of solids, 206–208; "subtle substance" (or "matter"), 207, 222
Dorna's demonstration of the parallelogram of forces, 138–139
δύναμιs (dynamis), *see* power

Egg, explanation of the resistance of an–, according to Pardies, 259
Elastic cord as a physical model of force and its action, according to V. Riccati, 124
Elastic curve, 274
Elasticity
  Descartes's explanation of– in terms of a "subtle matter", 222; Fabri's concept of– as "intermediate force", 254–255; Rossetti's atomistic

Elasticity (cont.)
explanation of–, 216, 217; Hooke's law of elasticity, 263–265; Mariotte's contribution, 266; Leibniz's contribution, 269; Jakob Bernoulli's stress-strain relation, 276; Newton's concept of– as "vis interna attrahens", 221–222; Marière's contribution to cartesian theory, 222; J. Riccati's hint of an energetical approach, 222–223; Boscovich's explanation of– by means of his general law of natural forces, 225–227; the molecular theory of elasticity, 227–232
ἐνέργεια (enèrgeia), see Act
Energy
Aristotelian root of– in the concept of act, 22–24; introduction to this term in modern science, 31–32, 89–90; conservation of–, 30–34
Enigma of "becoming" (in particular of movement)
Eleatic school's paradoxical thesis, 22; Aristotle's solution, 22–24, 77
Enigma of force, 3–4, 7–8, 12, 97
ἐντελέχεια (entelècheia), see Act
Epistemological reductionism, 7–8, 19, 97, 116–117, 129, 179–180
Experimental researches on the resistance of solids, 203–205, 280–283, 287–289

Fabri's
concept of elasticity as "intermediate force", 254–255; discovery of Hooke's law, 255; results on vibrating cords, 255; contributions to the theory of resistance, 256
Fabri–Pardies–Mariotte–Jakob Bernoulli's error on the bent beam, 256, 260, 267–268, 276
Fibers and filaments as a cause of resistance, 163–166, 204
Foncenex's
functional equation for the law of the lever, 63–64; fundamental lemma for the composition of forces, 127–130; demonstration of the parallelogram of forces, 130–133
Foncenex's and d'Alembert's functional equation, 132–133
Formal structure of the principles of statics, 15–16
Fossombroni's
"equation of moments" and "equation of forces", 98–99; demonstration of the principle of virtual velocities, 99–100; approach to deformable systems, 101
Foundations of mechanics, 7–14, 141–142
Fourier's
demonstration of Archimedes' lemma for the law of the lever, 66–67; concept of "moment", 101–103; extension of the principle of virtual velocities to the case of unilateral constraints, 102–105; identity criterion for mechanical systems, 103; demonstration of the principle of virtual velocities, 103–104
Fracture
Galileo's explanation of the causes of–, 163–166, 169–173; – as the effect of exceeding the maximum supportable weight or tension, according to Galileo, 179–181; Rossetti's explanation of– as the effect of exceeding the maximum supportable elongation, 217; Boscovich's explanation, 226–227; Mariotte's and Saint-Venant's criterions of–, 266

G*alilaeus ampliatus,* 238, 241

Galilaeus promotus, 236, 238
Galileo's
solution of the problem of the inclined plane by means of the equation of virtual velocities, 83–84; claim for the autonomy of science, 148–149, 156–157; concept of "sensata esperienza", 154–155; distinction between "logic" and "geometry", 155, 157; epistemological approach, 154–158; atomism, 152–154, 163–166, 173–175; first intimations of an atomistic theory of resistance, 168–173; concept of horror vacui, 166–168, 170; experiment for evaluating the power of the vacuum, 169; explanation of piston suction pump, 169–170; fundamental problem of the flexure of a horizontal beam, 176–197; concept of "absolute" and "relative" resistance, 178, 181–182; problem of solids of ultimate dimensions, 188–194; problem of solids of equal resistance, 194–197
Géostatique, 73
Girard's
experimental studies on the strength of materials, 283–288; contribution to the study of solids of equal resistance, 290–291
Gluey or viscous substance as a cause of resistance, 166, 206
Gregory's contribution to the problem of catenary, 273–274
Guidobaldo's approach to the principle of virtual velocities through his analysis of pulleys, 80–82

Hilbert's axiomatisation of geometry, 9–10; "Sixth Problem" of his *Mathematische Probleme,* 10, 12
Hooke's law of elasticity, 217, 255, 263–265, 266
Horror vacui
Aristotle's argument, 167; Medieval debate, 171; Galileo's concept of– as an explanation of resistance, 166–168; Galileo's experiment for giving– a quantitative estimate, 168–169; Baliani's refutation of–, 170, 200; Torricelli's experiment for disproving the existence of–, 198–200; Rey's and Mersenne's thesis against, 201, 204; Descartes's opinion, 206
Huygens'
theorem of "live" and "dead" forces, 6, 31; axioms for the equilibrium of planes, 58–59; demonstration of the law of the lever, 59–61; solution of the problem of catenary, 272; center of percussion, 279
Hylemorphism, 22, 152

Immobile mover, 23–24, 29; metamorphoses of the– in modern science, 30–34
Inclined plane, 4, 41, 83–84
Infiniti indivisibili, 175
Internal friction, 280
Irreversible displacements, 102–105

Lagrange's
demonstration of the law of the lever, 55, 66; first demonstration of the principle of virtual velocities, 95–98; "principle of pulleys", 95; attempt at eliminating the concept of force, 96–98; method of multipliers, 100, 106, 114–115; second demonstration of the principle of virtual velocities, 113–115; argument against the

Lagrange's (cont.)
geometrical demonstrations of the parallelogram of forces, 127
Leibniz's
arguments against atoms and Newton's attraction forces, 218–221; solution of "Galileo's problem", 268–271; foundation of the mathematical theory of elasticity, 271
Leonardo's
concept of force, 7; foretaste of the importance of funicolar equilibrium, 41; "discovery" of the principle of virtual work, 78; observations on the buckling of thin rods, 283
Lever, the law of the–, 4, 15, 34–35, 39–42, 43–76, 80, 87, 88, 91–92, 94, 95, 98, 103–105, 125, 126, 176–177, 241, 244, 249
Life of Galileo, 147–152
Live forces, 30–31, 88, 122–123

Magnus' functional equation, 16
Marchetti's
postulates for the concept of moment, 61–62; demonstration of the law of the lever, 63; axiomatic approach to the resistance of solids, 244–246
Mazière's attempt to explain elasticity, 222
Mariotte's
solution of "Galileo's problem", 265–268; criterion of rupture, 217, 266
Material body
Archimedean concept of weight as a measure of its matter, 43–44; Aristotelian theory of *hylemorphism,* 22, 152; Galileo's atomistic approach, 152–154; Descartes's theory of– as "extension in space", 206–207; Rossetti's explanation of– in terms of atoms endowed with poles, 212–214; Newton's atomism and Leibniz's monadology, 218–221;
Boscovich's concept of– as composed of unextended points, 224–226; the concept of– in the light of Noll's axiomatisation of continuum mechanics, 10–12
Mathematical induction, 137
Mécanique Physique, 228
Minimi quanti, 175
Molecular and astronomical attraction, 226
Molecular theory of elasticity, 228–232
Moment of resistance, 178, 180, 182–183, 188, 244–245
Momentum, 4, 16–20
Momentum ponderis, 17
Momentum temporis, 17
Musschenbroek's experimental work, 280–283; experimental law of the buckling of rods, 282

Natura non facit saltus, 140, 223
Navier's foundation of the molecular theory of elasticity, 228–229, 232
Newton's
mutual forces and third law of mechanics, 6, 11, 164, 224; demonstration of the parallelogram of forces through the composition of movements, 40; concept of action at a distance and of attraction, 26, 218; "*vis interna attrahens*", 221
Noll's approach to the foundations of mechanics, 11–12
Non-conservative aspect of material behavior, 280–283, 288
Non-Euclidean geometry, 70, 125–126, 141

Non-Euclidean statics, 125–126, 141
Non-mechanical forms of energy, 32–34
Non-theoretical terms of a physico-mathematical theory, 19

Ontological difference, 19

Pack twine [*susta*], 165
Paradoxes of the infinite, 171–172, 174–175
Parallelogram of forces (or principle of the composition of forces), 4, 15, 35–36, 40–42, 66–67, 88–91, 98, 101, 113, 115, 116–142
Pardies' theorem on catenary, 257–258; solution of the problem of the suspension bridge, 258–259; model for the bending of a beam, 259–261
Parvum clinamen, 212, 218
Philosophy of nature according to Cajetan and neo-scholasticism, 20
Poinsot's
  axioms for equilibrium, 105–106; general theorem on the equilibrium of a system of points, 106–107; demonstration of the principle of virtual velocities, 107–108
Poliplastes, 95–96, 113
Potentia, see Power
Potential energy, 32–33
Power (Greek: δύναμις; Latin: *Potentia*)
  Aristotelian concept of–, 21–22; Aristotelian explanation of motion in terms of– and act, 22–23; Aristotelian principle of conservation concerning a necessary proportion between– and act, 27–29; metamorphoses of– in modern science, 30–32, 68; – as synonym of force, 64, 78, 79, 80–83, 91–94, 95, 116, 118, 123–124
Power of a vacuum, *see Horror vacui*
Prime Mover, *see* Immobile mover
Principle of inertia, 6, 21
Principle of local action, 26
Principle of solidification, 53, 105–106, 258
Principle of the superimposition of equilibria, 56–61, 109; Ampère's argument against the–, 109–110, 113
Principle of the superimposition of figures in geometry, 61
Principle of virtual velocities (or of virtual work), 4, 15, 28–29, 39–40, 67, 77–115, 125
Pseudo-Euclid's
  axioms for the equilibrium of weights hanging from a horizontal plane, 56–57; demonstration of the law of the lever, 57–58
Pulley (or tackle), 4, 67, 80–84, 95–97, 113–115

Relative (or transversal) resistance, 204, 280; *see also* Moment of resistance
Reptio, 17
Retaining walls, 287
Riccati, J., explanation of elasticity, 222–223
Riccati, V.
  concept of action, 92–93; "universal principle of statics", 94; demonstration of the parallelogram of forces, 123–125
ῥοπή (*rhopé*), 17
Rossetti's
  concept of "poles", as centers of attractive or repulsive atomic forces, 211, 214–215; law of "appretence" and "abhorrence", 212–213; concept of "sphere of energy", 213; explanation of the resistance of solids, 215–216;
Rossetti's (cont.)

explanation of the elastic behavior of bodies, 216–217; intimation of a criterion of rupture, 217

Saccheri's
hypothesis that the center of weights is not situated at an infinite distance, 70; law of composition of impetus (velocities), 71; two axioms and one postulate for equilibrium, 71; new approach to, and criticism of the law of the lever, 71–73
šaḥaq, 17
Saint-Venant's criticism of force, 8; contribution to the molecular theory of elasticity in the light of Boscovich's theory, 231
Solids of equal resistance, 146, 194–197, 235–244, 250–252, 256, 271, 287, 290–291
Solids of ultimate dimensions, 187–194, 252–253
Stability of equilibrium, *see also* Buckling of a thin rod, 101, 104
Statical demonstration of the fundamental formulae of geometry, 126
Stevin's
approach to the parallelogram of forces, 4, 41, 101, 122, 248; improvement of Archimedes' demonstration of the law of the lever, 55; formulation of the principle of virtual velocities, 81–82
Suspension bridge, 258–259

Tackle, *see* Pulley
Theoretical terms of a physico-mathematical theory, 19
Torricelli's *extremum* principle, 85, 104; disproof of the force of the vacuum and barometric experiment, 199–200
Tractatus a vacuo, 171

Unilateral contact, 102–105
Ut tensio sic vis, 6, 263

Vacuum, *see Horror vacui*
Vailati's reconstruction of Archimedes' treatment on the center of gravity, 50–53
Varignon's
deduction of the parallelogram of forces from the composition of movements, 40; *Nouvelle mécanique* based on the principle of the composition of forces, 88; demonstration of the principle of virtual velocities, 89–91
Virtus, 18
Virtus continuativa, vel contiguativa, 253
Vis agglutinativa, adherescentia, 253
Viviani's
axiomatic approach to the resistance of solids, 247–250; contribution to the problem of solids of equal resistance, 250–252

Water rod (or column) eighteen braccia long, whose weight would correspond to the force of the vacuum, 170–171, 200